D1187240

Study and Management
of Large Mammals

Study and Management of Large Mammals

THANE RINEY

Visiting Fellow, Australian National University,
Canberra, A.C.T., Australia

A Wiley–Interscience Publication

1807 1982

JOHN WILEY & SONS

Chichester · New York · Brisbane · Toronto · Singapore

Copyright © 1982 by John Wiley & Sons Ltd.

All rights reserved.

No part of this book may be reproduced by any means, nor
transmitted, nor translated into a machine language
without the written permission of the publisher.

Library of Congress Cataloging in Publication Data

Riney, Thane.
  Study and management of large mammals.
  Includes index.
  1. Wildlife management. 2. Mammals.
  I. Title.

SK353.R56   639.9′79   81-11519

ISBN 0 471 10062 5   AACR2

British Library Cataloguing in Publication Data:

Riney, Thane
  Study and management of large mammals.
  1. Mammals—Ecology
  I. Title
  599.05   QL703

ISBN 0 471 10062 5

Typeset by Preface Ltd, Salisbury, Wilts. and
printed at Page Bros. (Norwich) Limited.

Contents

vi

Acknowledgments

I am strongly indebted to several organizations with which I have been associated and which have sponsored various phases of the work or have permitted further work on the book on my own time. I am particularly grateful to the University of California, the F. S. Hastings Natural History Reservation, New Zealand Wildlife Service, New Zealand Forest Service and the New Zealand Dominion Museum, U. S. Fulbright Scholar programme, National Museums of Zimbabwe, the Beit Trust, International Union for the Conservation of Nature and Natural Resources (IUCN), Food and Agriculture Organization of the United Nations (FAO), United Nations Environment Programme (UNEP), Department of Forestry and Natural Resources, University of Edinburgh, Smithsonian Institution, Rockefeller Brothers Fund and Centre for Continuing Education, Australian National University. Of these I am particularly indebted to FAO for the opportunity to visit most of the major regions of the world and, through numerous contacts with National Parks and Wildlife organizations in these regions, to further develop my understanding of the needs of organizations which are in the formative stages of development.

So many persons have helped over so many years it is impossible to mention them all and hard to know where to start. First, I must mention Abubakar of Eritrea, who taught me so much about observing animals. I am indebted to two game wardens in Zambia, the late J. Uys and B. Shenton, who manoeuvred me into promising to write a handbook. New Zealanders A. Douglas, P. Logan and J. Henderson helped more than they knew in providing the stimulus to start. I am particularly indebted to G. Caughley for critical review, advice and help with writing Chapter 7000; to Luna Leopold for the idea for the basic approach used in demonstrating man-induced accelerated erosion based on abnormal stream patterns and I have also, fortunately, had the benefit of his criticisms and comments on this section; to G. Moisan for helping with preparation of several sections for use in the West African training course handbook and to W. Dasmann for help with some of the techniques on plant utilization and plant response. I am indebted to J. Lockie for the phrase 'a way of looking' and the subsequent thoughts it stimulated, for assistance in the early stages of preparing this section and for much helpful advice. R. Noble and Graham Child were of assistance in sharpening up several parts of the manuscript; K. Miller made useful contributions to the section on national parks; I profited from the advice and assistance of Ian Grimwood, who helped in the preparation of sections dealing viii

with the organization of game departments and with the organization of hunting areas, and R. Smithers helped with Chapter 4000 and in other ways.

The late Sir F. F. Darling provided steady encouragement and proved an incomparable foil for testing various attitudes and approaches. The late C. P. McMeekan and B. Shorland provided a similar service in my early New Zealand trials of developing ways of improving methodologies for simultaneous studies of Animal and Environment.

In addition the following have helped in several important ways: M. Björkland, H. Buechner, R. Estes, R. Fontaine, P. Hemmingway, W. Low, B. Mitchell and B. Roux. P. A. Riney drew most of the illustrations and without her the book would never have been completed. K. Riney assisted in the final stages of preparation.

While all this assistance is gratefully acknowledged, I reserve the right to claim credit for any mistakes in concept, interpretation or presentation.

The practical matter of existing was occasionally a problem during the later stages of the preparation of the manuscript; a problem that was solved mainly through the generosity of the Rockefeller Brothers Fund in supplying a grant administered by the Smithsonian Institution, Washington, DC. The patience of Messrs Hyde and Moody of the former, and Drs Ripley, Challinor, Michaelson and Hurlbut of the latter organization was sorely tried. I hope they feel it was worth it in the end.

Preface

Progress in developing an integrated approach to the study and management of large mammals as presented in this book has been slow and corresponds with several phases of my own accumulating experience. Various observational techniques were developed in India, the Middle East and Africa in 1940 and 1941 and these and subsequent war years provided wonderful opportunity for becoming acquainted with a broad spectrum of habitats, including also the United Kingdom, parts of Europe and North America. There followed 4 years of intensive work with deer in the coastal ranges and the Sierra Nevada of California, where the ideas for the present approach first developed.

Early sections were prepared for training in field work with problem animals in New Zealand, where over a period of 8 years the general aim of combining evidence from both animal and environment in a synthesizing approach was first tested in practical ways and developed to a level that seemed to work with the simplified New Zealand environments. Then, for 4 years, the approach was further field-tested in southern and eastern Africa, in the most complex of all environments containing many species of large mammal. More sections were written and earlier sections modified during this period when the general content of the book first took shape.

Following these 18 years of intensive field experience, opportunity was taken, in the course of a 2-year Africa-wide survey involving twenty African countries, to further test and refine various extensive survey methods and synthesizing approaches to problem identification and analysis. Material for several sections was gathered and new techniques developed and tested in this period. During 9 years of largely administrative work at FAO headquarters in Rome, several sections were prepared and translated into French for use as an unpublished but working handbook for use in teaching at the FAO-sponsored school for training West African game wardens in Garoua, Cameroun. The sections prepared followed the general outline of the present book and several FAO officers assisted in modifying earlier sections, making them more suitable for this training purpose. Finally, on leaving FAO, a fresh look was taken of the entire field, over half of the earlier prepared sections being omitted in the light of new developments and the entire concept re-thought and re-written to its present form.

THANE RINEY
1981

Chapter 1000

Introduction

Many countries still have, as part of their national heritage, representative populations of large mammals, which may still be conserved and utilized for the continuing enjoyment and benefit of their citizens. The potential for improving earnings from tourism based on wildlife is being developed; in some regions the use of wildlife as a source of protein and profit is increasing. In all regions certain species are scarce and becoming scarcer. Rational management is perhaps the best, most certain way to protect the wildlife resource from large-scale depletion or extinction.

The need for training appears greatest in organizations in the early stages of developing an effective wildlife service. There appears a real need for strengthening and broadening in-service and regional training. The trend for universities to broaden Forestry, Natural History and Conservation courses to include appropriate training in wildlife management continues. But appropriate materials for teaching or for self-learning are still scarce.

Fortunately, there are still plenty of large wild mammals about, and field men throughout the world who are trying to do something about improving their lot by studying, surveying, managing, protecting, monitoring, improvising.

Field men often work in splendid isolation; free from the learned biologists who are up to date on the latest fashionable subjects to study; free from constant contact with technical over-specialization and the narrowness of view that often goes with it; free to use their own eyes, to be guided by their own experience; free to use their common sense. It is especially for these men the book has been written.

Historically, one traditional approach to gathering information as a basis for management has been simply to build up a fund of information directly or indirectly related to a problem until enough information has accumulated to make it apparent that some clear relationships exist. This has worked in the past and will continue to work as a research process. But it is a haphazard and luxurious strategy few organizations can afford. The main danger is that such an approach creates an environment which seems to incubate and hatch research workers who insist on studying animals or vegetation for their own sake, without regard for and, indeed, often avoiding practical considerations.

There are hundreds of techniques published and available to a field worker, thus some basis for selection had to be found. This has been done largely on the basis of practicality: simplicity of use and the least amount of expensive equipment required. One of the innovations of presentation is in the classification of techniques in terms of the kind and significance of the information

1

they produce. Every technique has limitations as well as uses and this aspect has been stressed. Much previously unpublished material has been included to exemplify a technique or approach. I have drawn heavily on my own experience for it seemed relevant and facilitated a consistency of approach.

The plan of the present book is simple. First, ways of forming questions and of planning an investigation are introduced; then a number of techniques for gathering information are described. This is followed by chapters on problem-solving and management of special areas as illustrations of how techniques described may be selected as appropriate, modified, or new techniques invented for a particular use.

Perhaps the principle of acceptable alternatives is one of the foundation stones of wildlife management. There is no one sustained yield possible, there are many acceptable sustained yields for a population in a given environment. In popular terms there is no one point at which 'nature' can be said to be in balance, there are many points. This results from understandable interactions between animals and the various habitat elements on which they depend: on the flexibility and elasticity of the living materials with which we have chosen to work.

I have faith that succeeding generations of field workers will continue to be as adaptable as the populations of plants and animals with which they work. Hopefully there will be increasing attention to carefully defining problems and objectives of study and management and more precision in selecting relevant approaches to deal with the issues at hand in a way that can be acted on by decision-taking and policy-making senior officers in Conservation Departments. This handbook has tried to present a range of choices concerning large mammals and their environment, in helping to decide what is the relevant next step to take in moving ahead towards more effective management.

Chapter 2000

Planning an Investigation

2100   HOW TO PLAN AN INVESTIGATION

This section is intended for several classes of readers: it should be useful to those who have cooperated with research workers or have carried out investigations of their own. It should help administrators not actively engaged in research to understand the nature of the work and the problems of their research colleagues. It should interest even those who have no connection with research, but are curious about the whole process.

It is appropriate first to define research and the scientific method by which it is accomplished as a background note for the sections which follow.

2110   The Basic Questions

The term 'research', as used in this book, is simply the process of working from the known to the unknown in order to find answers to the basic questions: who, where, what, when, how and why. Science tries to explain, in terms of general laws, the phenomena which we observe. It has been successful in providing a better knowledge of the world largely because of the scientific method which research employs.

2120   The Scientific Method (modified from Fairchild, M., 1926)

(*a*) The first step in any research project is to define the object and scope of the investigation; for example, to study the occurrence and distribution of large mammals in a national park, or the extent to which buffalo are damaging a pine plantation. Some research workers prefer to start with a hypothesis; for example, as it is applied in a certain way in a certain kind of country, fire may be suspected of being the basic cause of the increase of certain species of ungulate and the decrease of other species. This step is essential to sound research because it helps focus attention on the problem or problems to be solved, or the hypothesis to be tested and reduces expenditure of energy in other directions.

(*b*) Once the object and scope of the investigation is defined, the next step is to determine what sorts of data will most probably yield the information that is needed and to plan how data will be gathered and recorded. This plan should be tested and modified where necessary before starting the research.

In some kinds of research it will be useful to start with a quick observational reconnaissance in the field. This provides a means of clarifying the object and scope of the research and of judging better the kinds of facts required and available and how they can best be obtained.

(*c*) When these steps have been completed, the research worker gathers data within the selected field. Scientific data consist of numerous and accurate human *observations*, sometimes assisted and corrected by instruments of precision. The observations must be recorded in definite terms and in repeated measurements. While several, or even many, observers may collaborate in gathering such data, it must be remembered that all observations are not of equal value. As with other human activities, the ability to make sound observations requires care and training.

(*d*) Once the data are collected, they must be classified, that is they must be organized on the basis of similarities, variations, activities, processes, causes, results and the like. It is also necessary to distinguish between essential and superficial elements.

(*e*) Analysis follows classification. It is now necessary to use reasoning, discernment and known principles to formulate a reasonable generalization, a statement or a *hypothesis* (a working, or provisional explanation) that appears to solve the problem or explain the known facts in the selected field of investigation.

(*f*) This hypothesis must be verified or *tested*. The approach most favoured in this handbook is to form hypotheses as quickly as possible and then organize the later investigations to test (to disprove if you like) the explanations you yourself have just formed. The tests may be of different types; for example: (1) further observations of a more critical nature may be made to further explore the uses and limitations of the hypothesis; (2) different areas may be compared; (3) experiments may be made in which certain variable factors are controlled or (4) predictions of results may be tested for accuracy.

(*g*) Finally, when a hypothesis has been found to offer a satisfactory explanation of the observations, it may be considered as a *theory*, a law, or principle, depending on the degree of certainty attained in the results.

(*h*) The findings are then ordinarily described in a *report*, subjected to criticism and verification by other competent workers in the field, and finally *published*.

It is important to emphasize that although the gathering of data is essential, it is only one of several phases involved in the operation of the scientific method. These data have then to be analysed, interpreted, tested, subjected to criticism and published. The scientific method consequently necessitates: (1) intensive, systematic and persistent thinking; (2) care to avoid misunderstanding, superficiality and bias; and (3) complete loyalty to reality and truth.

2130 Project Working Plans

Preparing a working plan is a good way to get started on a research project, investigation or survey. There is no one way to plan work that will meet the needs of all individuals or departments. The following plan is not presented as an ideal solution, but as an example of a procedure that has enabled investigations to get started on a basis mutually beneficial to all concerned. Modifications can be made according to the special needs of the officer or department concerned.

From an organization's point of view, the scheme of work plans here outlined can serve as a focal point for discussion, review or criticism; a means of increasing understanding and improved communication within an organization; to assist in the training of young research officers and in co-ordinating the work of various individuals within a given project or between projects. Planning is normally essential to sound and efficient expenditure of time.

From an individual's point of view, by preparing a working plan along the lines described below, in a very real sense you are assisting in writing your own job and in contributing to the formation of policy in your organization. Once the detailed plan is approved you can proceed along the lines you have indicated. Working plans: simplify reporting (progress reports, annual reports, etc.); facilitate colleagues answering questions about a project while you are away; help avoid duplication; facilitate several kinds of communication about a project (with other organizations, public enquiries, etc.); result in considerable saving of time for the supervising officers and facilitate the assignment of priorities. They have proved equally useful as a means of closely integrating interdisciplinary teams where synthesized results are required.

Approach to its use

The following is an example of a normal procedure for using the working plans.

1. The administrator assigns a field officer to work on some general or specific problems with which the organization needs help.

6

2. The officer prepares a working plan or a series of such plans, directed towards accomplishing the objectives.
3. The plans are reviewed, discussed with colleagues and approved before the surveys or investigations start.
4. Periodic reviews are made to supervising officers during the course of any project taking 6 months or more to complete. This is perhaps the most flexible part of the procedure. The main consideration is for the officer to have the main parts of the working plan under frequent review.
5. The project or mission is finally completed and the organizational or published report is used as the basis for decision-making, or as a scientific contribution, depending on the nature of the organization and its objectives.

In modifying the above procedure, or in devising new methods of using working plans, the major purpose of having some procedure for their use should be kept clearly in mind: that is, at all times, both administrator, the officer concerned and those assisting or cooperating should know where the planning, management or research efforts are directed and why, and how the work is progressing and when it will be finished. Working plans greatly facilitate administration and supervision.

Giving a form to a ranger or biologist to fill out without specific instruction in my experience has not proved entirely satisfactory. Therefore, under the various headings of the following working plan specific instructions are given indicating how to prepare it. As experience with working plans in an organization develops instructions may be changed or modified for particular purposes. But be warned about compromising to the point where laziness may win. There are certain questions that should always be asked at the very early stages of planning investigations, including research projects. The specific plan presented here is a guide to asking these questions and is organized as follows: title, objectives, value, scope, relation to other projects, methods, personnel involved, equipment needed, costing and schedule.

Title

This should be brief. It should let a reader know, generally, what the project is all about. When starting to prepare the plan, if the title is obvious write it down on a working basis. If there is uncertainty, do not waste time but complete the objectives section first and then write an appropriate title. After each subsequent part of the plan is formulated, the preceding parts of the plan, including the title, should be again reviewed to ensure that it appropriately identifies the project and that the various statements are consistent throughout.

Objectives

List all objectives in terms of what the study, project or demonstration is to accomplish: To study . . . , To learn . . . , To describe . . . , To produce . . . ,

To review . . . , To solve . . . , To contribute to . . . , etc. Normally there is but a single major objective; however, there may be two or three closely related objectives in a single working plan. Check to see if two or more objectives may be consolidated; for major and subsidiary objectives; long-term or interim objectives, etc. If there are as many as five or more objectives one should consider the possibility of making two or more specific working plans, perhaps under the general umbrella of a larger-scale, or more generalized, working plan. The latter should always be a feature of cooperative interdisciplinary team planning. For example, if one wishes to study predation in a national park, it would be appropriate to prepare subordinate working plans dealing with each predator included in the over-all study plan.

In reviewing working plans, the thing to watch in this section is irrelevant statements having nothing to do with objectives.

Value

The ideas listed under this heading are arrived at in a very specific way. First the officer assumes that all the objectives have been accomplished to his complete satisfaction. He has in hand all the results of the specific survey or research project and they are written in an easily understood final form. Assume all this. Second, decide exactly what will be done with the results. How may they contribute to the over-all objectives of a general study, to the management of a special area. Of what use are the results to the head of the unit for which the investigator works, to the Service, the State, etc. After asking these and similar questions, the third step is to write a sentence or paragraph explaining the value.

Clarifying the usefulness of a project may be impossible without some reference to policy matters. This is to be encouraged for projects genuinely consistent with accepted policy should be considered seriously by the Directorate, whether accepted in the next budget period or accepted in principle and deferred to a later date. The greater the value to the organization, the better the chance of it being supported.

The assumption behind this section is: if investigators have clear objectives, they should be equally clear regarding the use that may be made of their results once the objectives are accomplished. This approach does not imply that all research should be applied. The information produced may be used as a stepping stone to further understanding. 'What is to be done with this information once it is obtained?' seems a perfectly legitimate question for any head of department to ask his biologists or other officers. It is a question of special concern to Ministers.

'Value' has proved the most difficult section for officers to state clearly in such a way that the administrator can also see the value of the particular work in question. Frequent phrases inserted here, which must be guarded against, are, for example: 'necessary as a basis for management', or 'to contribute to an understanding of the ecology of the species', or 'it is necessary to have this information for any animal species'. Avoid such clichés.

Be specific. For example: 'Average dressed weights of each sex and age group will be used to calculate the cash value of meat taken from the study area.'

Think ahead. For example: 'An understanding of home range (the study objective) is to be used as a basis for understanding habitat requirements, individual home ranges being used as a basis for comparing quality and quantity of habitat elements within these home ranges with a measure of animal response, using indices to physical condition.'

In reviewing working plans, a comparison between 'objectives' and 'value' as described above should make sense when considered together. If not, one or both sections of the plan should be modified accordingly.

Scope

Under this heading one lists the limitations in time, location or subject-matter that characterize the project. Thus, a particular food-habits study may be restricted to the end of the dry season and to the principal plants eaten by two competing species whose habitat overlaps.

Background (optional)

In some studies it helps to make a brief review of previous work and findings in the same field to give a better basis and perspective to the planned project. This may be difficult in some situations because of inadequate library facilities or because the study envisaged has little precedent that could confidently be used as a model. If this section is included it should be kept short and relevant. Background material can often appropriately be incorporated within the statement of 'value'.

Relation to other projects

If a particular plan is an integral part of a larger-scale, more general, project involving several or many subordinate plans, it is important to understand the relation of the various parts to the whole. This should be indicated in this section. Alternatively, the relationships may be between the project concerned and other national or regional projects, as a follow-up from past work or as a basis for future work. Be specific.

Check for consistency with above sections and modify accordingly.

Methods

This section is the place to indicate how the objectives are to be accomplished. This is normally the largest section of the working plan. Specific proposals for gathering data should be clearly stated, including the techniques to be adopted, modified or invented. Be specific.

A useful attitude for the author of the working plan to adopt is to recognize that his immediate supervisor or a critical directorate may be carefully looking at his 'Methods' section in the following way:

> In reviewing working plans this section telling 'how' should be carefully checked against the objectives. At this point it is useful to ask other questions such as: 'Is this the only way of accomplishing the objectives?' 'Is it the cheapest way?', 'Is it a method that can be used by the type of person who will continue with such work?', 'Is it the quickest way?', 'Is this method going to produce much irrelevant information in terms of achieving the specific objectives?', 'Can the methods suggested be modified to make them more appropriate for the specific objectives of this project?'

There are occasions when the most suitable methods are unknown during the planning stage. In such cases the investigator simply describes the various new, modified or old methods he hopes to try for a short (stated) period, to see which he will eventually adopt to accomplish the project objectives. When the methods are adopted, revise the working plan accordingly.

It is important to remember that the working plan is by no means unalterable. It is simply a written record, kept up to date, and which, for a certain project, at a certain date, records what the officer is trying to accomplish and why and how he is going about it. Obviously an investigator should be able to satisfy his department's questions on his activities for every project on which he is engaged. Working plans facilitate this.

Check for consistency with preceding sections and modify any part of the plan accordingly.

Personnel involved

If more than one person is involved in any stage of the project, the proposed part each person is to play should be made clear even early in the planning phase. It is essential that everyone participating knows exactly what they are doing and why and how it links in with the efforts of their colleagues. This is particularly important in an interdisciplinary project, where several specific project plans will contribute to an overall 'umbrella' plan. For example, A may be responsible for preparing, integrating, and finalizing Working Plan No. 6, assisted in specifically stated ways by B and C, while concurrently B may be responsible for Working Plan No. 7, assisted in this and that way by A and C. Be specific! Include cooperation with officers from other organizations. It is essential that everyone participating knows exactly what they are doing and why and how it relates to the work of others. This cannot be over-emphasized.

Check for consistency and relevance with above sections, particularly 'Methods'.

Equipment or services needed

Equipment needed on the particular project should be indicated along with such services as estimated hours of flying time, travel costs, etc.

Check for consistency and relevance with above sections, particularly 'Methods'.

Costing

The cost of the project must be anticipated by someone. This section is not elaborated here because it so very much depends on administrative procedures within a particular organization. However, even if the biologist may not make the final evaluation of the costs involved, it is useful to include this item in the working plan. In terms of a two-way communication between field man and administrator this is one way of making the officer aware that the costing of a project is an integral part of the planning exercise and of course one of the many realistic considerations on which priorities must be based. Quite apart from being essential for administration, it is good training for junior officers.

Check for consistency with above sections, particularly 'Equipment or services needed' and 'Methods'.

Note that restricted funds may force reconsideration of several previous sections of the working plan.

Schedule

This section of the plan indicates how long the project and its various phases will take to complete. Items listed under 'Schedule' will naturally vary depending on the character and circumstances of the project, except for one generalization. Always there is a start, work done along the way and a conclusion. The project is not complete unless the demonstration is made, the final report is submitted or published, or whatever else is consistent with realizing the objectives. Each phase of the work should be included in the schedule and a prediction made as to when it will be finished.

Sometimes the schedule is indicated diagrammatically, as with months aligned in vertical columns and the tasks to be completed indicated on horizontal lines. But the form of the schedule is entirely a matter of taste and custom within each organization.

Reports

Periodic reports on progress are required by many organizations. Such reports are of value both to the investigator and administrator as they allow everyone to keep up to date on the extent to which the objectives are being accomplished. Interim reports should be scheduled. They facilitate the preparation of later reports.

Revision

Working plans are working guidelines. They are, as their name implies, made to be used and they can be most useful if they are periodically consulted to keep the objectives clearly in mind and to review the extent to which the methods used are accomplishing their objectives. If a weakness is discovered, the plan should be revised. A normal practice is to schedule an early revision, for example, 2 months after work starts, and thereafter to modify as required.

Likewise such plans provide a concrete basis for discussion at periodic, scheduled revisions of progress by participants in an interdisciplinary project. For example, a modification in the schedule of one working plan may automatically mean consequent modifications in schedules of several other closely linked plans.

2140   The 'Team' Approach

As the body of knowledge in each field of science builds up, it becomes impossible for any one man to be aware of all the latest scientific developments. Teamwork between scientists is thus imperative. In a field as complex as wildlife management—involving wild and domestic plants and animals, soils, agriculture, forestry and people—investigations usually require the close cooperation of specialists in different disciplines studying various aspects of the environment and of man's impact on his environment. Often it is not enough to study the biology of an area; the economic and sociological aspects must also be considered. If specialists in the several fields work together in a team they will normally get better results (their results will reflect a better perspective) in solving a complex problem than if they work independently.

In developing countries where wildlife specialists are scarce and face many problems there is an especial need for teamwork, sometimes involving the close cooperation of the entire staff of a wildlife organisation, from the game guards to biologists and administrators. The field men can contribute considerably to the accumulation of essential data from their daily observations. It must be pointed out, however, that field men normally have regular jobs to do other than research; they should show genuine interest in contributing to the project before being asked to participate in an investigation. Otherwise, the information they gather may have to be thrown away because simple rules have not been met (certain measurements dropped, methods changed, schedules missed). It is normally the responsibility of the investigator to develop and maintain a genuine interest in field colleagues by briefing them at regular intervals in respect to the usefulness of information they are asked to provide, and to encourage them in specific 'observational' tasks. If field men really feel their work is valued they may be encouraged to gather food specimens, skulls, parasites, pathological organs or records of predator kills. They should be shown how to keep records of what they see and collect. This practice in accumulating recorded information and material will assist in the general management and maintenance of the wildlife areas once the study is completed and may provide a useful basis for future studies. Thus, the fullest possible use should be made of any volunteer field assistance available.

It is difficult to advise on ways and means of forming teams and integrating work on a team basis. The teams themselves can vary greatly in character. A biologist can form a team of technicians or volunteers cooperating in gathering data or two or more persons may voluntarily split responsibilities in a joint investigation. The most difficult type of team to manage involves large-scale surveys or problem analysis where it is obvious from the start that several disciplines must combine their efforts in making the study or analysis a success.

Placing projects on a work-plan basis (see preceding section) has proved a useful basis for integrated teamwork in several large-scale assistance projects mounted in developing countries. For such projects particular emphasis should be placed on planning the gathering of data from one discipline, e.g. the wildlife specialist, in such a way that it can be used as acceptable evidence by an officer representing another discipline, a hydrologist, agronomist or economist.

The fact that several persons are working on different parts of the same general problem at the same time does not mean they are part of an integrated team. Integration does not happen by accident. It must be carefully planned, and the logical place to start is in the earliest stages of project formulation. Thus, to refer again to section 2130, integrated teams must give extra attention to two parts of the working plans: (1) concerning the personnel involved, it is essential to make absolutely clear who is responsible for carrying out which action; (2) in listing 'values', include specific examples of how the information produced in one work plan can be used to further the objectives of another work plan. Precise planning for better integration and coordination sets the stage for closer collaboration and mutual assistance during the course of the project. It is not uncommon, under these circumstances, when the project draws to a close and final reports are being written, for the officers concerned to find difficulty in producing separate reports and to submit jointly authored contributions. This seems a trend worth encouraging.

2200   SYSTEMATIC RECORDING OF OBSERVATIONS

Field observations should never be entrusted to memory, and written notes should be taken in a systematic way. This section contains hints on note-taking, the design and use of forms and of checklists and reminder lists. Systematic recording of data must be emphasized, especially in investigations that involve two or more persons working in cooperation. Nearly every field worker has his own system for recording observations. The purpose of this section is not to force a different system on the reader but to describe some of the commonest systems and encourage adoption of those elements which appear appropriate.

When a biologist first starts producing research results it is usually surprising to him how much time is taken with converting his observations into an acceptable report or publication. Quite normally at least as much time is spent in preparing the report as in doing the field work required. This question has been discussed with senior biologists some of whom claim that the planning and writing-up aspects of their work take at least two-thirds of the total time spent on a project.

There is much room for improvement in this ratio and this section is intended to stimulate you to consider how you can improve the effectiveness of your own efforts in the field.

It is stressed that when two or more observers pool their observations in a study, the observations are of value only if the same criteria are used and in exactly the same way.

2210   Notebooks

There are two main types of notebooks: looseleaf and bound. Data cards and punch cards are considered modifications of the looseleaf notebook. No full discussion of notebooks is attempted here but emphasis is given on the need to use some form of notebook and on the use of field forms.

2220   Note-taking

Observing and recording go hand in hand. It can be recognized as the most basic of all techniques available to those wishing to understand more of wildlife and its environment and is prerequisite to accomplishing the objectives of a study. Attention should be given to reducing to a minimum the time between the data-gathering and the reporting stage. This implies thinking ahead towards classification of observations by the use of tables, forms or maps. Some field workers find that writing an early version of a report (or progress report) helps them to develop effective systems for classification of observations and points to gaps in knowledge.

If a looseleaf system is used, it facilitates classifying information. Separate series of pages may be devoted to each species and, when relevant, to each type of observation dealt with for a given species. It is also handy to have separate sets of pages for simple journal notes that include an itinerary, weather conditions, hours of observations and the like. It is best to write on only one side of each page; it will save you time in the long run.

When taking notes on behaviour of animals without any definite purpose in mind except to 'see what can be learned', include the date, time of day, nature of the immediate environment, specific behaviour, voice, reactions to danger, feeding habits, mating actions and so on. Exact distances and counts may prove to be important. Do not be afraid to duplicate observations and to repeat again and again for this may be important—especially when there is doubt as to the significance of an observation.

Keep up to date with notes

Each field trip should lead to an additional series of notes that should be completed at once. Do not let notes accumulate to write up tomorrow or later. The practice of copying notes from one bit of paper onto another is one of the worst time-killers and should be avoided. Each time a copy is made, additional errors have a chance to creep in. Original notes are a form of evidence and should be so organized that they can be readily relocated.

Use diagrams and sketches when these help to clarify an observation. Preferred style of writing may vary from a shorthand, telegraphic style to complete sentences or paragraphs written as closely as possible to the final report form. The latter approach is more consistent with the need to cut short the time elapsing between field records and final report.

A tape recorder can be a useful tool for recording field observations where otherwise two persons would be required; for example, when conducting an aerial survey or when driving a car. The tape recorder is invaluable for making a record of a trip during which one is not at liberty to stop for note-taking. But the material registered on a tape recorder should be transferred to a map or notebook or form as soon as possible.

2230   The Use of Forms

In many studies completed forms make up most of the original observational records. When the same observations have to be made over and over again, it is profitable to develop a form on which such observations are allowed for. The forms may fit in the notebook of the looseleaf type, or onto a clipboard, and are used largely to supplement written observations.

Forms are sometimes deliberately used as an intermediate stage in thinking, that is to provide evidence on which preliminary decisions may be made. For example, rough forms may be designed and used to test the suitability of two or more competing techniques in terms of contributing most relevantly to the objectives of your study.

2231   *The Design of Forms*

Some general guidelines to consider in developing and using forms are listed below:

1. Whenever possible, the forms should be the same size as the notebook, and consequently the notebook itself should not be too small. It is handy to have an inexpensive duplicator (such as a gelatine or a spirit duplicator) so that, once the form has been tried in the field, mistakes can be corrected, and the form altered accordingly and reduplicated.
2. It helps to arrange headings from left to right in the same order in which observations are made. This not only saves time but reduces error.

3. The use of different symbols in the same column should be avoided. For example, instead of one column for sex, use two columns, one for each sex. This saves time in classification and summarising.
4. Do not hesitate to include 'unknown' or '?' columns. Forms should facilitate the recording of honest observations. Thus, in the previous example if an animal was observed but its sex unknown it should be so recorded. Trials of forms in the field will quickly reveal practical ways of modifying a form for a particular purpose.
5. It is obvious that everyone using the form should clearly understand the precise details of observing and recording, otherwise the results of different observers at different times, or in different areas, will be impossible to compare.
6. Try to visualize and anticipate the final use of the information gathered. The closer your field form can correspond with the format of a table or other means of presenting data in your final report the more time will be saved when converting the original observations into a meaningful integral part of your final report or publication.

A number of different types of forms with suggestions for their use will be found in this Handbook (see, for example: Physical condition field form, Section 3864).

2240   Checklists

This is a list to remind you what kinds of information are needed in a particular study, in a particular area. It is a useful device for beginners as a means of directing their observations and it is useful for a biologist to use where several observers are working independently in the same area. The lists can be short and simple or more complex where longer-term studies are involved. They can simply consist of key observations to be made during a day, a season, a study, points to check in comparing two areas, observations to make on finding a dead animal, etc. When individuals of a team are observing independently, the joint preparation of checklists of observations required under certain study situations helped speed the accumulation of relevant useful information. (Examples: Species checklist, Section 3410; Checklist of vegetation types, Section 3510.)

2250   Use of Photography

Photography is another important tool for recording field observations. Among their many uses, good photographs will best depict the general description of an area, changes that have occurred following management practices, or examples of animal damage. They often stimulate the memory to recall many more relevant details than those recorded in the photograph. Photographs may be used to show animal specimens, their tracks, their food.

They may be used to study the behaviour of animals, the structure of herds, or to count the number of animals in large herds observed from an aircraft. They may also help to describe a method or an instrument. Videotape motion pictures of animal behaviour when played back again and again provide opportunity for repeated observations of the same incident and well may become indispensable for certain types of observations on animal behaviour. This method provides a preciseness of observation otherwise impossible. Both still and motion pictures can also be of considerable help in public relations work, such as illustrations for newspapers, or magazine articles or for public talks. In fact photography is indispensable for most of the activities mentioned above and many wildlife workers can hardly go in the field without a camera.

Photography can also become a real hindrance to research, an exciting tool for going off on such a tangent that pictures can become ends in themselves. This attitude is fine for professional photographers but the biologist must control the time spent on photography, keeping it within the requirements of his job and the scope of the particular project in hand.

My own experience has shown that photography can be a useful tool, not taking too much time away from the basic field observations but with the following reservations.

1. Either black-and-white or colour photographs can be taken. To attempt both seriously interferes with observations. Where both are required the use of colour negative film should be considered.
2. In general filming motion pictures should never be attempted unless you can give it your full attention.

The use of photographs as a data-gathering tool, as distinct from a means of illustrating a point, is dealt with in later sections of this Handbook.

The types of cameras and accessories are so great that it precludes their discussion here. The choice is mostly based on personal preference.

Before selecting a camera you may be able to save a great deal of money if you answer these questions. Precisely what part will photography play or be likely to play in your work? What will you do with the photographs once you have them? What is the maximum size enlargement you will need for publication? The kind of printing process normally used in scientific journals and most magazines and books does not reproduce the original photographic print in enough detail to show the difference between very expensive and more modestly priced cameras.

2300   METHODS FOR OBTAINING HISTORICAL RECORDS

When performing any kind of investigation in a given area, the wildlife worker should recognize that some historical knowledge is often essential for the proper interpretation of present conditions. In order to understand the complicated relationships between animals and their environment, the research worker should develop a good knowledge of the history of human activities in his study area, and of changes in habitats and in animal populations which may have occurred over a period of years prior to the study. Clearly existing records or early journals available should be consulted. However, knowledge of even the recent past is often difficult in developing countries because historical information is largely lacking in published form and few accurate records exist.

Sometimes the changes may not be man-induced. Volcanic activity, earthquakes, floods, changes in river courses, for example, can profoundly change an environment and influence the succession of changes in both plant and animal communities for many years to come.

Remember that the environment you see today is rarely the same as it was even 100 years ago. An understanding of past changes and their causes can be extremely important to the management decisions of today.

A technique used with success in Africa involves interviewing old men who have lived in a study area all or most of their lives. They provide an excellent source of information and a basis for cross-checking other historical records, or findings obtained by various field techniques. The objective of the interview is to learn something of the extent to which habitats for animals have been radically altered or large changes in animal population have occurred within a man's lifetime. The general minimum information to obtain is: estimated age of the man interviewed, time lived in the area, his occupation in his early and middle years, where his father lived (if his father lived in the same area, further questions may reveal information the father may have passed on to his son), changes in frequency of occurrence of water and its distribution, changes in well levels and river flow (especially in dry seasons), changes in vegetation (especially in proportions of trees, shrubs and grasses), pattern of stability or shifting of villages, changes in animal populations, occurrence of epidemic diseases in the area (both human and animal), die-offs of domestic or wild animals or humans (especially during bad drought years or years of crop failure).

Further questions may reveal the first dates of European settlement in the area or the early types of activities the settlers pursued. It is also a useful way to learn the former extent of forest clearing, frequency and pattern of burning, changes in numbers of domestic animals grazed, history of fencing, dates of construction of dams and boreholes, and the changes in density of wild animals associated with or following these events.

Experience has shown that the five most important techniques of questioning are as follows:

1. Ask only one question at a time. Questions regarding changes in animal numbers should be asked on a species-by-species basis. It is often useful to ask the same question in different ways but the basic question should be

the same and the only one considered until the answer given by the old person is understood.

2. Enough detailed questions are asked about the early youth of the person to get him truly talking about 'old times'.
3. Many answers have to be cross-checked with further questions, not immediately, but after a few minutes have elapsed and several other topics have been discussed.
4. Only one person is questioned at a time, even though this may take several hours.
5. Questions requiring 'yes' or 'no' answers should be avoided, as well as any other leading questions. It is better to ask the old person to recall the differences, now as compared with the days when he was a child.

Take your time. Remember that you may be talking with the only source of information that extends back over the past 70 years. Let these people help you at their own pace.

Occasionally, one encounters a person who readily answers according to what he imagines the questioner wants to hear. This is easily spotted either by the character of the answers, or by the contradictions that are expressed when cross-check questions are introduced. Such an interview should be terminated as tactfully as possible and disregarded. The information gathered in one interview should of course be used as a basis for cross-checking the information given by other old persons interviewed in the same area.

It will be readily appreciated that historical factors can have an extremely important bearing on the way in which research findings are interpreted. They can provide answers that yield great contributions to long-term objectives of the conservation of wildlife. Only a little time and effort spent in such interviews should convince you of the importance of historical records in your own area for your own purposes. You may have opportunity to interest your local conservation society, government departments, museums or other organizations in publishing and recording historical records. As a minimum requirement such records should be placed in the libraries of museums, game departments or historical societies so that future researchers may have access to them.

2400   OUTLINE FOR NATURAL HISTORY STUDIES

The following outline (modified from Taylor, W. P., 1948) may be useful for field workers wishing to study the life history of an animal species. It lists items descriptive of different aspects of the biology of a species. These items have unequal importance, some may apply to certain species, others may best be omitted. The outline may help point out gaps in knowledge. It can be used as a guideline, or modified to form checklists, to be adapted to specific cases and local circumstances.

1. Name of species and subspecies: scientific name, common name, vernacular or local names.
2. Description: weight, measurements, colour, sex and age variations in appearance.
3. General habits: voice or other means of communications (calling, singing, roaring, warning attitudes, glandular secretions, changes in colourations), hierarchy in groups, degree of sociability, types of shelter (nest, form, den, burrow), extent of home range, type of locomotion (running, jumping, flying, digging), speed and endurance, time of activity (diurnal, nocturnal, crepuscular).
4. Signs left by species: tracks (size, distance between footfalls), faeces (abundance, shape, size, colour, place of deposit), claw marks, wallows, beds, dust baths, runways, trails.
5. Distribution: locally and elsewhere.
6. Habitat relations: distribution in relation to soil, available water, vegetation types, other animals, reaction to climatic cycles or to unusual natural catastrophes (storms, floods, fires).
7. Food habitats: foods eaten, seasonal variations in food habitats, physical adaptations associated with food-getting, storage of food, dependence on water, frequency of drinking, other habits associated with feeding or drinking.
8. Breeding habits: dates of heat and copulation, length of gestation period, date of birth or young, number of young, courting behaviour, relation of the sexes (monogamy, polygamy, promiscuity), relations between parents and young (care of young, mode of feeding, weaning, mode of carrying).
9. Movements or migration, local or general migration, dates of movements, extent of direction of movements, relations to climate, or food supply, or breeding period.
10. Relation with other species: friends and foes, prey (modes of capture), competing species, parasites (external and internal), diseases (transmission, periodicity, rapidity or recovery), mutually advantageous relations.
11. Present and former status: past and present numbers, causes of increase and decrease, estimates and counts of animals per unit area, sex and age composition of groups, herds, or population, fluctuation in numbers from year to year.
12. Management: status of protection by law, trapping and hunting methods in local use, number harvested, price received for pelts or meat, relation to agriculture, cattle, public health, possibility of utilization as source of meat, or fur, or hide, or other products, needs for protection.

Although this list is by no means complete, traditionally these are the kinds of questions asked about animal populations by those who have studied mammals in their natural surroundings. On reflection it should be obvious that knowing the answer to such questions can be of unestimable value to wildlife managers. However, in terms of management based on an under-standing of cause–effect interrelationships between a population of animals and its surrounding habitat, some kinds of facts are generally much more valuable than others. Several such questions seem important enough to emphasize by elaborating in the chapter on data-gathering techniques (Sections 3000 to 3900 inclusive).

Chapter 3000

Data-gathering techniques

3100 INTRODUCTION

Most of the techniques described below are easy to apply but a given study cannot use them all. Limitations of time and expense alone will rule this out. The game then becomes one of selecting, modifying or inventing appropriate methods to accomplish your own or your department's research or management purpose.

It is hoped that it will become increasingly clear that there is no one way to achieve the objectives of your work plans. There are usually several acceptable ways and there will be even more ways in the future. What will not change, however, is the requirement to select only methods relevant to achieving the objectives in question.

Because of the large numbers of techniques available it is useful to present a plan for classifying them. The following plan stresses the constant interplay between animals and environment within and between three major levels of description possible in the area (ecosystem) under investigation. The purpose of this approach is to facilitate the selection of, or invention of, appropriate methods to achieve specific objectives of survey, research or management. The section emphasizes the relevancy of techniques for particular purposes and their inadequacy for other purposes, for every technique has limitations

21

22

as well as uses. It makes no pretence at being complete. Thus, investigational techniques may be grouped as follows:

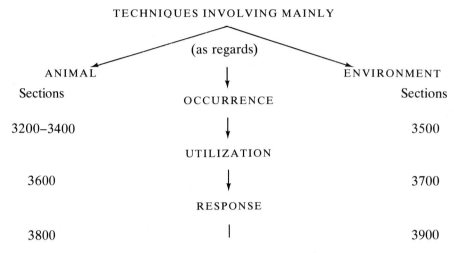

TECHNIQUES INVOLVING MAINLY

(as regards)

ANIMAL	OCCURRENCE	ENVIRONMENT
Sections		Sections
3200–3400		3500
	UTILIZATION	
3600		3700
	RESPONSE	
3800		3900

In this scheme the techniques dealing with animals, and the evidence taken from animals, are grouped separately from those related to the environment. The techniques are further divided into three classes: those dealing with occurrence, with utilization, and with response, a sequence that is related to the entire ecosystem. These three levels of description may be described as follows:

(a) Occurence

This is usually the simplest level of description. For animals this may involve listing the species present, determining areas where they are more or less numerous and/or obtaining some estimate of, or index to, their numbers. For the environment this may mean listing, locating and describing the available habitats.

(b) Utilization

This covers techniques that describe the use of the environment by animals (e.g. food habits studies), and the measurement of this utilization on the vegetation itself (e.g. on shrubs or grasses).

(c) Response

This is the most complex level of description. On the animal side techniques may, for example, measure animal response to habitat occurrence and utilization and be reflected in and assessed in terms of productivity, mortality, abundance and population trends, or in physical condition. On the other hand examples of environmental response may include measuring or assessing trends in the composition or in the quantity of vegetation, in changes in available habitats, or in demonstrating accelerated soil erosion.

Published studies that include evidence from all parts of the areas under study are the exception. Many studies deal entirely with the most simple level of description, the occurrence level. For example, a study of the distribution of a species on even the major types of habitat presently occurring, such as water, grasslands, forests, and so on, coupled with the distribution and density of animals throughout the year may in itself contribute significantly toward the understanding of habitat requirements for some animal species. Even when an investigation requires extracting from an area information from more than one level it is still possible to simplify by concentrating on relevancy in terms of objectives. For example, one may wish to reveal the extent that animals may be responsible for downgrading an environment. In this case obtaining a perspective of the extent and location of downgrading vegetation would be appropriate (an environmental response). At the same time it would be useful to map the distribution and comparative density of the various species of wild and domestic animals present (occurrence) as well as any changes in numbers or management practice over the past few years (historical evidence). If the information were done on an objective basis this would allow close comparisons and even statistical correlations to be made if required. In such a case, food habit studies (animal utilization), plant utilization studies, sex and age classifications (animal response), and studies of the physical condition of animals (also animal response) can be expected to provide additional supporting or cross-checking evidence. However, such supplementary studies may be unnecessary for the adequate solution of a research or management problem.

Thus, to summarize the approach suggested here: start simply and become more complex and sophisticated only insofar as is required to achieve your research or management objective.

3200 OCCURRENCE—ANIMAL . . . LIVING SPACE AND MOVEMENT BEHAVIOUR

3210 Introduction

One obvious characteristic of large mammals is that they move about. When first starting research on large mammals it is natural to be uncertain regarding what is really happening in a population. Individuals are often difficult to see and most wild species move about unrestricted by fenced enclosures.

In considering the occurrence of large mammals in an ecosystem we should recognize not only the way in which our study species moves about within its environment, but also its reasons for doing so. The following sections discuss basic living areas and several important kinds of movements, and suggest ways of learning more details of these for particular purposes. Chapter 5000 continues from this base in discussing habitat requirements.

3220 Basic Living Areas

3221 *Home Range*

Fortunately, for most kinds of large mammal, individual adult animals are normally more or less restricted to a definable area during their adult life.

This basic living area for an individual (or sometimes a group) is known as its *home range*. The term may be qualified, as by referring to summer or winter ranges; dry or wet season ranges, or to yearly or lifetime home ranges. But, regardless of how qualified, home range always refers to a more or less limited area within which an animal moves in conducting its normal activities.

An interesting characteristic of adult animals is that when captured and released outside their home ranges, they commonly find their way back to the same area and often to precisely the same home range. Although species vary, individuals are often so strongly attached to their home ranges they not only do not leave them during their adult life, they cannot be driven out of these areas except with the greatest of difficulty. When home ranges become completely inadequate and unsuitable individuals of many species die inside their familiar haunts rather than move to other areas.

I have seen mule deer of known ranges die of starvation on the side of a small stream, but 1–2 ft (0.3–0.6 m) deep and 20 ft (6 m) across, while on the opposite bank, but outside the home range, was an abundance of available browse. In Senegal, populations of kob, which depend on grassland, remained fixed to their home area even as the grassland on which they depended disappeared with the invasion of quick-growing woodland trees. In one such area, in Niokolo-koba National Park, kob in emaciated condition were living in an almost closed canopy woodland whose largest trees were 8–10 years old. Less than a kilometre away was another group in fair to good condition in an area where the invading trees had not yet reached and where grassland still remained.

3221A *Distinguishing between 'Home Range' and Closely Related Terms*

Home range is here used to mean the basic restricted area within which normal activities take place. *Territory* is applied only to those areas defended. Territories may be seasonal and thus smaller than the yearly home range. On the other hand they may be exactly the same area as the home range, in which case the only distinction is the fact that the 'territorial' area is defended. For many species areas of noticeably heavy use are found within the home ranges. Often home ranges of individuals overlap in these areas of heavier use as, for example, in seasonal feeding areas. When such areas of concentrated use occur they are referred to as *key areas* or *key utilization areas*. The existence of such areas might indicate some important *habitat element* within the home range. A description of the specific uses to which the various habitat elements are put is one way of understanding a species' *ecological niche* (its functional role). This is further discussed in Chapter 5000 (see especially Figure 120).

As elaborated in Chapter 5000 individual or group home ranges may be a result of the occurrence and interspersion of various essential elements of the habitat, or of various social factors, or of a combination of these. Knowledge of home ranges of large mammals is too scarce to allow much confidence in drawing broad generalizations. However, even now a few tendencies seem worth mentioning. Territorial behaviour (defence of all or part of a home range) seems more consistently observed with large predators than with other large mammals. A species living in areas of dissected topography seems to have more clearly defined and more consistently identified home range boundaries than the same species when found on more open, level, plains-type country. For example, mule deer in mountainous areas, in migrating from higher summer down to lower winter ranges, occupy exactly the same winter range of the previous year (Leopold, A. S. *et al.*, 1951). The same species, however, when migrating across a gently sloping plain, may have individuals returning to the same general area but with an individual's wintering area being several kilometres removed from its winter range of the previous season.

3221B *Importance to Management*

In several ways the home range concept is important to management:

1. It exemplifies the extent to which individuals within a population are localized: to what extent can one, for management purposes, be justified in considering different watersheds or parts of watersheds as holding essentially separate populations?
2. Knowledge of home ranges can contribute to defining a particular pattern of occupancy of a population in an environment.
3. By definition, all the essential requirements for an animal's existence are found within the boundaries of its home range. It thus provides the most precise unit for study of animal–habitat interactions and habitat requirements. For those species living in large groups or herds, the group-range will be the relevant unit of study (see also Chapter 5000).
4. Home ranges provide a basis for classifying those kinds of movements most important to management.
5. A mosaic of both home ranges and/or, for certain species, the defended ranges (territories) serves as a means of spacing out individuals of a population and thus contributes to maintaining stability of numbers.

3221C *How Home Range is Expressed*

Home ranges are usually defined only after repeated observations of recognized individuals have been recorded. It is interesting to review several methods of expressing home range, for their usefulness varies.

In the first authoritative book written on Game management, A. Leopold (1937) does not use the term home range but describes the individual's living space as the 'radius of mobility' and expresses the radius as average and maximum daily and yearly distances measured from the place of marking or banding. This means of expressing home range was still commonly in use 20 years later.

Although these early workers obviously visualized home range as an area, they actually expressed it as a linear distance. This proved of little use other than to indicate something of the extent to which the population was localized. However, under exceptional circumstances linear distances come close to being an accurate description of home range. For example, a particular band of baboons studied by J. H. Crook (1966) simply moved back and forth along a line of cliffs.

A logical next stage in presenting 'home range' was to use the average distance an animal was observed from the capture trap as the radius of its range, or the distance between the extreme observations as its diameter. Home ranges considered in this way were thought of as circular and on this assumption the size of the home range was often expressed in terms of acres, hectares, miles or kilometres (e.g. Leopold, A. S. *et al.*, 1951). This sounded a bit better, and in the 1940s and 1950s led to some fanciful calculations of carrying capacity (total study area divided by calculated average area of home range). Because of the assumption of circularity this was only a small advance over 'home range' when expressed as a line. However, in spite of its rarity, circular or nearly circular home ranges have been demonstrated, as, for example, for a lizard on a flat desert (Riney, T., 1953); and circular territories were descibed for a medium-sized ungulate, the Uganda kob, by H. K. Buechner (1961).

Most of the early work on home range was done with small mammals and often, in addition to expressions of distance or area, maps were shown which located traps and subsequent recaptures. This information was of course most useful and some of the maps interpolated the shapes of the ranges as well (e.g. Stickel, L. F., 1946). The shape of deer ranges was considered important by Riney (1950) as a basis for comparing the occurrence and spacing of habitat elements within different individual home ranges. As elaborated in Chapter 5000, the shapes of the individual ranges can be used as one basis for understanding habitat requirements.

Once the shapes of a few individual or group home ranges are known the stage is set for gathering many additional kinds of information. Most importantly, within a home range of known shape will be found areas of greater or lesser use. Elaborating more precisely the seasons and kinds of use made by an individual within its home range will lead to a definition of its ecological niche (see Figure 120) and 'key areas'. For some species the home range boundaries will include not only areas of special use, connected by trails, but the tradition will be so fixed that potentially usable areas will occur inside the range boundaries but remain entirely unused by the individual concerned.

In short, the most precise descriptions of home range will not only accurately plot the boundaries of the home range on a map but indicate areas of higher and lower use within the home range as well as the relation of animal use to specific habitat elements. If relevant, the trail system connecting these areas may also be plotted.

3222 *Territory and Territoriality*

Territory has been defined above as a defended area. Intruders are driven out. When the entire home range is defended, territory and home range are obviously the same. Territories were first recognized in bird populations during the breeding season. Likewise in large mammals territorial behaviour may be seen but usually only during a limited season, often the breeding season.

F. A. Pitelka (1959) feels that 'territory should be defined as an exclusive area, not merely a "defended" one' and this seems a point well taken in terms of recognizing the functional role of these areas (see Figure 1d).

Territoriality is apt to have a somewhat different meaning for every species studied. In recent years several fine studies of large predators have been made. With lions on African plains, territorial behaviour is so easily observed that some more subtle aspects of territoriality are revealed. For example, R. Schenkel (1966) in observing lions in a Kenya national park, concluded that only a central area within the lions' home range was actually defended. The degree of aggressiveness was influenced by: cubs belonging to the pride being in the vicinity; the numbers of adult members of the pride taking part in the aggression, and the comparative inferiority of the intruder in fighting. He concluded that, for his lions, territoriality is a dynamic system: 'not a stable independent factor, but the fluctuating result of a multitude of factors'.

In this same study Schenkel observed that lions distinguished their territories by combining several actions:

1. roaring;
2. scent marking with urine, (a) on bushes and (b) by 'squirting downwards while squatting in a crouched position and rubbing the feet in it, a behaviour pattern which results in scent marking the track'; and
3. 'the open, unconcerned manner of moving in their own territory and the "proud" posture in looking around.'

The term 'territorial' is often used carelessly in situations sometimes related to but not necessarily indicative of territoriality. Marking of home range or territory boundaries or trails with urine, faeces or scent glands may be fine for predators; but for large browsing and grazing animals such markings do not always function in maintaining an exclusive area.

The safest procedure would seem to be to stick to home range as the basic areal unit; to use 'territory' to indicate that home range or part of a home range which is defended. Aggressive behaviour not easily referable to territorial defence should simply be described in whatever terms seems appropriate.

I have observed that while several family groups of mule deer may be feeding in the same communal clearing, adult female family group leaders will not tolerate deer outside their group closer than a certain distance. If one had 'territoriality' on the brain one might describe this as an invisible, mobile, mini-territory that surrounded the adult as it moved, 'territoriality' being exhibited when strangers approach the edge of this mobile 'territory'. On the other hand one just might as easily describe the sensitive area surrounding an adult as a zone of intolerance or a threshold of tolerance and with no territorial implications whatever. At present this latter course is safer. In time and with increasing knowledge we may discover several other principles affecting movement behaviour of which we are yet unaware. Arbitrarily classing every observation into only those pigeonholes of which we know the names can only delay the discovery of new principles. When in doubt, describe.

3223 *Methods of Determining Territory and Home Range*

1. Plot on a map the observed location of naturally or artificially marked individuals. To define the yearly home range, observations should be made throughout the year. To learn the daily activity pattern within the home range, observations must be made at different times of the day. For small mammals home ranges were defined by capturing and marking individuals then plotting the sites of recapture on a map. This is normally impractical for large mammals and unnecessary when large animals can be visually located and recognized in the field.
2. Map the tracks and trails followed by individuals or groups of animals. Fresh tracks can be followed using a trained dog, and in certain conditions (e.g. in snow or mud) animal tracks can be followed for considerable distances. Following tracks and trails can be particularly valuable in defining seasonal shifts, or migrations and, for some species in fairly open habitats, this can be done by aeroplane.
3. As mentioned above, it is difficult or impossible to drive most woodland or forest animals from their home ranges. If the animal doubles back at a certain point this is worth noting and recording as a possible edge of its home range. Subsequent similar observations on other occasions and in other parts of the home range will further contribute to defining home range boundaries. Another way of understanding home range and pattern of occupancy is possible for those species whose individuals share feeding areas at certain times of the day and then return to their individual resting or bedding areas. When such groups are disturbed on a communal feeding area, as by a man or a vehicle, the individuals tend to run away in different directions. Some species tend quickly to reach the trail by which they entered the clearing and run rapidly back along that trail although it may result in some individuals running back towards the disturbing vehicle. In this respect warthog, zebra and several large ungulates in Africa behave

exactly as mule deer in the Western United States, or as several species of wallaby in Australia.

4. Radio tracking can be employed, as discussed later. Once a transmitter has been attached to an individual, subsequent radio locations may be plotted on a map in the normal way to assist in defining an animal's home range. Radio tracking is particularly useful for wary animals, nocturnal animals or those which live in situations that make observation difficult. Since relocation may be done from the air, this method has been successfully used in remote situations, on animals involved in lengthy migrations and on individuals within very large herds. The main disadvantage is the expense, the time involved and the expertise required for its use. For further details see Section 3313.

3230 Patterns of Occupancy

Although the boundaries of individual home ranges form the basic aerial units of a population, for management purposes equally important is a population's *pattern of occupancy*. While a knowledge of a few home ranges may contribute to describing these patterns, they can also be learned in other ways. Each species has its own range of patterns. For some species the range is so narrow there is but one pattern. For other species the possible patterns of occupancy are many. For a given species such patterns can vary with different environmental conditions (see Figure 1b).

When one considers groups of animals, as, for example, different family groupings or segregation by sex or age, then the possible patterns of occupancy seem practically unlimited, for within a given study or management area each species will have its own unique pattern of occupancy. A few examples (Figure 1) will illustrate something of the kinds of pattern one may expect.

Figure 1 (c) shows diagrammatically the relationship between billy and nanny-kid herds in an area occupied by feral goats (after Riney, T. and Caughley, G., 1959). Young males disperse from three nanny-kid herds (A, B and C) to a common billy herd (ABC). Shaded areas around nanny-kid herds represent the ABC billy herd's summer range, which includes and surrounds the yearly ranges of the three nanny-kid herds.

Figure 1 (d) indicates a situation where home range and territory are synonymous, as in certain territorial species of deer. Here little overlap occurs between territories.

Figure 1 (e-1) shows groups of female wildebeest moving from one male territory to another in a low-density population (Riney, T., unpublished manuscript). Figure 1 (e-2) shows tightly packed herds of female wildebeest and young with adult bulls circling the herd. Dotted lines indicate the defended territories. This pattern occurred in an area of very high wildebeest density (Talbot, L. M. and Talbot, M. H., 1963). An extra feature of this area (e-2) was that, especially when female herds were larger than 50, two or three

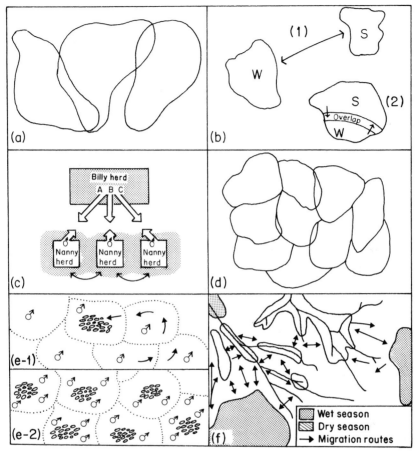

Figure 1 (a) An example of individual home ranges, including a smaller more restricted 'critical season range' (shaded). (b-1) Winter and summer ranges connected by a migration route. (b-2) A seasonal shift within a yearly range. (c) Diagrammatic representation of the relationship between billy and nanny-kid herds of feral goats. (d) A situation where home range and territory are synonymous, with little overlap between territories. (e-1) Groups of female wildebeest moving from one male territory to another in a low density area. (e-2) Tightly packed herds of female wildebeest and young with more than one adult cooperating, both in breeding and in defending the territory. (f) A pattern of seasonal distribution and movement of wildebeest, the animals migrating between wet and dry season areas of concentration and following frequently used routes (modified after Talbot, L. M. and Talbot, M. H., 1963)

bulls might share a territory and cooperate both in breeding and in defending the territorial boundaries against other males.

Figure 1(f) shows a pattern of seasonal distribution and movement of wildebeest in western Masailand, Kenya, the animals migrating between wet and dry season areas of concentration along frequently used routes. The movements on the left half of (f) have a maximum east–west spread of about

120 miles (192 km) and a north–south spread of about 140 miles (224 km). It is interesting to note that in other areas wildebeest are not migratory and remain within fairly small annual (group) home ranges throughout the year (Estes, R. D., 1968).

For species that must have water fairly frequently, it is common to find concentrations along running watercourses during the dry seasons. However, it is dangerous to assume this without checking. G. Caughley and his colleagues made periodic aerial observations in a Zambian national park and discovered the exact reverse of what some of the local game officers had predicted:

> The elephants move towards the alluvial zone about a month after the rains begin. By January the process is completed and the alluvium holds densities averaging $7/km^2$, densities in the remainder of the park having dropped to $2/km^2$. By March dispersal away from the alluvium has already begun and by May the elephants have taken up their dry-season pattern of distribution which comes close to an even density throughout the park. (Caughley, G., 1973).

In this area in the dry season small groups of elephants were able to use scattered permanent pools or to dig for water lying within a metre of the surface of the sandy stream beds.

3240   Kinds of Movement Within, Between and Outside Home Ranges

The basic movements within home ranges are those associated with normal daily activities. For example, movements may be directed towards feeding, water, bedding areas, refuge from hunters or other predators, seeking shelter from the elements, reactions to others of the same species or to other species, breeding, play and so on. Daily activities may be concentrated in certain parts of the home range during certain seasons.

3241   *Migration*

Migration is here used to refer to movements between seasonal home ranges. For deer of temperate climates the yearly home range normally involves a winter range and a summer range, connected by a migratory route. In Africa, migratory routes commonly extend between separate dry and wet season ranges (see Figure 1(b-1)). If no such separation occurs, or if there is an overlap (Figure 1(b-2)) such movements are more appropriately regarded as *seasonal shifts* within a yearly home range.

How far must the seasonal range be separated before the movements between them are called migrations? Theoretically if the seasonal ranges are separated even slightly and the animals remain throughout the appropriate season in each range, the movement could be called migratory. However, this is normally no problem because migratory movements are usually several to many times the length of individual seasonal ranges.

Large herds of plains game may exhibit a rather complex migratory pattern, as shown in Figure 1(f). Such seasonal shifts may involve several thousand individuals gradually moving in great circles or loops, or slowly drifting over a broad area between seasonal concentrations. In some semi-arid tropical regions of erratic and inconsistent rainfall specific movements of large herds may depend on the animals recognizing areas with rain at distances of several miles, different herds moving into these areas from several directions and assembling there in large concentrations.

Migratory movements in fairly open country are usually easily recognized; they will be known by local people. However, in woodland and forest areas seasonal movements of any kind may be more difficult to recognize. If this is the case a series of aerial counts, study of migration trails, mass marking or marking of individuals may be necessary.

Recognition of a population as migratory or non-migratory has several important practical implications for management. Before discussing these, however, it should be recognized that populations do not migrate without reason. If all the yearly requirements for existence are met within yearly home ranges there may be seasonal shifts but no migration. If such requirements are not met within a single range, then migrations must develop between seasonal ranges.

The most convincing evidence supporting this generalization is of two types. The first kind of evidence comes from introductions or reintroductions of large mammals. From the same liberation, as introduced red deer spread into certain valleys in the South Island of New Zealand they become migratory. In other valleys, colonized from the spread of the same liberation, the deer settle on a non-migratory basis and live within fairly small home ranges.

The second type of evidence is similar. Within the distributional range of a species of large mammal known as migratory will be certain areas or even regions where the same species is non-migratory.

These two kinds of evidence have important implications for management. First, a study comparing the habitat elements available in areas with non-migratory populations with those elements available in seasonal ranges of migratory populations may quickly lead to an improved understanding of habitat requirements essential to the species concerned. Second, the assumption that animals do not migrate unless they have to implies that in management terms the question of an area being a suitable ecologic unit to hold a population is a question of the moment; not necessarily an analysis of a permanent situation. Habitats may be altered to make areas more suitable, more self-sufficient ecologically. One frequently encountered objective of national park personnel in developing countries is to extend boundaries of a park to include migratory routes which normally take the animals beyond the park boundaries. Sometimes this is necessary; but in many such situations an alternative possibility exists: to manipulate habitat so all required elements are available within the existing park, thus eliminating the need for migration.[1] Comparative studies of migratory and non-migratory populations may thus provide the basis for improving the existing park environment. Viewed in this way the development of a park as an ecologically viable, self-sufficient management unit is often possible without the extension of park boundaries. This will of course not always be possible, but it should always be considered as a possibility where a park or reserve contains migratory populations which spend time outside the park and are in some kind of conflict with adjacent land uses.

3242   Dispersal

Dispersal, as used here, is the movement of a young animal away from the home range of its parents or family group to a home range of its own. In large mammals this takes place as the young animals reach breeding age and the new home range, once established, is maintained, normally with only slight yearly shifts, for the breeding life of the individual. Wandering of old individuals and their temporary association with first one, then, another family group is not included in the present definition.

3242A   *Patterns of Dispersal*

W. E. Howard (1960) introduced the idea that there were two basically different patterns of dispersal: environmental and innate. Howard's work was with mice.

In New Zealand, observations based on fifteen species of introduced ungulates entirely supports Howard's suggestion. Although evidence from large African mammals further confirms the existence of these two basic dispersal types, it is clear that, for certain African species at least, still other types may be recognized (Riney, T., 1966) and some of these are elaborated below. Regardless of what the future understanding may be regarding different patterns and aspects of dispersal, however, it seems useful now to be at least aware of these two basic patterns of dispersal. They provide convenient reference points for recognition of specific variations.

3242B   *Environmental Dispersal*

This refers to a pattern of dispersal by which young members of the population move into a suitable habitat within, adjacent to or near the parental

---

[1]For some slowly maturing species and where learning is important, a tradition for migration may be so firmly established that it may be difficult to break without removing a proportion of migrating adults, particularly the group leaders. A safe procedure is to think of the conversion to non-migratory populations in phases: the first stage would be improving habitat and observing the occupation of this habitat. Then, when a higher proportion of permanent residents has been established, one may consider the most appropriate means of breaking the tradition of migration.

home range. This type of dispersal probably occurs for all species of large mammal regardless of the presence or absence of other dispersal patterns.

Figure 2 shows diagrammatically the distinction between environmental and innate dispersal. In the species represented in Figure 2(a) only environmental dispersal is taking place. Dispersal is most frequently into adjacent suitable habitat and less frequently into less favourable habitats. Completely unsuitable habitat forms an effective barrier to environmental dispersal movements.

3242C   *Innate Dispersal*

As here used, this term indicates a type of dispersal occurring independently of the suitability of the environment. It consists of a certain proportion of young individuals, more commonly males, moving long distances across either suitable or unsuitable environments before finally establishing a new home range. As Howard (1960) notes:

> Possession of the innate dispersal trait implies that such an animal is predisposed at birth to leave home at puberty (as it comes to breeding age) and make one dispersal into surroundings beyond the confines of its parental home range. Such density-independent individuals have inherited an urge to leave home voluntarily. They often pass up available and suitable niches and venture across unfavourable habitats.

Figure 2(b) shows a population of a species from which a certain proportion of young are dispersing innately, although most young are still dispersing environmentally. Those individuals dispersing innately find unsuitable habitat no barrier to their dispersal movements and eventually settle in their own home ranges in more or less suitable habitat.

It is clear from the New Zealand observations that each species has its own characteristic pattern of dispersal and that, among the fifteen species introduced the pattern ranges from species like the Sika deer, whose dispersal pattern is almost entirely environmental, to chamois and red deer, who have strong innate dispersal characteristics in addition to the basic environmental type dispersal shared by all large mammals.

Judging by the infrequency of records coming to hand, the proportion of young that disperses innately varies greatly between species. For example, records of young chamois being found far from established chamois populations are much easier to find than similar records for Himalayan thar. Both species were released in the same kind of mountainous country in the South Island of New Zealand. This is interpreted as evidence that innate dispersal is stronger in chamois than in thar.

The most frequently encountered evidence of innate dispersal is the sudden appearance of a young animal in a place considerably removed from the near-

36

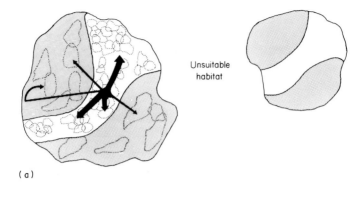

(a)

Unsuitable
habitat

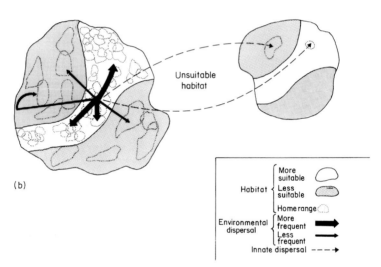

(b)

Unsuitable
habitat

Habitat {
More suitable
Less suitable
Home range
}

Environmental dispersal {
More frequent
Less frequent
}

Innate dispersal ----→

Figure 2   Diagrammatic representation showing the distinction between environmental and innate dispersal. For the species represented in (a), from the parental home range (shown in solid black) only environmental dispersal is taking place. For the species represented in (b), most young individuals still disperse environmentally (see text)

est known population. Young red deer have suddenly appeared in patches of forest 20–30 miles (30–50 km) from the nearest known populations. The movements of some of these young individuals become well known as various neighbours exchange news of their progress into unoccupied country.

A good example of species differences in this respect is the introduction of red deer and Sika deer at the same time and on the same farm in the North Island of New Zealand. The forest behind the farm consisted of podocarp and beech interspersed infrequently with small clearings bordering streams. In front of the farm was a very large area of grassland, as much as a mile wide in parts and extending for several miles in either direction. Red deer spread across this clearing within the first few years after introduction, then gradually spread northward, while the Sika deer took over 60 years to cross the same clearing to the opposite side. Dispersal of Sika was of the environmental type only. Dispersal was through the forest and eventually in this slow way around the clearing and back to a point opposite the initial liberation point. The large area of pure grassland was enough of a barrier to contain the Sika deer and innate dispersal was apparently lacking.

Clearly, if a proportion of individuals of a species were dispersing innately into already occupied habitat this aspect of dispersal would probably remain unknown unless individual animals were marked when very young.

3242D   *Rates of Dispersal*

Rates of dispersal have been recorded for only a few species of large mammal. Table 1 shows dispersal rates for several such species. The outer limits of most of these populations colonized new habitat at rates varying between 8.7 (5.4 miles) and 0.6 km (0.4 miles) per year. These rates are essentially rates of environmental dispersal. The records usually involve the rate of advance of a moving front of a population as it gradually increases its range by dispersing into adjacent suitable habitat.

Within one species the rate of dispersal may be affected by the size and shape of the occupied habitats. For example, animals introduced into and living along a narrow grassland ridge-top situation bordered by forest would be expected to disperse along the ridges more rapidly than in another situation where environmental dispersal could take place equally in all directions as with a forest-dweller. Similarly, environmental dispersal should be slower in the best habitats which can sustain the highest densities. Innate dispersal would probably be more noticeable in large rather than small populations.

Table 1   Rates of dispersal of some large mammals (modified from Caughley, G., 1977)

Species	km/year	miles/year
Horse	45	29.8
Chamois	8.7	5.4
Grey kangaroo	4.8	3.0
Himalayan thar	3.2	2.0
Red deer	1.6	1.0
Sika deer	1.6	1.0
Whitetailed deer	1.0	0.6
Fallow deer	0.8	0.5
Rusa deer	0.8	0.5
Sambar deer	0.6	0.4
Wapiti	0.6	0.4

However, there is little evidence from large mammals demonstrating differences in dispersal rates as associated either with population density or with qualitative or quantitative differences in habitat. Many more relevant observations need recording before a broader perspective of dispersal can emerge.

3242E   Comparisons between Antelope and Deer

Some information is in hand for about thirty African ungulates (largely bovidae). This provides a basis for comparison with the fourteen species of large mammals, largely deer, that were introduced into New Zealand from both temperate and tropical regions.

In African introductions (and reintroductions) of large mammals one can find the same range of dispersal patterns as is present in deer. Records of innately dispersing kudu, for example, are obtained with about the same frequency in Zimbabwe, Zambia and the Transvaal as are records of dispersing red deer in New Zealand. Buchbuck, on the other hand, resemble more the Sika deer, who disperse almost exclusively environmentally. In Africa, the main difference was that certain species of antelope were characterized by an essentially different pattern of dispersal, which one may call group dispersal. To my knowledge group dispersal is unknown in temperate region deer.

3242F   Group Dispersal

While dispersal as thus far discussed involves the separation of a young individual from its family group and the subsequent establishment of a home range of its own, in Africa certain species disperse in mixed groups: a kind of mass dispersal.

I have recorded observations of groups of young adults of several African species suddenly appearing in various areas where they were previously unknown. To mention a few observations made in Zambia and Zimbabwe, groups consisted of twenty waterbuck, over a hundred wildebeest, sixty to seventy giraffe, and seventeen sable. Dispersal of each of these groups was considered as a form of innate dispersal as the newly located animals were many miles from the nearest resident populations of the same species. For each of these species. I would also expect (but cannot prove) group dispersal to take place on an environmental basis.

It has been suggested (Riney, T., 1967b) that group dispersal of certain African ungulates may be part of the explanation for the present-day survival of a Pleistocene African large mammal fauna. Although there are bones of ancestral deer in various parts of Africa south of the Sahara, in the drier areas without dependable water supplies present-day species of deer might temporarily establish themselves but would ultimately die. It is suggested that deer lack the two main mechanisms that have allowed African ungulates to survive: either a physiological adaptation to allow survival for months without access to surface water, or a mechanism for dispersal that permits at least small groups to disperse to areas where water is still available.

3242G   Other 'Special' Types of Dispersal

Although the above discussion describes three basic types of dispersal in large mammals there are other exceptional ways of achieving dispersal of a species. No discussion of dispersal in large mammals would be complete without mention of the spectacular mass movements of springbok (*Antidorcas marsupialis*) on the dry central plateau of southern Africa.

These mass movements, superficially reminiscent of the mass movements of locusts or lemmings, were first reported in 1894 by observers who called them 'migrations'. S. C. Conwright-Schreiner (1925) showed that this was not the case; that movements were not seasonal, as there were only four significant peaks in the Preska district of the Northern Cape (South Africa) between 1887 and 1896. More recent records describe similar springbok movements that occurred in 1946 and in 1950 (Child, G. and Le Riche, J. D., 1969).

Such multitudes of animals were involved that most observers avoided attempts at estimating numbers. For example, two observers accustomed to judging numbers of sheep and using binoculars, estimated 50,000 moving springbok while standing on one vantage point. This vast herd covered an area some 130 by 15 miles (208 by 24 km). Davis is quoted by Child, G. and Le Riche, J. D. (1969) as estimating the size of one mass movement as one hundred million springbok. He drove 47 miles (75 km) through massed springbok. The most recent large trek reported by the same authors took place over a 200 mile (320 km) front and took 3 days to pass an observation point.

These mass movements were locally known as 'treks'. Springbok on 'trek' moved at a steady walk and without their customary alertness so that thousands were killed by the press of animals behind them, for example, while trying to drink from streams. They moved forward until, in some instances, they met the sea, where one observer recorded large numbers dead along 30 miles of seashore.

As with locusts, once numbers exceed a certain level the behaviour of the individuals within the massive aggregations changes and the mass movement assumed a character of its own. Several observers recorded the springbok's lack of fear on trek. Animals were clubbed to death in streets and gardens of country towns. Horses and waggons passed through masses of springbok which scarcely allowed room for them to pass. Herdsmen caught up in the treks were trampled to death. Flocks of sheep and other groups of wild animals, including kudu, hartebeest, wildebeest, blesbok, zebra, eland and other antelope were caught up and swept along by the treks into areas where these species had become extinct: a kind of involuntary but natural reintroduction.

Both the above-mentioned references noted the immense damage caused by such treks, the country being left virtually denuded of plants.

That these springbok treks disperse populations is clear. G. Child and J. D. Le Riche (1969) were satisfied that there was a marked reduction in the resident springbok population in and around the Kalahari Gemsbok National

Park following treks in 1946 and 1950 and that numbers remained low. Likewise T. Riney and W. L. Kettlitz (1964), referring to an area in the Transvaal, noted that numbers were still low in 1964 although the last trek had taken place over 30 years before.

Such treks do not fit either definition of dispersal as discussed above. They are not a form of 'environmental dispersal' for once movement starts it takes place over any traversable habitat and has involved distances in terms of hundreds of kilometres. Neither do the treks seem 'innate' in the sense described above. The movements bear no resemblance to the movement of some proportion of juveniles or largely juvenile groups, leaving the home ranges of their birth and travelling many kilometres to a new and suitable location. One feature of these treks is that all sex and age groups are involved. Another characteristic is that treks are known to have originated only from areas holding unusually high concentrations of springbok.

T. Riney and W. L. Kettlitz (1964), in looking for a basic pattern or cause of the treks, suggested that the history of human occupancy and consequent changes in land use may be responsible. European hunters first operated in the Transvaal area mentioned above, in 1794. The areas were settled by Europeans who moved up from the south with their cattle and sheep in 1834. When the early settlers arrived there were very few springbok. The great springbok treks took place in this area between 1920 and 1933, that is starting 86 years after the arrival of the first settlers. In other more southern areas, colonized earlier, springbok treks were first known from 50 to 90 years after being settled by Europeans. As burning and eventually overgrazing was characteristic of this settlement period, this was a period of marked change in habitats. While becoming less suitable for domestic stock, large areas became more suitable for springbok. The suggestion is that eventually the springbok erupted to such high numbers that the special 'trekking' mechanism was released and off the springbok went on their historic treks.

G. Child and J. D. Le Riche concluded in 1969 that the springbok treks were spectacular dispersal movements *from areas of high population density in which all sex and age classes participate*. They noted that springbok have considerably extended their range of distribution by means of such sporadic treks over the previous 50–60 years.

Springbok occur in areas subject to marked climatic fluctuations; but the details of changes in weather and habitat conditions in the months and years preceding such treks is largely unknown. Obviously much more research is needed before this type of movement is fully understood. Springbok treks certainly represent an extreme way of dispersing a population of large mammals: a kind of movement behaviour that comes into play only when numbers exceed a certain level of overpopulation.

Finally, as noted above, still another kind of dispersal—an involuntary, forced dispersal of groups of other species swept along within the body of a vast springbok trek—must be included in a list of known dispersal types.

One must conclude that a pattern of dispersal is as characteristic of a species as its more easily recognized physical attributes; however, such patterns are known for comparatively few species of large mammal. It is an interesting subject for future study.

3242H *How to Obtain Information on Dispersal*

Information on dispersal in large mammals is rare, partly because biologists have been slow in appreciating its significance in management but also because of the difficulty in obtaining such information. There are three good ways of obtaining information to assist in defining dispersal patterns: by tagging individuals, through historical records of animals moving into previously unoccupied areas, and exploiting situations associated with animals introduced by man.

Knowledge of dispersal may result from a long-term programme of tagging very young animals or otherwise recognizing them as individuals and mapping subsequent trappings or observations. Particularly for forest or woodland species where thresholds of visibility may be high, this may require many marked individuals. For example, in a study involving mule deer (Riney, T., 1950), approximately fifty deer were individually marked each year for 3 years. Marking and subsequent observations were mainly in a restricted wintering area where deer were concentrated. Each year twenty-five deer were marked before even one of these was seen outside the traps. After 4 years and 150 marked deer, only three instances of innate dispersal movements were recorded, each involving a young male in its first breeding season. If the tendency for innate dispersal had been weaker, innate dispersal may have been identified only after many more animals had been marked and several years had elapsed.

A second method is to alert any cooperative and interested persons working outdoors in an area, to your interest in knowing details of individual or groups of animals that suddenly appear in areas where they were previously unknown. The dates, exact locations and as accurate an assessment as possible of the sex and age class of the animal are the relevant details required. In my experience it is common to obtain ten or more such records before one can be certain of the age class (even in terms of yearling or 2-year-old) so most of the incoming hearsay information will not be useful. It is important to alert the 'network of eyes in the field' to the need for notifying the biologist or warden so the estimate of age can be confirmed. Even with this limitation this method seems to produce more usable information on dispersal quicker than by marking animals, as mentioned above.

The third and easily the most productive method for learning about dispersal is to follow the increase and expansion of introduced, reintroduced or self-introduced populations. Under these circumstances not only can dispersal patterns be readily identified but, in addition, such refinements as differences in rates of dispersal, associated both with different types of habitat and with different species, may be recognized. Observers should be particularly alert to the nature of the individuals and groups of colonizing animals on the fringes of expanding populations.

3242I   *Practical Implications for Management*

A knowledge of dispersal patterns has obvious practical implications for management, as, for example, when cropping schemes are used in order to reduce the spread of unwanted ungulates into nearby farming or ranching areas. Selective shooting at the right time of the year, and emphasizing certain sex and age groups, can help contain populations. On the other hand, there is some evidence both from New Zealand deer control operations and from African tsetse fly control operations that when dispersal patterns are ignored and shooting takes place at the time of the year when young are naturally dispersing, this accelerates the rates of dispersal.

Howard (1960) observes that all dispersal of whatever type is a mechanism for mixing genes and permitting species to occupy a new habitat as it becomes available through changing climates or large-scale activities of man. Exploitation of situations surrounding introductions allows us to see this mechanism in operation quickly and clearly without long-term, expensive marking campaigns. For example, consider the record of a single male thar suddenly appearing and taking up residence in a mountain valley, several valleys removed from the nearest New Zealand thar population, and living his entire adult life alone on the small home range he established. In time, after thar have spread slowly and mainly environmentally into this valley, only expensive marking studies could permit a similar observation.

Since a particular pattern and rate of dispersal is characteristic for each species this is an obviously important kind of requisite knowledge preceding the introduction of a species into another area. For example, species with little or no capacity for innate dispersal may be safely contained in an area surrounded by unsuitable habitat, as if on an island. Or knowledge of a characteristic dispersal pattern can thus provide forewarning of the speed and pattern of occupancy of the new species as it starts on its initial, colonizing, eruptive oscillation (See Figure 126).

3243   *Wandering*

Another kind of movement is wandering of old individuals and their temporary association with first one then another family group or other established group. It is difficult to make such observations as it requires recognition of habitually associated groups as well as the even more difficult task of recognizing old individuals in the field.

3244 *Relation of Life Phase to Kinds and Location of Movement*

Between birth and reaching sexual maturity, young animals usually remain within the mothers' home range. They may be attached to the mother or mothers' family group, or they may be part of a small family group which includes young of the previous year, or juveniles of some species may form calf (nursery) herds attended by one or more adults. However, species vary enormously in terms of segregated family, sex or age groups, or as they exhibit various kinds of mixed groupings.[1]

Dispersal takes place as the young animal reaches sexual maturity. Although dispersal normally involves a young animal leaving a family group and locating the home area of its adult life, in some species dispersal involves mass movements of groups of young as noted above.

Once the adult home range is established movements associated with normal activities take place within this home area throughout the adult breeding life of the individual.

Figure 3 shows diagrammatically the main relationships between life phases and kinds and locations of animal movements. Although wandering old indi-

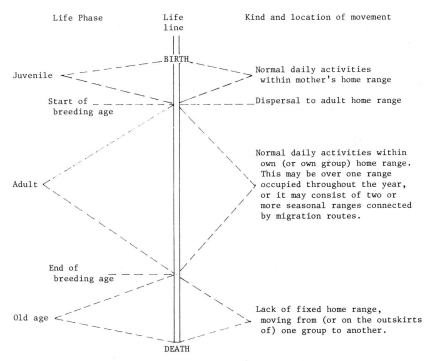

Figure 3 Relationships between life phases and kinds and location of animal movements

[1]These thoughts have been nicely elaborated in a stimulating paper on the Social Organization of the African Bovidae by R. D. Estes (1974).

44

viduals are difficult to recognize in the field this is not important for management, as their numbers and effect on the rest of the population are probably insignificant. For those species showing mass dispersal, difficulty in distinguishing between dispersal and migration may be more serious. Dispersing groups will consist largely of young animals, while migrating herds will consist of individuals of various ages.

3300 OCCURRENCE—ANIMAL . . . RECOGNITION OF INDIVIDUALS AND OTHER WAYS OF LEARNING OF MOVEMENTS OR THE EXTENT OF LOCALIZATION

Detailed understanding of a population is not always necessary in management, but when it is it is usually desirable to know at least something of the individuals that compose it.

Recognizing individual animals may be useful in several ways: in describing patterns of occupancy, a home range or a territory, in more precisely defining an individual's ecological niche, in assisting in recognizing dispersal patterns, in following migrations, in observing seasonal changes in physical condition within a defined home range, in the study of various social interactions with a population. Recapture of known individuals allows growth curves to be defined, criteria for ageing to be developed and various physical characteristics or its state of health to be assessed at periodic intervals.

Almost any imaginable way of marking animals has been tried by someone. Literally hundreds of different techniques are available, either for recognizing individuals or for determining localization. Thus in this section a thorough review is not attempted. Instead a selection of different approaches is presented, with a few principles, guidelines and examples. As with other techniques, before selecting a method the objectives of study or management

46

should be carefully reviewed and the method of marking or other means of determining localization selected or developed with these objectives in mind. In this context the basic questions to ask are: 'Why do you want to identify individuals in the first place?', 'What will you do with these identifications after they have been made?' and 'Is there a simpler, quicker, less expensive way of achieving the same objective?'

Often an initial objective is simply to improve one's understanding of the extent to which a population is localized. Knowing the size and shape of even a few home ranges will allow one to judge, for example, the extent to which populations in adjacent valleys, or within a certain distance of scattered watering points, may be regarded as essentially separate populations for management purposes. Learning something of the size and shape of a few home ranges is in my opinion essential to developing a feeling for interactions between a population and its environment.

3310   Techniques for Recognition

Techniques for recognizing individual animals are of two broad types: recognition of individuals without marking, and recognition of artificially marked individuals.

3311   *Natural Markings*

With the exception of identical twins, which occur rarely in large mammals, every individual recognizably differs from every other one of its kind. If enough time and opportunity for detailed observation is available, as with animals in a zoo or other enclosure, individual recognition is normally not difficult. However, for free-ranging wild animals special equipment and procedures may be required and of course for some animals recognition based on natural markings may be possible but not practical in the field.

Ideally recognition should be quick and accurate and the use of binoculars is recommended. For most large mammals magnifications of between eight and ten power are usually sufficient. Optical equipment with more powerful magnification is slower to use, more difficult to hold steady and on sunny days heat waves make precise observation difficult or impossible. Where temperatures are not too high and animals may be observed from considerable distances, telescopes held on a firm rest may be useful.

3311A   *Recording Observations*

A basic method with many possible variations is to maintain separate notebook pages for each individually recognized individual, whether the individual markings are natural or are artificially applied. The first record sheet should contain all the identification data, each item being entered in the same location on the page for future ease of reference. This information will include: the date, the number and sex of the individual (for example, the fifth male might be designated M5; the fifth female, F5) and an estimate of age class. Distinguishing marks will be clearly described and the exact location of the animals recorded as well as time of day observed and other animals with which it was associated. Subsequent observations will record the date, exact location, time of day, activity, associated individuals and any other observations relevant to the study in hand.

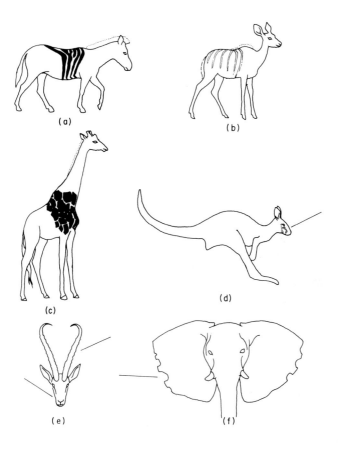

Figure 4   Six profiles on which can be entered individual markings: individual stripes or patches from selected 'key' areas, as in the zebra or giraffe, or the entire stripe pattern as in the kudu (b); or special attention to other variables associated with antlers or horns or facial markings, as in the red kangaroo (d) or Soemmering's gazelle (e); or distinctive tears on the ears plus tusk characteristics as in the African elephant (f). Outlines (a), (b), (c) and (e) were modified from drawings by P. Dandelot © 1970 (Dorst, J. and Dandelot, P., *A Field Guide to the Larger Mammals of Africa*, William Collins Sons & Co. Limited, 1970); (d) and (f) were drawn from photographs

In recording distinguishing marks a common practice is to have a series of blank sheets with profiles drawn and to note distinguishing marks on these front, side or rear profiles. When starting, a few attempts at identification are usually enough to determine which kind of silhouette will be most useful. Examples of several such profiles are shown in Figure 4. If an artist is not available useful profiles may be copied from handbooks for identifying animals, or from appropriate photographs.

3311B *Naturally Conspicuous Individuals*

In the early stages of a study do not attempt to identify every individual in an area or group. This may be possible, but for management purposes is usually not worth the extra time and effort required. Instead, initially concentrate on individuals easily recognized by some special features.

The easiest type of individuals to recognize are those with uniquely distinguishing characteristics: a deer with deformed antlers, an antelope with an unusually twisted or broken horn, an albino or partial albino. Conspicuous scars from past predation may readily identify an animal, as can a partly missing leg, a missing or shortened tail, or a limp from an old wound. Torn or notched ears may result from negotiating barbed-wire fences or from catching on thorns or sharp branches during flight; as a result of fights or from an encounter with a predator. Any such marks may help in recognizing individuals.

3311C *Colour Patterns*

For animals like zebra and giraffe, whose patterns of stripes or reticulations are distinctive for each animal, photographs provide an obvious way of trying to record individuals. But photographs alone seldom provide a requisite rapid re-identification method. This is partly due to the difficulty of getting clear photographs in just the right light to show the pattern to best advantage, but mainly because, even with a good photograph, one still has to recognize the distinctive pattern. Even with animals like zebra and giraffe this can be difficult. The difficulty may be minimized by concentrating on key areas. For giraffe this may involve the neck or shoulder area; for zebra this may involve the colour of stripes, the presence or absence of 'shadow stripes' (additional stripes of lighter colour). Key areas for special attention may be the stripes on one side, particularly in areas where different stripe patterns meet or in sections of the fore or hind leg. I have found it practical to draw in a key area on a zebra outline in a notebook, making appropriate additional notes on colour of stripes, presence of shadow stripes or other special features. Several workers have developed formulas which they find more convenient for rapid re-identification. As long as it can be duplicated by others the method is not important; use any system which suits you and which serves your purpose best.

3312   *Artificial Marks*

3312A   *Without Capture*

3312A(1)   *Uses*

The biggest advantage is the ease of quickly, and comparatively inexpensively, marking individuals. This means that mass marking can be considered, as when many animals may be marked a short time before the start of a seasonal movement or migration whose direction or destination is unknown.

Sometimes recognition of an individual may not be as important as marking a number of animals in a certain location—for example, within a small feeding area or around a watering point. Animals in one area may be marked with one colour. Other animals in nearby areas may be marked with another colour. This is a good and simple way of learning something of the extent that populations may be localized and of obtaining some details of occurrence away from the marking site. Mass marking used in this way can quickly increase understanding of a species' seasonal 'pattern of occupancy' in a study area.

Another use for temporary markings without capture is that this often paves the way for subsequent recognition of naturally marked individuals, who may otherwise have remained unrecognized. For example, in a Zimbabwean study in 1958 several species of large mammal were being marked with paint delivered by capsules fitted to arrows fired from a 55 lb (25 kg) longbow. One species, the steenbuck, had eluded all efforts at marking. The reactions of this small, uniformly coloured antelope were so quick that even when facing away from the marking team, the noise of the arrow leaving the bow would stimulate it to look up and leap out of the arrow's path, even at distances as close as 20 yards (18 m). Finally, my wife, a much better shot than I, successfully anticipated the direction of the jump, shot to one side and as the steenbuck leapt into the path of the arrow we had one marked steenbuck. Before this marking we had estimated between two and three steenbuck as habitually living along a 2 mile (3 km) stretch of a certain tourist road. However, within 2 weeks after the single marking, eleven different individuals were recognized along this same stretch of road. Once certain recognition of one individual was possible, closer examination revealed previously unsuspected individual markings for each of the eleven steenbuck. In this example in recognizing individuals, two variables were combined: differing dark lines on the ear and slightly different facial markings.

3312A(2)   *Limitations*

Where permanent markings are required paints and dyes are inappropriate. They will not last beyond the following moult, which gives it a maximum potential life of recognition of about 6 months. Paint applied to horns can of course last much longer, especially if applied to those parts of the horn base where rubbing is minimal.

3312A(3) *Paints and dyes*

Captured animals may be unavailable or too difficult or expensive to obtain. Then consideration may be given to marking animals without capture. Paints and dyes are suited to both mass-marking and automatic marking techniques and can be most useful when information on movements is needed in a hurry.

Paints and dyes can also provide a useful extra means of marking captured animals. This usually involves painting large numbers on easily observed parts of the animal's body.

Paints

Paints are in general the least effective means of marking without capture as in my experience even the best paints tend to rub off fairly quickly. However, some paints are good for several weeks and where mass-marking is desired and when appropriate dyes are not available, paint may be worth a try.

Different kinds of paint have been tried with various degrees of success. The most successful paints I have used are those designed to mark highways.

Suitable paint may be located by contacting the commercial firm supplying road-marking paint to the local roads department. Luminescent road paint is particularly good. It can usually be obtained in at least two colours but white road paint is always available and it can be coloured by the addition of other suitable paints recommended as suitable by the paint supplier. The latter should be made fully aware of your intended use of his paint since improvements in the durability and visibility of paints are constantly being made.

Dyes combined with other substances

Dyes have been well known as a system of marking wildlife since the early 1940s (e.g. Moffit, J., 1942) when various alcohol-soluble dyes were applied to the feathers of wild birds. Applying these dyes to mammals was less satisfactory although dying mammalian hair proved adequate for various short-term purposes. For example, alcohol-soluble dyes were used on bighorn sheep at a watering point by one worker (Hansen, C. G., 1964) who was making 3–12-day counts and wanted to avoid duplication. The following proportions were used by Hansen to make 1 gallon (3.8 litres) of solution:

Pink Dissolve 2 grams of Rhodamine B Extra in $\frac{1}{2}$ gallon (1.9 litres) of 99 per cent isopropyl alcohol; then add $\frac{1}{2}$ gallon of water.

Green Dissolve 4 grams of malachite green in $\frac{1}{2}$ gallon of 99 per cent isopropyl alcohol; add $\frac{1}{2}$ gallon of water.

Yellow About 25 grams of picric acid in $\frac{1}{2}$ gallon of alcohol; add $\frac{1}{2}$ gallon of water.[1] This was the most permanent colour used by Hansen.

[1]Picric acid is dangerous. Labels should be carefully read and safety precautions strictly adhered to.

Violet   A violet colour was obtained by mixing the green and pink solutions to the desired shade.

Commercial fur dyes have been used with some success when combined with hydrogen peroxide for fixing. According to R. D. Taber and I. McT. Cowan (1969), Nyanzol 4R (reddish brown) and Nyanzol D (black) are good examples, the latter remaining visible for between 30 and 240 days. Other fur dyes are said to last from 4 to 6 months.

Selecting dyes

One practical approach to the selection and application of dyes is to proceed as follows:

1. Review study or management objectives to determine if temporarily marking a number of animals would usefully contribute to achieving those objectives.
2. If dyes or paints are to be tried, determine how durable a dye is required for the purpose in mind. In the above-mentioned example involving bighorn sheep 3–12 days was sufficient. In a migration study the dye should last for several weeks. In defining home ranges or in migration studies dyes should be as permanent as possible; i.e. until the following moult.
3. What minimum size mark and colour range dye is required for identification at what distances in the field?
4. The paints or dyes are then tested under relevant field conditions.

This procedure should help in setting standards against which the effectiveness of various dye and dyes combined with other substances may be tested.

Pre-testing under field conditions (an example)

The following example illustrates how this approach may work in practice. In a Zimbabwean study it was desired to mark numbers of animals of several species to quickly learn something of local movements at the latter part of a dry season. For testing the dyes, fresh skins were taken from recent predator kills of kudu, impala, wildebeest and grysbok. The colours of these skins were dark greyish-brown (wildebeest), fawn-grey (kudu), rich rufous-fawn (impala) and rufous intermingled with white hairs giving a grizzled appearance (grysbok). Since these skins included both dark and fairly light colours it was felt that tests made on them would be useful for other species as well.

The durability of the dyes was tested in the following way. A teaspoon of dye was splashed on a vertically held patch of skin from a distance of about 3 ft (1 m). This was done in the afternoon and the dye was normally dry within a few minutes. The entire patch was then completely covered with water, held

52

Table 2   List of various dye combinations passing a durability test[a]

Species	Dye combination	Visibility rating
Impala and Sharpe's grysbok	1, picric acid; 2, water; 2, weak hydrogen peroxide	Fair to good
Impala and Sharpe's grysbok	1, picric acid; 2, water; 3, weak hydrogen peroxide; 1, e-black	Fair
Impala and Sharpe's grysbok	1, picric acid; 3, water; 1, e-blue; 1, e-black	Fair
Impala and Sharpe's grysbok	1, picric acid; 4, 40% hydrogen peroxide; 3, e-blue	Good
Impala and Sharpe's grysbok	1, picric acid; 2, 40% hydrogen peroxide; 3, e-blue	Fair to good
Impala and Sharpe's grysbok	1, picric acid; 2, 40% hydrogen peroxide; 4, e-blue	Fair to good
Kudu and wildebeest	picric acid	Good
Kudu and wildebeest	1, hydrogen proxide; 1, ammonia; 1, mercurichrome	Fair
Wildebeest	2, picric acid; $\frac{1}{4}$, ammonia; $\frac{1}{4}$, mercuri-chrome	Good
Wildebeest	2, picric acid; 2, ammonia; 2, brilliant green	Fair

[a]See text for description of test.

beneath the surface with a stone and soaked for 24 hours. On removal the skin was placed on a flat surface and scrubbed with a stiff floor scrubbing brush for 5 minutes.

Eleven of the thirty-six combinations thus tested were classed as not durable. The twenty-four combinations passing this 'durability' test were then allowed to dry and were subjectively assessed for visibility at a distance of 100 yards (90 m), using 10× binoculars. In this way the following arbitrary and entirely subjective assessments were made; not satisfactory, fair, fairly good and good. A further fifteen combinations were eliminated in this way as they were not readily distinguished in the field. Table 2 lists only those ten combinations considered as acceptable. The visibility rating might differ if other species were tested. The prefix 'e' in the dye combinations refers to a concentrated commercial dye for human hair obtainable from suppliers to hairdressers. The brand used was Enecto, but other brands may be equally or even more effective. All but one of the dye combinations used in this trial included picric acid, which was combined with hydrogen peroxide, ammonia, or both.

Projecting the substances to the animal

A number of projecting methods have been tried, including shooting balls or capsules of paint from specially designed guns (useful only at short ranges of from 10 to 15 yards (3–5 m)) to pelting elephants with blown eggs filled with paint (the wrong kind of paint was used which lasted only a few days) (Hanks, J., 1969), to automatic marking devices as mentioned below.

The most successful ways of throwing paint or dyes onto a free-ranging large mammal have involved bows and arrows and there have been two main approaches used.

In the late 1940s the State of Colorado marked deer by shooting them with bow and arrow; ordinary target arrows were used, tipped with a sponge rubber ball and dipped in common sheep-marking paint of various colours. The method was also used on elk and bighorn sheep. In a later study enamel paints were said to make clearer marks than oils (Manville, R. H., 1949). The disadvantages were that animals had to be at fairly close range, a good shot with a bow was needed and the paints did not last long.

The second method of projecting paints or dyes with a bow and arrow is to use a moulded candle-wax capsule filled with paint or dye and fitting snugly on a specially turned light wood or plastic head, which in turn is fitted to and replaces the head of a target arrow. The moulds are easily made and many 'loaded' arrows may be prepared on the evening before a day's marking activities. Figure 5 shows the mould, resulting candle-wax capsule and formed

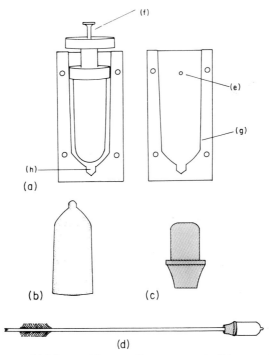

Figure 5   Brass mould (a), resulting candle-wax capsule (b), turned head for fitting over end of arrow (c) and head of arrow with dye-filled capsule ready for firing (d). Note overflow hole for excess hot wax (e) and nail or wire plug in place for pouring (f) (its removal is necessary to remove capsule after cooling), the slope to the sides of the capsule (g) and solid wax nipple (h)

head for the arrow. By trial and error it was found that maximum splash effect was obtained only after the sides of the capsule inclined slightly (Fig. 5(g)) and a solid wax nipple formed its apex (Figure 5(h)).

As much care should be taken in recording artificial marks as in recording natural markings (see Section 3311A).

3312A(4) *Self-marking devices*

These have been of three main types: devices using paints or dyes, marking by baiting, or various kinds of self-affixing collars.

Dyes and paints

These have been self-applied by using a trip mechanism to propel or release a spray along trails or at artificial feeding sites, in bedding areas, natural feeding areas or at watering points. A certain proportion of self-marked animals may come to be recognized as individuals even after the marks wear off. But this method is used mainly to contribute to understanding the role of a particular habitat element in the life of the population. In a sense this is a way of testing the drawing power of the habitat element. For example, for animals that must drink daily a watering point is an essential part of their home ranges. Self-marking of numbers of individuals can thus contribute to an understanding of the size and shapes of home ranges and the extent to which individuals may overlap in their use of key habitat elements.

One self-marking system uses a short length of $\frac{1}{2}$ inch (12.7 mm) pipe and a specially loaded shotgun cartridge. One end of the short pipe is open to eject the dye or paint. The other end is closed with a metal nipple that screws firmly into the pipe. The nipple is drilled so that a nail—the firing pin—may slide through it. A shotgun cartridge loaded with about 1.77 grams of powder is inserted into the pipe and the nipple screwed on. The colouring material is then inserted through the open end of the pipe. The striking bar of a rat trap forms the hammer that hits the nail. When an animal presses the trip wire or string attached to the treadle of the rat trap the striking bar of the trap hits the nail which in turn explodes the cartridge and the marking substance is propelled for several feet as a spray (Taber, R. D. and Cowan, I. McT., 1969).

The main disadvantage of explosive marking devices is that they must be reloaded after each discharge.

Simpler and more commonly used systems involve the use of compressed air to discharge a spray of dye. An early version simply used a soda-syphon bottle charged with compressed air cartridges (Moffit, J., 1942). Release may be arranged using a self-tripping device, or an observer at a distance may pull a cord attached to a discharge lever, or the device may be triggered by a solenoid activated by a concealed treadle or by a photoelectric cell. Other more recent and more ambitious devices use larger pressure spray tanks that depend on pumping and compressed air. For example, C. G. Hansen (1964) used a standard spray tank with built-in pressure pump, fastened the nozzle to a pole adjacent to a watering point and the spray was released by an observer concealed 80–100 ft (24–30 m) away and pulling a string. Almost any kind of pressure tank and spraying device can be used. They simply require firm mounting and a tripping wire, solenoid or other device to release the spray lever.

Two more tripping devices devised by Taber, R. D. *et al*. (1956) seem worth mentioning. The first device consists of a trip cord stretched between two trees, leading up and to the centre of a plank of wood fixed between the two trees. On the plank is affixed a rat trap and on the trap arm a razor blade. On tripping, the razor blade cuts through a plastic bag releasing dye or paint on the animal below (see Figure 6(a)).

Another system consisted simply of an arrangement whereby the trip cord pulled a wooden or metal arm anchored by a pivot in such a way that when the wire is tripped coloured material is thrown over the animal's neck and shoulders (Figure 6(b)).

I observed a simple but effective automatic marking device made by two keen national park wardens in Zambia. Observing that zebra habitually

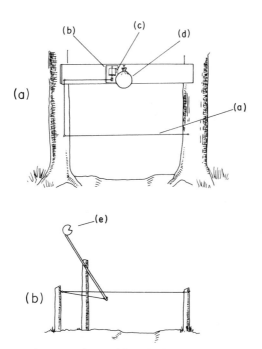

Figure 6 Two types of automatic marking devices developed in Ontario (6a) and Wyoming (6b) respectively (from Taber, R. D. *et al*., 1956). Trip wire (a) is attached to treadle of rat trap (b). When sprung, razor blade (c), affixed to the striking arm of the trap, cuts into the plastic bag containing the paint or dye (d), thus releasing the marking fluid. In 6b (e) shows location of specially cut rubber ball

rubbed against a certain pole or tree after dusting, they buried a similar pole about 5 in. (2 cm) in diameter and about 3 ft (1 m) deep in hard soil near a dusting area. At the top of the pole, well above zebra height, was nailed a tin can full of road paint. The rubbing caused the paint to splash on the zebra.

Many other less generally applicable devices for self-marking have been used. These have, for example, included allowing elk to mark themselves by placing freshly painted ropes around hay stacks. The elk, thrusting their heads between the ropes, thus marked themselves.

When conditions are right ungulates can mark themselves at waterholes and thus be recognized at varying distances from the waterhole. Circumstances especially suitable for this kind of marking are pools with muddy edges. The hoofs of large ungulates sink several inches in the mud leaving a hole partially filled with water. Ideally the holes should hold water to a depth of 10 in. (25 cm) or more. Dye powder such as basic fuchsin is then placed in several of these small holes and subsequently animals stepping in the holes mark a lower leg with the dye.

Marking by baiting

The droppings of animals may be coloured by including powdered dye in food or salt. Faeces have been dyed experimentally for sheep, wapiti, red deer, Australian possum (*Trichosurus vulpecula*) and several African ungulates. Methylene blue, crystal violet, basic fuchsin and aniline blue are satisfactory for staining faeces blue-green, violet, red and blue respectively. These dyes were used on sheep and wapiti by F. Kindel (1960) by mixing the powdered dye with salt. I have used basic fuchsin powder in gelatin capsules (obtainable from chemists) inserted in carrots and apples to bait wild red deer and Australian possums.

In powder form 1 gram is sufficient to dye the faeces. The dye appears about a day after it is taken and continues to colour the faeces 2–4 days afterwards. I found that faeces dropped as much as a week after ingestion could still be recognized as dyed, providing the pellet was broken open.

For most purposes dyed faeces will be less useful than naturally or artificially marked individuals. However, in situations where bait or salt is readily taken it is a good way of mass-marking a site. Each animal that ingests 1 gram of dye will be leaving coloured droppings at the rate of ten to fifteen groups a day for several days. Since a higher proportion of droppings are usually associated with feeding areas (ingestion stimulates defaecation) an initial trial might be to locate the bait at a bedding area and observe various nearby feeding areas for coloured droppings, or to bait at one feeding area and test other nearby areas to see to what extent the 'marked' animals were also using other areas.

The main disadvantage of dyed faeces is that, for most species, droppings are not usually distributed over all parts of its home range. However, they are usually most numerous in those areas of highest use. Because it is an inexpensive and simple technique to use, dyed faeces thus seem a useful supplementary and indirect means of obtaining information on localization of movements and the differential feeding use of animals in relation to different available habitats.

Self-locking collars

A different type of self-marking device is the self-locking collar, which fastens to the neck of an animal as it moves forward along a trail. Figure 7 shows one such collar. R. D. Taber and I. McT. Cowan (1969) have indicated that several problems must be overcome before it can be used with confidence:

> it must lock tightly enough to stay on but not choke the animal, even if the neck swells and yet not be so loose that the animal will stick a foreleg through it when feeding on the ground; it must lock in place before the animal becomes frightened and withdraws its head; it must come free shortly after locking so that the animal does not become injured in its struggles to escape.

The first problem implies that this device is better suited to small ungulates which quickly grow to a standard size and where both sexes are approximately the same size. It would be least appropriate for larger mammals where one could expect various-sized animals to be moving along the same trail with the strong chance that the size collar set would be inappropriate for the animal entering the snare.

Another kind of self-collaring device involves a rubber collar which is held open on a wooden frame, mounted over a watering trough or artificial feeding

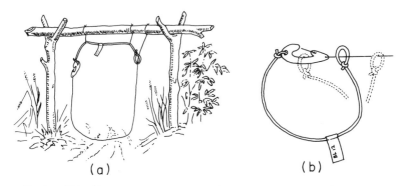

(a)   (b)

Figure 7   A self-locking collar which acts on the principle of a snare: (a) shows a method of suspending the collar above the trail with wire thin enough to break once the collar locks around the animals' neck; (b) shows details of the locking device (drawing modified from Romanov, A. N., 1956)

station. On drinking, the animal trips a release which releases the rubber to collapse around the animal's neck (Beale, D. M., 1966).

3312B *Capturing and Marking*

Mass capture techniques have been described elsewhere as a means of capturing animals for subsequent release or as part of a harvesting programme (Sections 7530–7540). For the purposes of marking, however, animals are normally captured either individually or in small numbers. Capture may be accomplished either by mechanical or by chemical means.

3312B(1) *Mechanical capture*

In spite of recent advances in the capture of animals through the use of chemicals, some form of mechanical capture still remains most appropriate for many species. Four basic techniques—chasing, netting, snaring and trapping—are suggested for consideration.

Selecting a precise technique will usually involve a compromise that takes into account management or study objectives, money, manpower, time and equipment available, and the proven skills and ingenuity of those doing the capture. Selection must also consider two aspects of the habits of the animals to be captured: (1) the timing and location of their daily movements, and (2) the nature of their reaction to various kinds of disturbance.

With the above in mind it is obviously impossible to describe a universal technique that will work for all large mammals. It is hoped that the following examples may serve as a guideline for developing appropriate capture techniques for large mammals not mentioned. They are taken mainly from two sources: Riney, T. and Kettlitz, W. L., 1964 and Taber, R. D. and Cowan, I. McT., 1969.

Chasing

Several species of large mammal may be captured by chasing with a horse or vehicle. The animals should be in open country and an intimate knowledge of the terrain is essential both for success in capture and for minimizing damage both to the animal and the operators. Each species must be handled somewhat differently, as the following examples illustrate.

When *blue wildebeest* are captured from horseback after a short chase and from alongside, a horn or tail is grasped and the animal is thrown much as a cowboy will 'bulldog' a calf. Blue wildebeest can be safely held down by standing on a horn. The animal is then blindfolded and tied up until it can be taken away, or until it is marked and released. If it is to be transported, at least three men are needed for handling it by the horns and tail as they manoeuvre it into a transport crate.

When capturing from a vehicle a common practice is to fix a noose on a long pole and manoeuvre it over the running animal's head. With the rope held tight the animal is grabbed by the head and tail and thrown as above.

They are occasionally aggressive when released.

Black wildebeest. Horses are sometimes used in separating an individual from a herd but at a certain point of fatigue this species turns and fights its chaser, which precludes the use of horses for capturing. This habit is, however, utilized in capturing black wildebeest from a truck or other suitable vehicle. The procedure is for two men to ride on the bonnet of the truck; one man grabs the horns of the wildebeest as it attacks the truck, while the other man leaps off and grabs it by the tail. It is then held and handled as for blue wildebeest. Black wildebeest are very quick-moving and must always be considered dangerous.

Zebra are captured either from a vehicle or on horseback. They are chased, caught using a noose as for wildebeest or by the tail, and a collar or halter is placed on them after which they are either blindfolded, tied up and marked, then released, or led to trucks for transporting. Casualties with fully grown zebra appear to be less than with young individuals, although the same technique is used for any size.

Impala may be captured by hand at night. They are located by two powerful (blinding) spotlights on a four-wheel-drive vehicle specially fitted with a firm seat for a man on the bonnet. For capturing, an open background has been found unsatisfactory; the best situation being for the impala to be fairly close to shrubs or trees. The approach by vehicle is slow until almost upon the animal, then men leap from the vehicle and grab it by one hind leg, lifting only that one leg off the ground. Another man is required to cover and hold the head for marking, or if transporting, to assist in leading to a communal crate on the back of another vehicle following the capturing vehicle. For transporting the Transvaal Department of Nature Conservation captures only young animals, 3–5 months of age; 35–40 lb (16–20 kg) weight (Riney, T. and Kettlitz, W. L., 1964). Care must be taken not to injure their backs. Impala are sensitive but not aggressive.

Giraffe are the largest animals to be regularly and successfully caught by chasing. This was done in the Transvaal by three men on horseback. Compared to capture of other animals by chasing this is a highly skilled operation and not to be generally recommended. It was developed before the use of immobilizing drugs, which are now recognized as more suitable for capturing the largest mammals.

Driving

Driving animals into nets is a very old but still an effective way of capturing many species. I have seen it used as a normal way of obtaining meat by a hill tribe in the Western Ghat range of India. It is most effective for smaller mammals, but ungulates with a shoulder height of under 3 ft (1 m) may be readily captured by driving into nets. It is particularly effective in scrub or forest areas where the net may easily be propped up by vegetation and camouflaged.

A common way of arranging such a drive is to quietly set a net in a longbow shape with the net held or propped up to a height of about 5 ft (1.5 m) in such a way that it will collapse on, and entangle, the animal when it is struck. Before the drive men are stationed out of sight and at intervals behind the net that they may quickly rush in once an animal becomes entangled. A line of beaters, making conspicuous noises, gradually move the animals into the net. To initiate an effective drive of this kind something should be known of the pattern of daily movements of the species concerned as well as an estimate of the average size of the home ranges and the population density.

An obvious disadvantage of this technique, as with any 'drive', is that animals with small home ranges tend to lie quietly in thickets or to double back rather than leave their familiar home areas. Thus the animals caught are more apt to be those whose normal home ranges include the area where the net is erected.

Capture using nets

Nets can be effective in capturing individuals of some species. They are either suspended above a trail or fastened on the ground along one side and extending up at an angle above the trail. The net is 'sprung' either by a self-tripping device or by a cord pulled by a camouflaged observer from a distance. The latter is preferable, for as soon as possible after becoming entangled the animal should be held, any relevant measurements taken and recorded, marked and then released.

Snares

Self-stopping snares, camouflaged and set along well-used trails, have been used in capturing such animals as zebra, wildebeest, kudu, waterbuck and eland by A. S. Mossman and associates (1963). They used a 15 ft (5 m) steel cable, prepared as shown in Figure 8. Following their procedure a 'hard eye' is placed at one end and the cable runs through it to form a noose. A 'soft eye', 8–10 in. (20–25 cm) in circumference is formed at the other end and a clamp, as shown in Figure 8, is fastened to the cable within the noose and 24–25 in. (73–76 cm) from the opening in the 'hard eye'. This gives a loop a fixed circumference of about 24 in (73 cm). Different species will require different settings. To prevent slipping, any preserving grease is washed off from the area of the clamp with petrol and the clamp nuts securely tightened using locking washers. The snare is anchored by encircling a tree limb and passing the snare through the 'soft eye'. Mossman found that the cable should be anchored high by encircling a tree limb about 8 ft (2.5 m) high. Snares in semi-arid habitats caught better when set near water and locations frequently changed. Moonless nights were more successful. Careful preparation along

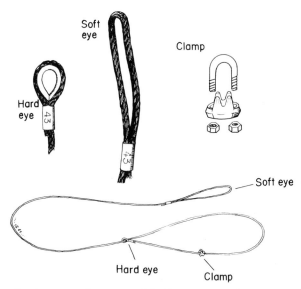

Figure 8   Details of a stopped snare used for live capture of several species of African ungulates (modified from Mossman, A. S. *et al.*, 1963)

constricted paths and close attendance are essential for the successful use of this technique.

Another approach to snaring is to go for the legs rather than the neck. Ashcraft and Reese (1957) snared sixty-two mule deer by setting on the ground two double-loop snares which were attached to a strong rubber band made from discarded truck tyre tubes. When a trip mechanism was released the snares caught the lower legs of the deer, holding them fast. Figure 9 illustrates the method of setting.

The authors claim there is less chance of injuring the animal if it is held by the legs rather than by the neck. Once caught the animal is prevented from applying its full force to the restraining ropes by the elasticity of the rubber which activates the device. A bell attached to the restraining rope signals the operator, who remains in the vicinity servicing a number of snares. If bells are used as a marking device the snare bell should have a different pitch and quality than the marking bells in order to avoid confusion.

The snared animal, if approached cautiously, keeping the anchor point between the operator and the animal, will usually throw itself. Care should be exercised to prevent the accumulation of slack line, which the animal would use to develop momentum, resulting in damage to the animal and the device. Once the deer has thrown itself and been held down, or rendered immobile with a small net, it can then be examined, marked and released.

The same principle but using a spring rather than rubber, has been incorporated into a snare trap patented in New Zealand and available commercially.[1]

[1]From Kerridge Odeon Export Ltd, Auckland, New Zealand.

62

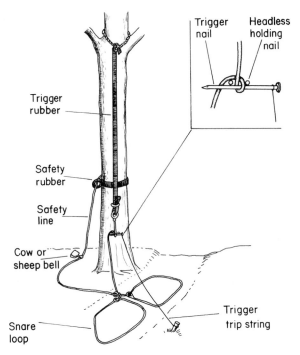

Figure 9   Modified drawing of a leg snare devised and successfully used by Ashcraft
and Reese (1957) on medium-sized deer

I have seen two prototypes, one suitable for smaller rabbit or possum-sized
animals; the other suitable for deer.

Trapping

Individual traps

The simplest, most commonly used individual traps are of the box type. These
are constructed either with sides of solid wood or with heavy wire netting.
When set, doors at both ends are raised. They are normally set on trails and
baited with an attractive bait. (Attractive means to a particular species in a
particular area at a particular time of year.) On taking the bait the animal
triggers a release mechanism that allows the doors to fall shut thus capturing
the animal. The captured animal is then either moved into a portable crush or,
more simply and economically, run into a heavy net held at the entrance of
the trap, then held on the ground while being examined and marked.

Figure 10 shows one such trap in operation as a deer is being released into a
strong rope net for handling.

Figure 10   The operation of a simple box-trap as a deer is being released into a net for
handling

Corral traps

Various kinds of corral traps will be mentioned in the section dealing with
methods of harvesting (Sections 7530 to 7541B). Animals caught by any of
these means may be run into an enclosed space, held securely, marked and
released. Where such measures are appropriate and practical, and the objectives can justify the expense, this is probably one of the best ways of capturing
numbers of animals for permanent marking.

3312B(2)   *Chemical capture of large mammals*

Chemical capture as we know it today was first tried on White-tailed deer in
the 1950s, first using the drug Flaxedil (Hall, T. C. *et al.*, 1953), then nicotine
salicylate (Crockford, J. A. *et al.*, 1957). During the following 10 years many
workers experimented on many species with a wide range of different drugs.
In this period mortality rates were apt to be high. Drugs were not something
one could confidently recommend for use on a rare species, for some embar-
rassing incidents had occurred. On the other hand the technique was becoming used by more and more wildlife biologists for some species with good
results and with lowering mortality. Many of the drugs and dosages developed
in this decade can be found in the *Wildlife Management Techniques* handbook
of 1969 (Taber, R. D. and Cowan, I. McT., 1969).

During 1963 two drugs (etorphine (or M99) and fentanyl) began to be used and, according to Pienaar (1973), this set the stage for the first really successful field trials on a wide spectrum of wild large mammals. Using these drugs great reductions in mortality rates during field application were observed, much smaller dosages were required, and remarkable tractability achieved. These drugs soon became favoured for immobilizing large mammals.

Drugs such as succinylcholine chloride are still in use today and species such as Uganda kob, impala, yearling wildebeest and others may be captured with a reasonable degree of safety with this quick-acting drug. However, for marking and most other procedures the duration of the effects of this drug is too short and now preference is given to safer and more easily controlled (i.e. reversible) drug combinations.

In recent years efforts have been directed at perfecting dosage rates for particular species and to discovering the most suitable tranquillizing drugs for combining with either M99 or fentanyl. Some species react better to certain drug combinations than others.

Immobilizing techniques for most large African mammals, and many more, have now been worked out and most of these techniques are quite safe if properly used. Unfortunately it is becoming more complicated. Over fifty drugs or drug combinations are now in use for one aspect or another of capturing animals through the use of chemicals.

Much of the pioneering work has been done by skilled veterinarians. Certainly when initiating chemical capture, a qualified veterinarian should be involved. Often inexperienced persons (sometimes including experienced biologists or wildlife managers) think they can capture game animals successfully, when equipped solely with drugs, syringes, something to project them with and a published table indicating a few dosage rates. Unfortunately heavy mortalities still occur—as E. Young (1973) has noted—'through lack of appreciation of the complexity of the task and the necessity of a sound knowledge of not only anaesthesiology but also physiology, animal management and even of disease'. In the conclusion of this same book it was suggested that 'a profound knowledge of the drug reactions encountered in each species is the only real guarantee of ultimate success. Study of the interspecies differences alone would occupy the lifetime of many workers.'

An early draft of this handbook included a separate chapter on the chemical capture of animals and a long table of dosages. However this section has been omitted for two very good reasons. First is the publication of a book, *The Capture and Care of Wild Animals*, edited by E. Young (1973) which is an excellent reference dealing with all aspects of chemical capture, transport accommodation and nutrition as well as post-capture complications and treatment. The second even more convincing reason is the publication of another book entitled *The Chemical Capture of Animals*, by A. M. Harthoorn (1976). This is easily the most comprehensive guide to the chemical restraint of both wild and captive animals. The book is so complete that, where chemical capture of large mammals is contemplated, the best advice I can give is to obtain a copy before starting, then, with the assistance of a qualified veterinarian, proceed to obtain the most appropriate equipment and drugs for your particular requirements. Initial tests should always be made in close cooperation with a qualified veterinarian.

K. van der Walt (1973), in discussing limitations of chemical capture techniques, notes that in many countries there are still various obstacles to overcome, such as estimating distances and body weights, most suitable drug mixtures, the best area to hit on the animal and ways of eliminating or reducing initial disturbance. The extreme danger associated with the use of M99 remains a problem. Finally, losses are still great through failure to track an animal from where it is darted to where it becomes immobilized. He concludes with a statement worth keeping in mind for other techniques as well: 'immobilization of wildlife is simply a technique and must be recognized as such. If this technique is not employed in a gainful manner it will become a gimmick, which should be condemned in the strongest possible terms.'

3312C *Marking Individuals*

Introduction

Although there are many methods of marking or tagging these can be grouped into general categories as follows: (1) mutilation, branding, clipping notching; (2) tattooing; (3) tags, bands, streamers, rings; (4) marking the coat with dyes or paints; (5) collars and neckbands; (6) radioactive materials in droppings; (7) internal markers for later recovery; (8) bells; and (9) radiotracking.

Once captured marking an animal is usually a simple procedure. However, since there are many marking methods it is useful to keep the reasons for marking firmly in mind as a guideline for selection.

B. J. Neal (1959) has reviewed the characteristics desired in a good tag or mark. With its practical use in mind, ideally it should: (1) be quickly and easily applied; (2) be retained indefinitely; (3) be readily observed at the distances and under the conditions required for the particular study or management problem; and (4) identify separate individuals, years of tagging, marking locations or herds as appropriate.

On the other hand the mark should not affect mortality rate, behaviour, or the ease with which the animal may be observed, or subsequently recaptured.

Mutilation

Mutilation is a quick and inexpensive way of producing a permanent, readily recognized mark on a number of species of large mammal.

Ears may be marked in the same way cattle are marked in certain ranching areas. With a sharp knife the point of the ear is removed to form a square rather than the naturally rounded or pointed tip; or V-shaped notches may be cut from either the tip of the ear or from its top or underside. Cuts should be made only in the region of the outer half of the ear, thus avoiding the cartilage. Ear markings of this type have been made on most hoofed animals studied and on other species of large mammals including predatory animals such as the hyaena (Kruuk, H., 1966).

The main advantage is to draw attention to marked individuals from great distances. Where numbers of animals are marked closer inspection is usually needed to recognize individual markings of some other type.

As noted above, individuals with tails shortened by some natural means (as by a lion) can be recognized with binoculars at distances well in excess of 100 yd (300 m); and shortened tails are equally conspicuous if done on purpose as part of a marking scheme. Another form of mutilation has been the clipping of claws or removal of a toe joint on a predator to form a distinctive pug mark.

Clipping

For animals with manes, such as zebra, close clipping over a section of the mane remains conspicuous for over 2 months and can thus provide short-term information on animal movements.

Branding

Making permanent scars in the shape of numbers, letters or symbols is an obvious technique to try on large mammals because of its widespread use on cattle. It is not advisable for animals with long hair but for many species of large mammal branding can provide inexpensive and permanent marking.

The most commonly used technique involves a copper-headed soldering iron heated by a burning jet of propane gas from a portable cylinder. Brands 8 in. (20 cm) high can be seen at distances of up to 150 yd (138 m) (Hanks, J., 1969) and individual brands have been recognized several years after branding.

Two biologists working on African mammals (Orr, R. T. and Moore-Gilbert, S. M., 1964) found that certain letters (as M, H, W, F, O and D) were confused at a distance and should be avoided. Closed letters, like A, B should also be avoided because of the tendency for the enclosed skin to slough off. In my experience various symbols, such as bars, crosses, triangles, etc. are more satisfactory. It is suggested that before branding various letters, numbers or shapes be tested at distances of 100 yd (90 m) or over and under field conditions, using skins taken from dead animals (predator kills, for example).

The main difficulty in branding is that open wounds may be created and become infected. If used: (1) care must be taken to leave the brand on the skin only long enough to do the job of preventing the hair from subsequently growing and (2) each branding operation should be carried out as quickly and cleanly as possible. This will involve previously clipping the area to be branded, disinfecting with alcohol, branding and daubing with an antiseptic grease or tar.

For any form of mutilation where the skin is broken an antiseptic cream should be applied to prevent infection. This is particularly important for predatory animals.

J. Hanks (1969) has noted that several workers have experienced difficulties in branding elephants, either from the skin peeling off about the wound and obliterating the number scar, or from the brand wound becoming septic. The same author also notes that chemical-type brands tried on several African species were not successful.

Freeze-branding offers a good pain-free solution but is expensive. Freeze-branding has resulted in a growth of white hair at the branding site. Apparently, this is due to the local destruction of pigment-producing cells, leaving an area of pigment-free skin which then grows white hairs. The individual can immediately be recognized for the brand remains visible until the white hairs start to grow. This method is said to be permanent when used on cats, dogs and cattle (Farrel *et al.*, 1966).

According to Hanks (1969) the two most commonly used refrigerants are crushed solid carbon dioxide in ethyl alcohol (-70 °C) and liquid nitrogen (-190 °C). First an area of skin is prepared by closely clipping and moistening with 95 per cent ethyl alcohol. A copper instrument is then immersed in the refrigerant until boiling ceases and the release rate of bubbles is constant. The brand is then applied to the skin for 30 seconds.

The above is only a rough description. The appropriate application time may vary greatly from species to species; the optimum times must thus be established for each species. At present freeze-branding is expensive, dangerous and little used. However, when facilities and trained personnel become available it may be worth considering.

Tattooing

Tattooing is normally not a practical way of marking for field identification but it can provide a permanent mark for identification of recaptured individuals. This is usually done on the ear by using tattooing devices available for domestic animals.

Tagging

This was the earliest method to be developed for marking large mammals and is still widely used.

Ear tags can be either used alone as a means of individual identification or, in combination with collars or other markers, as a kind of reserve marker if other marks disappear. Three types of ear tags have been commonly used: metal bands clipped to the upper stiff cartilage of the ear; circular tags, usually fixed with a self-riveting device through a hole punched in the ear, or streamers inserted through a slit in the ear near its upper edge.

Both the metal bands and circular 'button' rivets are used for marking domestic animals and are usually serially numbered. To facilitate recognizing the animals at distances it is advantageous to arrange for the tag to hold a distinctively shaped piece of coloured vinyl or other suitably coloured marker.

The following method of marking used in a study of a medium-sized deer in California (Leopold, A. S. *et al.*, 1951) may serve as a practical example of one way in which tags have been successfully used. Each animal was marked for future recognition by a combination of ear cropping and riveting coloured plastic (vinyl) tags into one ear. By varying colour combinations and the shapes of the plastic tags, and by using various combinations of ear cuts (cropping, swallowfork, underbit, etc.) 247 deer were each differently marked.

The plastic tags, $1\frac{1}{2}$ in. (3.81 cm) square, were cut into four shapes designating different sexes and ages as follows: male, square; female, triangle; male fawn, rectangle; female fawn, cross. Shapes other than the square were backed with plastic of a contrasting colour. Each tag was fastened to the ear by an aluminium rivet stamped with a serial number and a return address. Tags were placed on both sides of the ear. Individual marking was derived from combinations of ear markings such as mentioned above and by placing the tag either in the right or in the left ear.

This type of marking allowed deer to be recognized with $10\times$ binoculars in forested country at distances of between 50 and 100 yd (46 and 91 m) and occasionally at distances of 300 yd (274 m).

Ear streamers consisting of strips of soft plastic (vinyl) may be doubled and the doubled part then slipped through a slit in the ear below the cartilage to form a loop. Both ends are then passed through this loop and knotted against the cartilage to form an excellent mark. Such streamers have been used on deer and on a number of African mammals and have been seen in the field at distances of up to 800 yd (731 m) in good light (Hanks, J., 1969).

3313   *Radio Telemetry*

In radio telemetry a radio transmitter is used to carry information from its source to a distant observation point, where suitable radio receivers are located for monitoring the transmitted signal. Some forms of radio telemetry have been in use for biological purposes since 1921 (Winters, 1921) when the US Army Signal Corps developed radio telemetry equipment for transmission of heart sounds from ships to shore-based receivers. In 1959 Le Munyan and three colleagues initiated the use of a miniature radio transmitter for use in animal studies (Le Munyan, C. D. *et al.*, 1959). It was used in studying the movements of the woodchuck or marmot, a large rodent (*Marmota monax*). Such an obviously useful technique developed rapidly and in the last 20 years the technique has been used on dozens of species and in all the major regions of the world.

Two books may be consulted for information on the broad field of radio telemetry, including telemetry developed for medical and physiological studies (see Slater, L. E., 1963 and Mackay, R. S., 1970). The latter contains details of the electronics involved and should prove most useful for those who wish to build their own equipment.

In locating individuals by radio telemetry an individual is equipped (by collar, harness or surgical implant) with a miniature signal-emitting radio transmitter. The animal carrying the transmitter can then be located by receivers fitted with directional antennae. The receivers can either be fixed in one place or used as portable receivers. In either case the directional antenna indicates the direction from which the transmitter signals are coming by emitting louder signals when the antenna is pointed in the right direction.

If receivers are stationary at least two receivers with rotating antennae are spaced at some distance apart in the study area. They automatically take bearings on the transmitter at frequent intervals or continuously. The animal carrying the transmitter is located by triangulation. Several transmitters can operate simultaneously on different animals in a study area, the individuals being separated by different pulsing rates or different frequencies or pitches. Information can be stored on tape or cards and the data analysed automatically by computer. Bearings can be taken on about ten different individuals every minute (Adams, L., 1965).

Used in this way the technique requires an expense and degree of expertise that is beyond the capacity of most game departments in developing countries. A very careful assessment of the likely benefits to the organization of the costs of equipment and highly skilled manpower needed should be made before seriously considering fixed receiving stations.

Portable receivers with directional antennae present a simpler and much less costly tool. These can be mounted on a motor vehicle or airplane (Weeks, R. W. et al., 1977) or carried by hand. Because of the short antennae, very high frequency receivers can be operated from inside an aeroplane (Newgrain, K. and Horwitz, C. M., 1979). The animals located may then be recorded on a map at the location where the two bearings intersect. Black bear and elk have even been located by satellite (Craighead, J. J. et al., 1971 and Craighead, F. C. Jr et al., 1972).

Radio tracking can thus provide a very fine means of building up information on home range. For very shy animals in country with good cover it may prove to be the only practical way. However, as suggested elsewhere, dots on a map are not necessarily the answer to practical management problems. Ideally one would combine field observations and radio tracking to obtain more relevant information. Fortunately there are many good studies which demonstrate the usefulness of this technique as an integral part of a field study.

Two examples of a fine combination of radio telemetry with other field

techniques are the studies of M. G. Hornocker (1970) who studied and analysed mountain lion predation on two species of deer in the Western United
States, and Seidensticker, J. C. IV *et al.*, 1973 who studied the social organization of the mountain lion. Without radio telemetry, studies of this excellent
calibre would have been impossible. In a study of grizzly bear, F. C. Craighead Jr and J. J. Craighead located hibernation den sites by placing transmitters on a number of bears and using portable direction-finders. In 2 days they
were able to monitor approximately 1500 square miles (3885 km^2) of country
using a transmitter with a 4 mile (6.4 km) range. They feel that one of the
greatest assets of a tracking system is the advantage of getting an observer
quickly to a position where he can observe a particular instrumented animal
and reason about and interpret what he observes on the spot. Even in very
large forested or wooded areas man can arrange to meet with an animal and
observe it undetected. 'This ability to locate, move in, or follow and then
observe is in itself a tangible research reward' (Craighead, F. C. Jr and Craighead, J. J., 1965).

Radio telemetry has been used as a means of learning the cause of death, as
the transmitter still functions after the animal dies (Zwank, P. J., 1977).
Because of the difficulty in locating dead animals in the field soon after death
occurs, this technique seems likely to contribute much to our future understanding of natural mortality.

Using radio telemetry, D. H. M. Cumming (1971) studied warthog and
learned rather precise sizes of home ranges (varying between 160 and 488
acres—65 and 198 hectares), that the home ranges were permanent, their
extent of overlap (which was great) and that dispersal normally took place in
clans rather than individually. In radio tracking the white rhinoceros R. N.
Owen-Smith (1971) obtained information on the pattern of occupancy of
cows, water-dependency, birth and sub-adult dispersal. I. Douglas-Hamilton
(1971), through radio tracking, was able to considerably improve our understanding of home ranges in an elephant population. Even though hemmed in
by human settlement on three sides and a lake on the fourth side, the
elephants did not use all the land that was available to them.

An innovative extension of radio telemetry has been developed in New
Zealand by attaching transmitters fitted with small microphones to collars on
the possum (*Trichosurus vulpecula*), a small marsupial introduced from
Australia (Greager, D. C. and Jenness, C. A., 1979). This enabled these
observers not only to relocate the animal but to record various characteristic
sound such as licking the fur, threat calls, feeding, chewing, biting, scratching,
breathing while asleep and the call of a native owl.

L. Adams (1965), in investigating the cost of relocating small mammals by
radio telemetry, found it to be only about one-hundredth of the cost of
obtaining the same information by live-trapping. His comparison did not
include the expense of developing the equipment and technique, which is
great. But when prorated over future use he concluded that radio telemetry is

an economical way of getting a great deal of useful information on movement of small mammals.

For large mammals, particularly ungulates, this is not always the case, for marked individuals of many species may be repeatedly observed under natural conditions without much difficulty. On the other hand, for wary ungulates, nocturnal animals or some of the secretive large predators radio telemetry is perhaps the most appropriate technique to use in learning about home range, migration and other kinds of movement behaviour.

The biggest disadvantage in using radio telemetry is the initial cost and the expertise needed to develop a practical workable system of relocating animals. Because of its sophistication normally this technique would not be recommended for new or recently formed wildlife organizations in developing countries. However, since transmitters and receivers of various kinds are now being produced commercially, this brings the technique within reach of many wildlife organizations which otherwise could not formerly seriously consider its use. Even with transmitter and receiver equipment reasonably available prospective users are strongly advised to consult someone competent in this type of electronics at as early a stage as possible to determine to what extent commercially available equipment may be suitable for the particular study or management purpose in mind.

With this in mind it may be useful to summarize some of the difficulties that may be associated with radio tracking. If only a general location is required then difficulties are few. However, if a precise location is desired then several factors can combine to make this difficult. The problem of sampling error has been elaborated by J. R. Tester (1971) and J. T. Springer (1979). The problem concerning these workers may be regarded as one based around sampling error. When relocating animals by radio, an error in sampling is not an error in the location of a point. The error is an area, which they have called the 'error polygon'.

Briefly, when using a directional antenna attached to a receiver, a series of measurements of direction will not be exactly the same. The variation in directional readings forms, for each set of readings, an arc of error. The location of the transmitter-carrying animal can then be defined as the intersect of two or more error arcs. These arcs thus form an error polygon (Springer, J. T., 1979) as shown in Figure 11. In practice this lack of precision is usually not great enough to prevent the use of the technique. It is mentioned here simply to illustrate the main problem faced by field men, of trying to make the 'error polygon' as small as is consistent with study or management objectives.

J. R. Tester (1971) considers that the three main sources of error are: (1) error in the system, that is inherent in the mechanics and method of operating both transmitters and receivers with their directional antennae; (2) reflection or refraction of signals by physical features of the environment can cause error; and (3) when an animal is moving and bearings from two observation

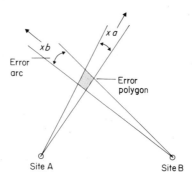

Figure 11 An example of a location obtained by radio triangulation. Xa and Xb indicate the spread of readings obtained at sites A and B respectively. The maximum differences between readings at each site form the error arc. The shaded area formed by the intersecting error arcs forms the error polygon. This is an area somewhere within which should be the transmitter-carrying animal (modified from Springer, J. T., 1979)

locations (for example, A and B in Figure 11) are taken at two different times.

Tester also mentions practical considerations which are useful to keep in mind:

1. The most precise locations are taken with bearings intersecting at 90°.
2. The user must know his exact location on a map before the proper bearing can be plotted on a map.
3. With two fixed receiving stations the distance between the two antennae used in triangulation is critical to the accuracy of locating an animal.
4. The distance from the animal being tracked to the field observation (triangulation) stations has a marked influence on the accuracy of location. The greater the distance, the lower the level of accuracy.
5. If animals are moving, 10 minute intervals between different sets of readings will improve accuracy.
6. Electric power lines, dense stands of trees, buildings or an animal being in an underground den may affect accuracy of location, as in these circumstances the signal may be subject to refracting.

Finally, the more sophisticated or costly the technique the greater the chances of losing time and money if it fails. In the early phases of a study (survey or management programme) one must emphasize against the ideas presented earlier in Sections 2100 to 2130. If radio telemetry is the only way, the least expensive way or the best way to accomplish a particular clearly defined study or management objective, and if it is possible to do so then by all means use it to good advantage. However, if the same objectives can be accomplished by less expensive means or by means more suitable to the expertise available within the existing wildlife organization, or quicker, then the more locally suitable techniques might well be favoured.

Although elaborated in Chapter 8000, it is well to remember that in the technique-choosing (method formulation) stage of a project some techniques may be as valuable for publicity or public relations purposes as for learning about animals. The capture of live animals and radio telemetry sometimes fall in this category. This is by no means a criticism of such techniques. It is simply another consideration that must be kept in mind in developing methods to achieve the study objectives within the policies and objectives of the organization concerned.

3320   Other Ways of Learning of Movements or of the Extent of Localization

Even the largest of the large mammals can, when environmental circumstances are right, live consistently within a surprisingly small area; and for all but a very few species the boundaries of their home ranges are remarkably fixed, especially during the period of their breeding life.

Once individuals have been captured, marked and released, a picture of the way in which a species occupies an area may be built up from recognizing and recording the same individual in its natural habitat over a period of time. It is not always necessary, however, to mark individuals to improve understanding of, for example, the pattern of occupancy, migration routes or the extent to which a population is localized.

Migration routes can often be determined by observing trail patterns from the air during or immediately after the migratory movement. For example, trails of some of the large African mammals can be as easily traced from the air as can routes along which numbers of cattle have been herded. Patterns of game trails of various kinds that lead to water and that disperse from water at various distances (and according to various specific patterns) are easily recognized.

One of the commonest reasons for marking animals is to see the extent to which individuals are localized, and this in turn can be used simply to see if, for example, one population can be considered as essentially distinct from another for purposes of management. This question commonly arises when there is no obvious barrier to movement between the two areas, that is in two adjacent or nearby areas which are clearly part of a continuous population. For management purposes can the two arbitrarily divided areas be considered as separate or must they both be managed as a single unit? The answer to this question will be important in assessing the intensity of plant and/or animal manipulation that will be required in a given management area.

If the habitat is essentially the same in both areas then it may be necessary to determine the home ranges of a few naturally or artificially marked individuals to get a feeling for the extent the population may be localized. However, if between the two areas there are obvious differences in character or in proportions of the various habitat elements available, and if these differences are important for the species, then *the population will have responded accordingly*. Such habitat differences may thus be reflected in differences between the two animal populations. Such differences may be demonstrated by differences in size, productivity or physical condition, the latter being the most sensitive evidence of change, as noted in Section 3800.

For example, in two such adjacent areas if there is a demonstrable difference in physical condition at a certain time of the year (usually taken near the end of the most critical season) then this fact alone indicates that, in spite of the population being apparently continuous, individual animals could not be moving back and forth freely throughout the two areas considered or there would have been no such demonstrable difference. Individuals may have crossed the arbitrary dividing line but home ranges were small enough and fixed enough to allow the two areas to be recognized as holding substantially different populations.

This is crude evidence but, nevertheless, good indirect evidence of localization. It can be made somewhat more precise by narrowing the distance between samples. A real advantage of such evidence is that it may be quickly obtained. Both for deer in New Zealand and several ungulates in Africa I have demonstrated significant differences in physical condition (assessed visually as shown in Section 3800) by carefully assessing the condition of twenty adult females in each obviously different habitat, the entire exercise taking less than half a day. Because it is so easily obtained this kind of observation can prove useful in selecting specific areas for marking or otherwise identifying a few individuals whose known movements would then act as a direct cross-check against the indirect evidence. This approach is further elaborated in Section 3800 and Chapter 5000.

3400   OCCURRENCE—ANIMAL . . . MAINLY CONCERNING POPULATIONS

Any study requiring knowledge of the species of animals present, their past and present abundance, their distribution, together with a description of location and character of available habitats, should start with a survey of existing literature. A review should be made of relevant documents pertaining to the area, including previous studies, official reports, government files, travellers' diaries and journals, and of course books on the fauna and flora. Checklists of mammals and birds exist for almost all continents or zoological regions. The natural history museums can also be asked for information on the localities in which specimens have been collected in the region concerned. In national or regional studies it is useful to maintain a map for each animal species, on which a dot is made for every authentic record, cross-referenced to a register giving details of the record. If this map is supplemented and kept up to date with records from all other sources such as local naturalists and hunters, a distribution picture soon builds up which greatly reduces the area it will be necessary to survey to determine the present status of the species concerned.

Such reviews are particularly useful at national or regional levels. However, in developing countries, for a given park, reserve or in your particular study area, it is rare to find such information available in the detail you require.

The initial step taken in most wildlife surveys is to determine what animals occur in the study area, where and when are they most numerous and, for some purposes, how many. Consistent observations well recorded will result

76

in a list of all species actually seen and of other species not seen but known to occur on the basis of tracks or other signs observed in the field. One good way of building up this type of information is the 'Species Checklist'.

3410 The Species Checklist

This is a list of species encountered in the field day by day. It is convenient to construct a field form so that a space is avaiiable for each day of the month (Figure 12). Such a list is of special value for field workers stationed in parks and reserves. It can be a valuable cross-check against journal-type and other notes. It also helps in preparing the weekly or monthly summary of field activities, in compiling the frequency with which different animals are observed over a standard route or in recording the first and last seasonal observations of a species. Separate sheets should be kept for birds and mammals.

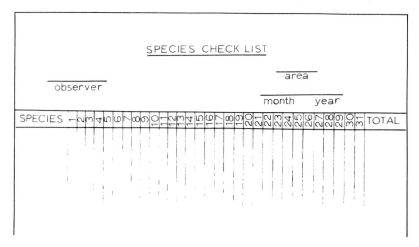

Figure 12   An example of a species checklist

If use conscientiously, the species checklist will supply information on the species present, and their relative frequency. If a day is spent exclusively in savannah dominated by mopane, one would expect a different association of animals than if observations were made in a teak forest. One hundred such days in each area, for example, would produce checklists that will differ not only with respect to species composition but also with respect to the relative frequency with which any one species was seen. Even a simple form such as this can, when consistently and thoughtfully used, be expected to raise important questions in the investigator's mind concerning the distribution and habitat requirements of the animals observed.

3420   Wildlife Population Surveys

The following will describe some of the wildlife population survey methods most commonly used. It will cover briefly the merits and weaknesses of each, with emphasis on those which have application in developing countries, and especially in Africa where some were tested. One may wish to use these techniques for various purposes. Some are used to obtain *total counts* or *population estimates*, that is a count or estimate of the actual number of animals present in an area. Others aim at obtaining *population indices*, which are used to detect upward or downward trends in a population, or to find out how two populations compare one to another. These two approaches are considered separately.

3421   *Population Counts and Estimates*

The methods available for determining the number of animals present in an area may be divided into two categories: the true census, where all animals are actually counted, and the sampling estimate, which derives from counts made on sample plots. Actually most counting techniques are really estimates because total counts are rarely feasible and usually costly. The techniques may also be divided into three classes depending on whether they are based on (1) direct enumeration of animals, (2) ratios or kill data, or (3) indirect enumeration. That is the presentation adopted below.

3421A   *Total Counts*

Total counts are normally attempted either from the air, or by intensive drives on the ground. Game species of the open plains, or species living in large herds, or confined within an open enclosure, are those most satisfactorily censused by total counts. Total counts have been made both aerially and from the ground of some species on areas varying from a few acres to a few thousand square miles. A comparison made by the College of African Wildlife Management between simultaneous aerial and ground counts of medium-size game in sparse bush cover (about 3 per cent of the ground covered by trees and shrubs exceeding 6 ft (2 m) in height) of Tsavo National Park gave a ground count more than double that of the aerial count. Yet the persons counting from the air felt they had counted all or nearly all animals. The species involved were hartebeest, Grant's gazelle, gerenuk, oryx and impala. It was concluded that aerial censuses of these species in areas covered only lightly with woody vegetation were unsatisfactory. However, at the same college it was found that aerial counts of larger species such as zebra, wildebeest, eland, buffalo, giraffe, hippopotamus, rhinoceros and elephant can closely compare with or exceed ground counts in cover even heavier than 3 per cent. In this East African example flying was done at 200–1000 ft (60–300 m) depending on the density of cover. Since, obviously, some aerial observers are more proficient than others at seeing game animals from an aeroplane it follows that hard and fast rules about what species can or cannot be counted totally under different circumstances are not feasible.

Total counts on large areas are subject to vegetation cover limitations and high cost. When total counts are attempted it is essential that the counting be done systematically by small sub-units bounded by identifiable ground features such as roads, streams, ridges, cover types and the like.

Certain wildlife schools, such as the one mentioned above, consider organized drives a method of making a total census even where there is substantial woody cover. In conducting such drives, they instruct that participants should be so spaced that no animals can pass back through the line without detection. Further, animals moving ahead or to the side of the drive should be tallied by observers stationed on the perimeter or enumerated through a count of fresh tracks leaving the area.

In this connection it is useful to remember that some species of deer and small ungulates are quite capable of remaining motionless in a thicket while beaters pass within a few feet. Graham Child (1968) noted the ability of several African animals to remain motionless, lying on the ground even in the face of lines of beaters making noise and throwing sticks and stones. Other species will slip back through the line even though the beaters are only a few yards apart: this is particularly true for animals with small home ranges. Still others are so wary that most of them will quietly leave the area as the participants in the drive are taking their places preparing to start. For many species in areas of good cover it is simply impractical to consider total counts by driving as a practical proposition. Before attempting total counts by any means, check again your objectives to see if an estimate or an index to numbers would not serve your purpose equally well.

The use of tracks in arriving at a total count also presents difficulties dealt with in Section 3423A.

Regardless of the difficulty in driving animals for the purpose of obtaining total counts, such drives can nevertheless be most useful in learning more about the species in question. Organized drives are particularly useful in in-service training courses as a means of quickly gaining experience in some of the difficulties in obtaining census estimates depending on seeing and counting animals. Such drives often result in a rough but useful estimate of the size of home ranges; an increased awareness of the remarkable tenacity with which many species refuse to be driven from their home ranges.

3421B *Sample Counts*

Most censusing of ungulates is based on incomplete counts of animals or sample counts. The results from counting on randomly or systematically located strips or other type plots are then applied to the entire area. As with attempts at total counts, sample counts can be made from both the ground and the air, but aerial counts involve a rather complex statistical treatment. The sampling design of aerial surveys, as well as the results, should be analysed by one familiar with statistics; otherwise the conclusions may be in error (see Norton-Griffiths, M., 1975, or Bleazard, S. R., 1973).

One problem which must be met when sampling by transect or strip is the determination of the effective width of the strip; that is the width of the strip on which *all* animals can be seen. King's method (Mosby, H. S. *et al.*, 1963) is believed to have merit in censusing a variety of game species in plains country to open savannah woodland. The formula is:

$$P = \frac{AZ}{2XY}$$

where P = population of area censused
A = total area
Z = number of animals seen (each species treated separately)
X = mean sighting or flushing distance
Y = length of census line.

The method necessarily involves measuring and recording the distance between observer and spot where each animal is sighted. These measurements are averaged after the strip count is completed. It is reasoned that this value reflects the distance at which all individuals of a species are being seen under prevailing conditions of visibility. Inasmuch as the average is derived from sightings on both sides of the line of travel, the distance is doubled to give the effective width of strip. The method has been tested for accuracy by randomly placing a specified number of wooden blocks in an area of grassland or open bush and running census lines through it. It has also been verified on an area of 35 square miles (92.5 km^2) in Tsavo National Park by the students of the College of African Wildlife Management. This test involved a complete aerial count and a ground count of elephants by King's method. On the ground, each of seven parallel transects, 1.2 miles (2 km) apart, were surveyed by a crew of three students. The locations of all elephants seen by the ground and the aerial crews were plotted to ensure that none had been overlooked by the aerial crew. The sighting distances were computed as follows: a compass bearing was taken on each animal or group at the moment of sighting; the right-angle distance from the point of sighting to the census line was then measured and recorded. Sighting angles were computed by subtracting the bearing to the animals from the bearing of the census line. The right-angle distances were then converted to sighting distances through division by the cosines of the sighting angles. It was found that 137 elephants were seen both from the 'plane and from the ground. Another similar verification was made with wildebeest and zebra, and results again were almost the same.

These results indicate that King's method, under some conditions, is reliable for certain African game species. It is necessary, however, to point out limitations and precautions to observe in its use:

1. The method is not accurate when the cover is sufficiently dense to permit some animals to move away from the observer's range of vision without being seen, or to move for a distance before being observed. It is probable that the method can be used on most game species where there is no more than a 15 per cent cover of shrubs or trees exceeding the height of the animals. For some species of animals it may be used even in denser cover when the trees shed their leaves seasonally.
2. The method is not effective for many of the predators because of their ability to move away without detection.
3. Each species must be estimated separately since sighting distances vary by species.
4. Unless observers are skilled in estimating distances, or unless distances are measured, it is recommended that sighting distances be computed as outlined above rather than estimated. Experience has shown that estimates may vary considerably with individual observers.
5. If the area of census includes several clearly defined vegetative types in which visibility conditions differ, each type should be counted and the populations should be computed separately. This requires of course a good vegetation map or set of recent air photos from which such a map may be produced. Without such a map the method has little merit as a means of estimating numbers, although, as mentioned below, it would still remain useful as an index.
6. Census transects should be far enough apart to avoid the frightening of animals from one transect to where they may be seen from another transect. Minimal spacing is perhaps 1.24 miles (2 km) in open bush savannah of 5 per cent or less cover to 0.62 miles (1 km) in cover up to 10–15 per cent.

Some wildlife biologists have used a variation of the King method to census animals. They have used the average of right-angle distances from the census line to observed animals to compute strip width. It is obvious, however, that if such a measurement is used for the mean sighting distance, the result will be different from that of the King method. If the same number of animals is counted by both methods, the average sighting distance used with the King method will provide a lower population estimate, since the computed width of strip will be greater.

Another variant of the transect method of sampling is the use of a strip of predetermined width. It may be of constant width throughout in a uniform habitat, or it might be variable to fit visibility conditions along the transect. If visibility and species density vary by vegetative type, errors will arise when counts in all types are lumped collectively. For best results, the width should be varied by species and cover conditions. A strip sufficiently narrow to allow all species to be observed in all types results in inefficient use of effort. There is also a temptation to use a wider strip than can be adequately covered in the denser cover types, resulting in low population estimates.

Under certain conditions, as in intensive long-term study areas where one can afford the time to build and use more precise standards of observation, it has sometimes proved useful to develop ratios of conspicuousness, based on a similar idea suggested by Howell (1951). This can only be done if for some specific area you have knowledge of the exact numbers of animals against which you can compare the number you see on a walking transect or on a roadside count from a motor vehicle. For example, in a Zimbabwean national park a certain stretch of road was known to have at least eleven steenbuck as they were recognized from differences in their facial and ear markings. However, over a period of 3 years, never more than three of these eleven were seen during any one observation run. Expressing this as a ratio (3/11), if one travelled a certain distance in similar country, observing under similar conditions (vegetation, season, time of day, etc.) and saw fifteen steenbuck, this would indicate a population of about fifty-five steenbuck along the count strip:

$$\left(\frac{3}{11} = \frac{15}{x} \; ; \quad x = 55 \right).$$

The strip count method also lends itself well to walking transects to calculate numbers of dead animals in an area in times of high mortality.

Another variation is the use of a car or other vehicle along a road, the road being the census transect. In using vehicles the same limitations and precautions mentioned above apply, but because of the longer distances covered it is especially important to carefully locate each observation in the appropriate vegetation type for separate calculations in proportion to the occurrence of the different vegetative types present in the study area.

Extra precautions to keep in mind in association with vehicles are:

1. The same vehicle or type of vehicle should be used. Counts differ when counts from Landrovers were compared with counts from other cars of different colours or shapes.
2. The counts should be made in the same way; e.g.:
   (a) Counts made by standing in the back of a Landrover or truck differ from those made from the front seat.
   (b) Driving should be at the same speed. In a test made in Wankie Game Reserve, Zimbabwe, 15 miles (24 km) per hour was set as the standard speed. Slower or faster runs altered results.
   (c) In this same area it was found that a deliberate systematic turning of the head from left to right produced more visual contacts with animals than by darting glances here and there at random. The area was one in which the range of visibility varied greatly second by second from 25 yards (23 m) to 300 yards (273 m) in the dry season and in the wet

season from 5 yards (4.5 m) to 200 yards (182 m). These are fine distinctions serving to emphasize the need to be absolutely consistent in the use of this technique.

(d) A consistent time of day should be used, preferably before other cars have used the road. The numbers of previous cars and the length of time elapsed since the passing of previous cars will greatly affect your counts of certain species.

(e) As mentioned above season and time of day are important; but if you are in a national park or game reserve which is open to the public only for a certain season then the time of your counts in relation to the opening date of the tourist season will affect the counts. For example, in Wankie Game Reserce, in 1958 and 1959, at the opening of the tourist season, warthogs ran away from cars at distances of over 100 yards (91 m). Three months later cars could stop on the road and photograph warthogs from distances as close as 15 feet (4–6 m). Thus wariness affects flushing distance which in turn affects the threshold of visibility (see Section 3426B).

The constant width strip for a given species in a given cover type is the only feasible means of using the transect method in aerial census. Various techniques may be used to measure the width of the strip observed from the aircraft. A common method is to use strips of tape on the wing strut, so placed that the angle between them and the viewer's eye intercepts a 100 (91 m) or 200-yard (182 m) or wider marked strip on the ground from a fixed flight level. It is important with this method that observers maintain a constant posture. Slumping into the seat, or straightening up, can result in sizeable difference in counts.

Other factors influencing the accuracy of censusing are the uneven distribution of certain species, variations in visibility of the animals, variations in human ability to see and count, variation in altitude and attitude, and the difficulty of maintaining a straight flight line. For a guide to publications on air survey techniques and practical examples of the use of aircraft on a specific project see Norton-Griffiths, M. (1975) and Caughley, G. (1979). Table 3 exemplifies variations in technique depending on species and flight objective.

Another variation on estimating numbers of animals from the air is to photograph them and count the individuals later from the photograph. This is sometimes useful for large animals occurring in very large herds and conspicuous enough to be recognized from a photograph or projected photographic slide.

If this method is applied, because of what seems an inherent human tendency to photograph every photographable animal, special care must be taken to ensure that some standardized system of fixing and aiming the camera is used to include only those animals within the strip to be counted. Shutter speeds must be fast, 1/1000 second or over if possible but negatives need not be large. Enlargement from either 35 mm or $2\frac{1}{4} \times 2\frac{1}{4}$ in. are perfectly adequate for this purpose.

Table 3   Flight plan requirements of a research biologist[1]

Purpose	Height	Strip/W metres	Strip/W yards	True air speed (mph)	Headings	Flap	Aircraft
Hippopotamus census	400	N/A		<60	As reqd	15°	S/Cub
Elephant densities across park boundary	200	100	110	70	As reqd	15°	S/Cub
Census of ungulates	200	50	55	70	As reqd	15°	S/Cub
Aerial v. ground counts	200	100	110	70	As reqd	15°	S/Cub
Census of rhino in GMAs	100	100	110	75	As reqd	Nil	S/Cub
Elephant visibility v. time of day	200	100	110	70	As reqd	15°	S/Cub
Survey of elephant track density	200	200	220	70	160°/030°/240° Magnetic	15°	S/Cub
Census of tuskless cow elephants	200	N/A		70	As reqd	15°	S/Cub
Sampling national park/game management areas	200	∞		110	As reqd	Nil	C185
Census of elephant	200	100	110	93	130°–140°/310°–320° Magnetic	10°	C185

[1] After Bleazard, S. R., 1973.
N/A = not applicable.

3421C *Ratio and Kill Data*

There are numerous methods providing population estimates from capture–recapture data, or changes in sex and age ratios created by known
selective kills, or tagged–non-tagged ratios, but these are not discussed here
in detail since they have had little application to management of wild herbivores under prevailing conditions in most countries.

In North America and in Europe this has proven a valuable technique for
estimating very large populations of ducks and geese where large numbers
can be tagged and returns of tags from successful hunters are well organized;
but for large mammals population estimates based on tagged animals seems
as impractical as attempting total counts. Those interested in a more detailed
assessment of this, and several related techniques, including a more elaborate
discussion of limitations to its uses because of the difficulty in realizing several
prerequisite conditions may consult Mosby, H. S. (1963) and Giles, R. H.
(1969).

Cropping rates are sometimes used to estimate numbers when comparable
figures of animals removed annually are available for several years. In Tanzania, for instance, approximately 2750 elephants were killed annually for
several years and the herd size was believed by biologists to have remained
constant. Annual recruitment less mortality was estimated at 5 per cent in a
stable herd. A minimum herd estimate for Tanzania (exclusive of non-hunted
area) would thus seem to be 55,000 (since 0.05 of 55,000 = 2750). If poaching accidents and disease are responsible for some losses, say 20 per cent of all
mortality, then 2750 would be closer to 4 per cent (0.05 less 20 per cent) and
the estimated population would be about 69,000. This method is a bit rough
but sometimes suffices for management purposes (see Chapter 7000 (yield)).

3422 *Indirect Enumeration*

Indirect estimates of game numbers have been achieved through track and
defaecation counts and quantities of forage consumed. Track counts have
been used to count animals going to and from water where tracks crossing a
path close to the only available water are counted and brushed out daily.
Migratory herds of deer have been censused by much the same method by
daily counting and obliterating tracks of deer that had crossed a dirt road
traversing the migration route at right angles. The method is not usable for
species like wildebeest where large numbers may migrate on the same trail. It
appears to have limited value in Africa except for species where movement is
essentially individually or in small family groups (see Section 3423).

Counts of faecal droppings can be used as a census method if the following
conditions hold:

1. The defaecation rate for the species concerned is known.
2. The number of days a specified area has been occupied by the species is
   known.

3. There has been no loss of droppings during the interval, and those dropped during the specified interval are identifiable from older ones.

Then the population estimate is derived from the formula:

$$\text{Population} = \frac{\text{Area} \times \text{Droppings per unit area}}{\text{number of days} \times \text{defecation rate}}.$$

One factor that works against use of this method in tropical areas is the action of invertebrates in destroying animal faeces. Another is the habit of several game species to defecate repeatedly in the same spot; other species cover their droppings. Also, with non-migratory species, one cannot determine the age of droppings except by removing old droppings from permanent plots, or spraying them with paint at the beginning of the count period. Defecation rates for most large wild species are unknown. Not much has been published on the defecation rates of African game mammals, for example, although limited information for the elephant indicates a rate of about twenty per day. Elephant is one species which might be censused through faecal counts since it defecates more or less randomly and the droppings are less destructible than those from smaller game (Robinette, W. L., 1968).

At present measuring quantity of forage consumed appears a method of little value in censusing because of the problem of separating use by different animal species.

In summary, it is stressed that this is a brief statement of some methods for censusing game animals. Only the more common methods have been discussed.

For those wishing additional information on animal censusing, I recommend four publications: Overton, W. S. and Davis, D. E. (1969); Robinette, W. L. (1968); Norton-Griffiths, M. (1975) and Caughley, G. (1979).

3423 *Population Indices*

Many wildlife biologists and administrators feel that, since population estimates are difficult to obtain, the results are often debatable, and the costs are high, they prefer indices. Indices to population trends are easier to obtain and are normally sufficient for management purposes and for many research purposes as well. Many wildlife agencies base hunting quotas or other management measures upon analysis of range conditions, hunting kills and population trends. For the latter purpose a reliable index to numbers is as useful as a census and much cheaper and simpler to obtain.

Although each of the above-mentioned methods has serious disadvantages when used to count and estimate populations, most of these disadvantages disappear when they are used simply as indices to monitor changes in numbers of animals present in an area. For example, instead of using the counts of animal droppings as a basis for calculating animal numbers, the same counts may be used as an index to numbers, expressed in terms of so many droppings per 100 sample stations. In this way you can compare observations and recognize the data as simply an arbitrary number which, if taken in the same way in the same season year after year, will let you know if the population is becoming more or less numerous or is remaining about the same. Likewise the other census methods mentioned above may be more accurately recognized as indices rather than censuses. Of course once indices are available for two or more years they can be used as a basis for estimating populations or allowable harvest and this is suggested as the safest procedure for a young game department to follow.

3423A   *General Hints on the Use of All Indices*

In adopting, modifying or inventing a technique to obtain an index to animal numbers there is one overriding rule: be consistent. This means in general doing exactly the same thing in the same way; usually in the same season of the year. In other sections some of the changes in technique which have produced different results have already been mentioned (see e.g. Section 3421B); but this phenomenon of changing results due to changing technique is common to all techniques. Differences between results taken by different observers are in many cases due to slight differences in the application of the technique. Particularly where more than one observer is involved it is considered essential to prepare written 'ground rules', thoroughly understood by all observers.

Often a situation arises where someone has obtained an index to numbers several years past and, in considering getting another check on the population, you have recognized deficiencies in the early method which you want to improve. The normal procedure in this case would be to make your improvements by all means, but also take the observations in the same way they were taken earlier so the results of your changes can be compared with the old way. This will help to produce a better interpretation of the past figures. If you do not do this it will be difficult, and maybe impossible, to interpret your new figures in comparison with those taken by the previous method.

It is misleading to the point of being unacceptable to simply take someone's past population estimates and compare them with your own present estimates without giving very careful attention indeed to the question of the methods used in each case and whether these figures can in fact really be compared.

Confidence in the use of various techniques and in interpreting the results naturally comes with increasing experience in their use. The best method is to be cautious of accepting anyone's advice without checking, testing, seeing in some way for yourself if the advice is sound, if the technique is relevant and usable for your purposes and under your conditions. This applies also to any advice given in this handbook.

Sometimes circumstances are such that counts are so easy to do, they are made, perhaps because they are interesting and fun to do, without any notion of using them for any practical purpose. Even so, if they are done in the same way it is useful to construct a form including the date, time of day and the number of animals seen, for example, in a single sweep of the same power telescope or binoculars from a certain spot. If the counts reach any degree of consistency, and they often do, they can be very useful indeed either in providing clues to answering some of the questions relating to your study objectives or providing more questions which you otherwise may not have considered.

For example, a field crew trapping deer in a California study (Leopold *et al.*, 1951) drove each morning from a forest station up a logging road to a ridge along which several traps were located to trap and mark deer on their winter range. The first traps were at or near the edge of a large clearing over 400 m long and about 300 m wide. The crew formed the habit of stopping before the edge of the clearing as they arrived first thing in the morning, walking quietly to a large rock from the top of which most of the clearing could be seen and then looking for marked deer and counting all deer seen. Counts were so consistent that they were used as an index to numbers of deer in and adjacent to the clearing. This was later useful in determining the dates the deer arrived, increased, and finally levelled off in numbers in early winter as they migrated into the area, and also the dates and rate of migration away from the area in springtime. Although the presence of lynx, coyote or bear had no effect on the threshold of visibility for these counts, counts dropped sharply when the ridge was visited by mountain lion. When the herd suffered heavy mortality in an exceptionally severe winter, counts from the rock were consistently 80 per cent lower in the following year, although similar counts from other observation points showed little change. The question of differential mortality thus raised was eventually associated with inadequate shelter rather than food, as had previously been assumed.

This question of spot-counting has been elaborated not because it is considered a particularly fine technique, but because as an index, it could hardly be more simple—almost too simple to be considered seriously—yet the information it produced added importantly to the kind of understanding at which the study was directed, *because it was done by the same observers in a consistent way*.

3423B *Animal Counts*

Trends are often obtained by counting animals over specified routes at the same hours of day, the same rate of travel and same time of year. The system is often not very sensitive, however, in indicating actual population changes. Various studies have shown that numbers counted vary with the person and with various factors of weather, as well as other conditions affecting visibility (see Section 3426B below). These limitations, however, also exist with population estimates.

3423C   *Track Counts*

Tracks counts have been used in monitoring changes or trends in game numbers. Observations are made on roads or lanes or randomized plots from which vegetation has been removed and the old tracks brushed out. The technique may yield satisfactory results in some areas but has limited value for general use.

The biggest problem with this technique is not only the difficulty in attributing various sets of tracks to different animals, but in observing a similar proportion of tracks under different tracking conditions, which would be required to make tracks useful as an index. A group of fifty animals strung out in single file would be almost impossible to assess as different from a file of five to ten animals under many tracking conditions. Before using tracks as a technique for any counting purpose you would be well advised to watch a group of five or more animals cross a road or path together, then move up and see how many of these tracks you would have identified as from different animals if you had not seen them cross.

Tracks and trails are useful aids in identifying feeding areas, bedding areas, daily movements, migratory movements and so on, and it almost goes without saying that it is important for any field man to be able to identify the tracks of the animals with which he is concerned (see Section 3425, Identification of Tracks).

3423D   *Defecation Counts*

Dropping counts can be used for assessing annual trends in animal numbers if the disintegration rate for droppings is reasonably constant and droppings from different species are distinguishable. Faecal or pellet group counts can also be used to indicate relative numbers of large mammals, wildlife and livestock associated with different habitats.

One advantage of dropping-count indices is that they lead themselves to being closely integrated with evidence of occurrence of habitat elements in that both can be incorporated along the same transect. They also provide one simply obtained index to numbers which is not influenced by the *threshold of visibility* (Section 3426B) as are those methods which depend on an observer seeing an individual animal. The following three illustrations are given as examples of how observations taken in this way help form a clear idea of to what extent several species may or may not be competing, based on evidence of occurrence alone (Figure 13); how indices may be useful in recording seasonal shifts in animal concentrations (Figure 14); and finally how they may indicate changes in numbers following a hunting season (Figure 15).

Normally droppings are made either on cleared plots or on evenly spaced plots along some kind of transect (as described in Section 8341). If some fraction of an acre is preferred, mil-acre plots involve a circle with a 44.7 in radius. If meters are preferred, a circle with an area of 4 m^2 has a radius of about 1.13 m. Cut a stick of appropriate length and use it to determine if faecal pellets are inside or outside the circular plots.

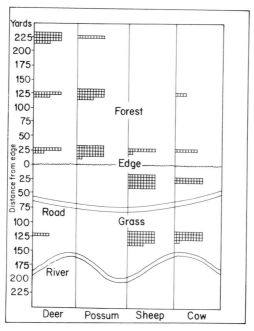

Figure 13 Differences in the occurrence of four species within 205 m (225 yd) of a forest edge in the South Island of New Zealand (modified from Riney, T., 1957a)

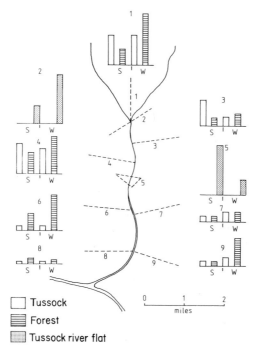

Figure 14 Distribution of total deer defecations classed as 'fresh' and 'medium' along nine transects counted in summer (April) and winter (November) from Riney, T., 1957a)

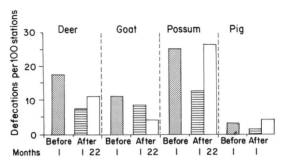

Figure 15   Number of defecations per 100 plots for six transects in a New Zealand watershed, taken 1 month before a 3-month Government-conducted hunt and 1 month and 22 months following the hunt (after Riney, T., 1957a)

Transects will be preferred by most departments because less time and effort is normally involved in getting a representative sample. After the length of transect and number of stations per transect is determined (see Section 3426A) they will be located appropriately depending on the study or management objectives. In Figure 13, for example, the object was to see to what extent the different species were associated with either forest or grassland. Transects were thus located along the edge of the forest and others were run parallel at varying distances inside the forest and away from the edge into the grassland.

In Figure 14, where an idea of the nature of seasonal movements within a watershed was desired, transects were run in different seasons, along the same lines in representative lower, middle and upper levels of a watershed, taking care to locate each transect in such a way that it sampled all habitats available at a particular level in about the proportion they were present in that area.

Transects producing data for Figure 15 were taken in the same way as for Figure 14, with the precaution that the hunters did not know the location of the pre- and post-hunt transects. They were simply told that the area had been sampled in various parts and that, if their hunting was effective, it would be reflected in changes in numbers of droppings later counted. The distribution of animals killed was in fact fairly evenly distributed throughout this drainage.

A defecation was counted if half or more of the heap was inside the mil-acre circle and each set of pellets strung out through the plot was counted. Arbitrary age classes are sometimes assigned to each pellet group, as, for example, 'fresh', within the first week following defaecation; 'medium', between 1 week and 3 months; and 'old' as probably between 3 months and 1 year. Known-age defaecations are used in setting up criteria for assigning age classes to defaecations. A certain amount of practice is usually required before accuracy can be attained in assigning age classes to faeces, so, before an observer is permitted to run a line alone, it is suggested he practice with an experienced technician until he can closely duplicate (to within 5 per cent) the results of the more experienced man.

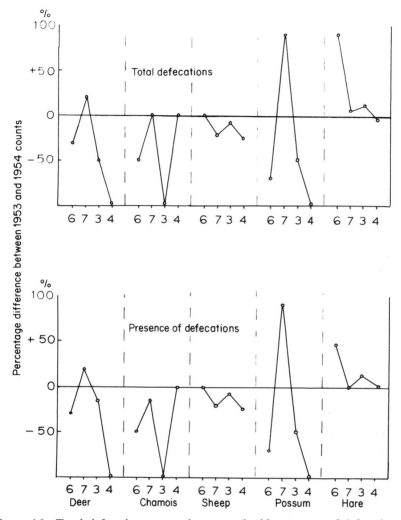

Figure 16 Total defecations counted compared with presence of defecations (expressed as a percentage of plots containing faeces) for five species on four transects in two consecutive years. Average number of plots for each of the numbered transects is 177 (modified after Riney, T., 1957a)

Results of the faecal counts are expressed in terms either of numbers of defaecations per 100 stations, or of their presence or absence per 100 stations.

If pellets are classed into age groups some arbitrary decisions must be taken to set criteria. For example, in temperate regions the individual pellet when fresh is often soft and slippery, with a surface slime. Normally, in about 1 week the pellet has become firm on the outside even in humid situations. 'Old' pellets may be distinguished from 'medium' pellets on the basis of the extent to which they are judged to be decomposed. Pellets are broken open before assigning them to an age class, and so long as the inside of the pellet appears firm and not decomposed, it is classed as 'medium'. Droppings are judged 'old' when the inside of the pellet shows obvious signs of decay, and pellets that have started to crumble are disregarded. Assigning an arbitrary age to pellet groups represents the most complex way of recording faecal pellets.

At the other extreme, it is possible to simply record presence or absence of faecal pellets inside the circular plots. When the purpose is to record changes in numbers from year to year this much simpler type of observation will give the same results as when total defecations are counted (see Figure 16). However, normally one must make more observations to obtain a comparable degree of consistency with presence or absence compared with total counts.

3423E *Call and Drumming Counts*

Call and drumming counts are used for population trend determinations of a variety of game birds. There seems little opportunity for use of such methods with large herbivores.

For those large animals making vocal calls, such as several species of deer, elephant, large predators, the recognition of distinctive calls can assist in locating animals. Thus Jenkins (1972) in making a short survey of the location of the sparse surviving population of Red Deer in Sardinia knew of their definite presence in several valleys filled with dense cover for he heard them roaring there. This would obviously not be a suitable technique for assessing numbers or even an index to numbers.

3424 *Identification of Animal Droppings*

In any work concerned with distribution, abundance and movements of game animals, it is necessary to be able to identify the droppings which are observed on the ground.

Each field worker should learn to identify, *in the general area where he works*, the droppings of all mammal species, domestic species included, and of the main bird species. Instead of relying on second-hand information, it is better to build a local collection of droppings, taken from places where animals are seen defecating. These droppings may be stored dry in labelled containers for future reference. The investigator should commit to memory the detailed characteristics of each kind of dropping by thorough, patient and perseverant review. When in doubt consult the reference collection.

In time it may be clear that droppings of a range of species are not only consistently identifiable but show negligible variation over very large areas. In these circumstances drawings may be prepared for use in local park, reserve or departmental handbooks. Examples of characteristic defecations of eleven common species taken from one such series made in a national park in Zim-

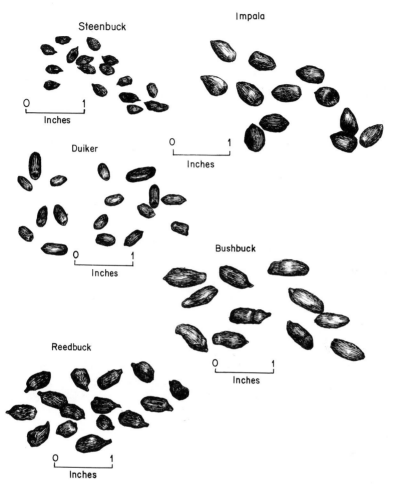

Figure 17   Differences in size and appearance of defecations of five small to medium-sized ungulates

babwe are shown in Figures 17 and 18. Differences in size and shape of adult droppings may be characteristic, as shown in Figure 17. Predator droppings once identified are usually readily distinguished (Figure 18a and b). However, drawings will probably never take the place of a positively identified collection. Using a reference collection to distinguish sable from kudu from giraffe droppings is not difficult. However, it would be difficult if only drawings were used as a guideline (Figure 18d, e, f).

Before using either a reference collection or drawings in another region, local samples must be taken for comparison. For example, in several West African countries waterbuck droppings, even when fresh, are voided as discrete pellets rather than tending to clump as shown in Figure 18c. A word of caution. Do not make your collection by labelling droppings identified by

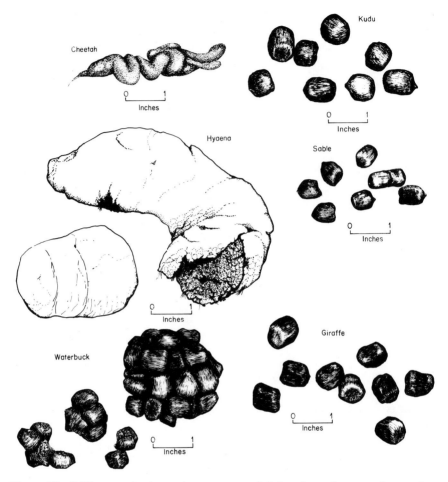

Figure 18 Differences in size and appearance of defecations of two predators, three large ungulates and the giraffe

game wardens or game scouts, no matter how reliable or how experienced these men may be. In this matter the only absolute authority is the animal itself. Observe them being dropped, collect and label. In one study area, I had to throw away nearly a year's information based on droppings because I foolishly failed to follow this common-sense precaution.

3425   *Identification of Tracks*

Tracking animals by footprints in dust, mud, sand or snow is probably the oldest known method used to identify animals present in an area, to trace their movements, to reconstruct events by sign, and to determine location and direction of crossings of individual species at regular intervals of time. Here again, as for droppings, identification of tracks can be learned through study and experience. Here too it is recommended that the track be positively identified immediately after watching the animal make it. This will diminish errors.

Mud or wet sand around a waterhole or at a stream edge, or snow will give excellent impressions of footprints. Clear footprints can be obtained by establishing strips of fine sand or soil which are swept regularly to obliterate old tracks (Bider, R. J., 1962 and 1968).

The tracks left by animals differ by species, and within the same species; they vary between the front and hind feet; between sex and age of animals. Many field books on mammals show drawings of the tracks of many species, but here again the best way to learn is to build a collection for an area by means of plaster casts. The equipment needed is very simple and can be easily carried in the field. It consists of a basin, bottle of water, several pounds of plaster of Paris and a few sheets of newspaper.

Plaster cast technique

When a well-imprinted footprint is found:

1. Put a ring of metal or bark, sand or soil, about 1 in (2.5 cm) high, around the footprint, to hold the plaster;
2. Mix about two parts of water to one part (in volume) of plaster of Paris until it has the consistency of thick cream and pours easily. *Work quickly*, as the plaster starts to set within a minute or two;
3. Pour the mixture carefully into the spoor making sure the track is fully covered;
4. Leave the plaster for about 20 minutes until it is set but still soft; then remove it and wrap in newspaper;
5. When the plaster is well set, unwrap the cast and put it in the sun to dry;
6. After an hour, wash off all the soil and debris adhering to the cast, using slow running water;
7. After 24 hours, clean the cast with a soft brush and write on the back of the cast the notes taken in the field at the time of collection. It can then be coated with a thin layer of lacquer.

A collection of plaster casts of tracks makes an interesting exhibit in information centres in parks or reserves; and in areas where animals are rather scarce, enabling a visitor to identify tracks, if even a few species, greatly improves the quality of his experience in the area.

Another way of recording tracks has been developed by S. R. Choudhary in India on tiger, and is described by H. S. Panwar (1979). It consists of carefully placing a glass plate over the track but just above it (e.g. on four small bits of wood) so as not to disturb the impression. The outline of the track is then drawn on the glass with a glass-marking pencil, labelled, protected and later the drawing is permanently transferred to paper. With the tiger, distinctive lines, scars, etc. were also recorded with the result that it was in this way possible to identify individual animals, in some instances each adult in a forest. Using this evidence, the restricted area of its home range, within which it habitually lived, was defined.

Some trials made on elephant suggest that the pattern of lines across an elephant's foot is as distinctive a means of identifying individuals as a fingerprint is for humans, or a footprint for tigers.

Such a technique would obviously be unsuited to many situations, particularly where elephant populations are high; but where animals are few and difficult to see, if some understanding is required of the daily movements of individuals and the extent to which they are attached to a certain area, the glass plate technique is simple and may be worth further use.

3426   *Special Problems in Sampling and Interpretation*

3426A   *How Can You Tell When You Have Enough Observations?*

How much information do you need to have confidence in your figures as a basis for comparing the same figures taken in the same way in different areas, or between different areas? This kind of question can lead to complicated mathematical discussions which are outside the scope of this handbook. The approach presented here is certainly not the only way, but it is a simple test to let you see for yourself when you have enough data to learn what you need to know for either investigation or management.

In taking a series of observations as a sample one must make two decisions. First, where to sample, then how many observations are needed to have a representative sample, for example, of each habitat.

Where to sample

If you know little of the habitat requirements of a species then samples should be made in all available habitat types. If you already know the habitat requirements then samples should be taken within each habitat of significance to your study animal. If 'key' areas can be identified, that is areas which are known to be critical to the species throughout the year or at special seasons (e.g. the winter season for deer in temperate regions, or the dry season for many ungulates in semi-arid areas), then it is often more useful to sample thoroughly these 'key' or critical areas than to take random samples over the whole area. Hints on how to recognize such key areas are found in Section 3732.

How many observations?

The second decision involves deciding how many observations you need. One practical approach is to ask the question in a slightly different way: is there some simple way I can tell when to stop? If 100 or 200 observations (or half that) will give nearly the same results as 3000 observations, why should I do all that extra work? If this attitude is appropriate to your case then you may wish to use a cumulative means (accumulating averages) test.

The test can be applied to many kinds of observation; for example, numbers of animal droppings per 100 mil-acre plots, ratios of males to females or young per 100 adult females, or observations made along transects such as percentage of ground cover, or average height of grass. You may want your observations (always within a given habitat type) to become so consistent that if you were to make twice or ten times the number of observations you would not change your conclusions nor add to the understanding possible with a smaller sample. If you are dealing with indices you may want some average, or ratio or final arbitrary number you can use as a guideline for further comparison.

The procedure is as follows: first, make 100 or 200 observations. Then divide these observations into groups of ten or some other size series if you prefer. Then calculate the arithmetical average for the first ten observations, then the average for the first twenty, then the first thirty and so on until all of your observations have been included in the calculation of the final cumulative mean (average). It is then a simple matter to take each mean and note the differences from the final mean (see Table 4). When plotted on a graph you will get results as shown in Figure 19. In this graph each point represents the deviation from the final mean. When averages get consistent within 5 per cent you will recognize that when your index shows changes of the order of 15–20 per cent or more between areas or in the same area in comparable seasons in different years then the change is a real one.

It is an easy method to use when you are not certain as to the best way of applying some techniques. For example, Figure 19 shows that, in one part of Africa, to obtain a comparable degree of consistency of results in describing the percentage of ground cover it required twice as many observations in the wet season as in the dry season (Riney, T., 1963a). (It also required over twice as much time to run a transect in the wet season.)

Table 4   Working table for calculating deviation from final average (example)

Sample point groupings	Sample points with litter	Accumulating total	Cumulative average	Deviation from final average
0–10	8	8	80	9
11–20	9	17	85	14
21–30	9	26	87	16
31–40	4	30	75	4
41–50	10	40	80	9
51–60	6	46	77	6
61–70	5	51	73	2
71–80	6	57	71	0
81–90	7	64	71	0
91–100	7	71	71	0

98

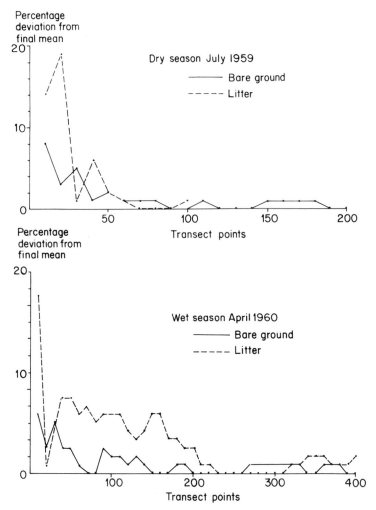

Figure 19   Technique for determining the number of transect points necessary to obtain consistent results (for explanation see text). It is clear from this test that, for these two kinds of observation, to achieve a comparable degree of consistency, well over twice the number of transect observations are required in the wet season than in the dry season (after Riney, T., 1963a)

As soon as you start using this test you will discover other things for yourself through experience, such as: the more uniform the habitat the sooner your results will become consistent.

3426B   *Threshold of Visibility*

It is important to draw attention to a commonly occurring phenomenon which should be considered before selecting a census method and before interpreting your own or another person's census estimates or indices. Any census method that depends on observing and recording numbers of animals seen is potentially in danger of being misinterpreted because of a relationship called here the *threshold of visibility*.

The threshold concept implies that the animals seen are not necessarily always in consistent proportion to the density of vegetative cover. More specifically and practically it is possible for sudden changes in numbers of animals to be observed and recorded without there being a correspondingly real change in population numbers.

Diagrammatically, Figure 20a shows what we often think we see when we assume that we are seeing less animals than are actually present but, nevertheless, in the proportion that they occur in the area. This assumption, though common, is misleading for it can easily lead to entirely unjustified conclusions as it ignores the possible operation of a threshold of visibility which, for decreasing and increasing populations, may be as shown in Figure 20b and 20c, respectively.

The reason for a threshold occurring at a certain level, at a certain point in time, almost always involves three self-evident relationships.

1. For a given species the proportion of animals actually observed will vary *as the character of the vegetation varies*.
2. The proportion of animals seen will also vary *as the wariness of the species changes*—as, for example, with the sudden opening of a hunting season or tourist viewing season.
3. If one is observing more than one species in a given area, because of species differences in habitat preferred and due to inherent differences in wariness or secretiveness between species, it is only natural that in a given habitat *one would expect to see a higher proportion of one species than another*.

   Although this principle is important for most species in most area, it is obviously less important for large mammals in completely open areas of bare ground, areas with low grass or with scarce vegetation. But even in these open conditions animals can become so wary that on sighting a man or a vehicle, even from a great distance, they will quickly leave the area.

Some practical examples will illustrate the way these three relationships may affect our assessment of animal occurrence, both in terms of where and how many animals were seen and the proportion seen as compared to those present as judged by other means.

One example involved a small valley in New Zealand, in which red deer were annually shot by hunters paid to eliminate the deer from the area. In the same area a biologist was trying to determine the effects of this hunting on the deer populations and if seasonal changes occurred in the pattern of deer use within the watershed. The topography was broken with many side-streams and gullies, and, broadly speaking, in this watershed, the major habitats con-

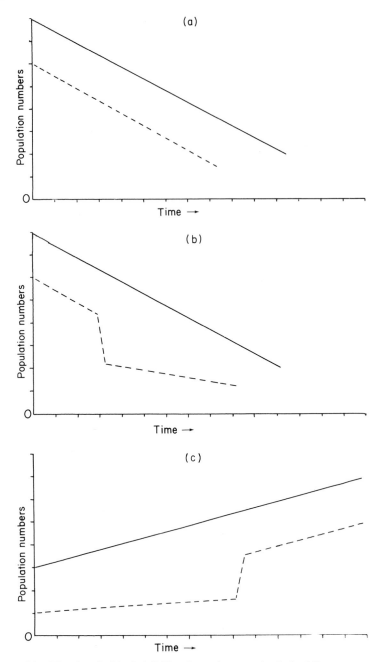

Figure 20 The threshold of visibility. In each example dashed lines represent actual field observations; solid lines represent changes in population numbers: (a) represents what we often think we see, that is our observations are assumed to bear a consistent relation to actual population numbers; (b) and (c) indicate how a threshold of visibility might operate in a declining and increasing population respectively

sisted of grassy flats in places along the river, forested slopes and grassy tops above the forest line. The difference in elevation between the stream bed and the adjacent ridges was about 2000 ft (610 m). Fresh and 'medium' aged droppings were defined, classified and counted along nine transects, all but one starting at the stream and continuing to the ridge top. Lower, middle and upper parts of the watershed were sampled in this way, as well as the area between the uppermost major forks of the stream. Figure 14 (taken from Riney, T., 1957a) shows the distribution of and frequency of occurrence of these droppings both at the end of summer, during which hunting took place, and at the end of winter.

The team of hunters involved was said to be the best that existed at that time in the South Island of New Zealand. The hunters, as well as their field supervisor, had many years field experience; they were interested in deer. Their opinions—based on considerable practical experience—when compared with the results shown in Figure 14 have several practical implications.

At the time the counts were made this drainage was regarded by government shooters as one suitable for conducting a drive. In such a drive, shooters normally start in the lower reaches and move towards the head of the valley, where it is usual to kill the greatest number of deer. The higher kill in the valley headwaters was interpreted by them as a direct result of having driven the deer there. However, the evidence shown in Figure 14 suggests an equally acceptable explanation; in this drainage, more deer can be shot in the upper reaches simply because that is where the greatest number of deer habitually live in summer when the shooting took place. The higher elevations also held a higher proportion of grassland and low shrubs where deer were more easily observed.

The hunters also considered that after a team of four men had shot the area for 4 or 5 days the deer had been virtually eliminated from the valley and that most of those few deer still alive had moved to adjacent valleys. When first entering the valley each year, a hunter would contact deer about twenty-five times daily, shooting seven to ten deer. By the end of 4 days he would be fortunate to contact deer three or four times and would be lucky to average one per day. However, as an experiment, at the end of 4 days' hunting, observers were stationed at sunrise on hillsides across the small valley from the most experienced hunter and kept him under observation throughout the day. Twenty-three times deer were seen to observe the hunter and quietly move out of the way. At this time a shot heard distinctly by the deer, even from distances of 2–3 miles (3–5 km), would send most of the deer into denser cover. The hunter shot one deer and was aware of contacting deer only four times.

Another belief held by the hunters was that winter shooting was not appropriate here, for in winter it was claimed that deer migrated over the main mountain range into other areas. In fact the counts of droppings showed no such decrease in the winter months in spite of the fact that winter was the season with the most rigorous weather. Evidence from the dropping counts suggests that deer use forest and shrub areas more in winter than in summer and are therefore harder to see. The discrepancy between how effective and devastating hunters think they are as hunters and how ineffective they have been proven by careful investigations is well known to biologists, for it is commonly noted in the literature. Yet in many instances where careful study has been made to evaluate the effect of hunting, differences are so small as to have little effect on the population (Leopold, A. S. *et al.*, 1951). The principle not only applies to hunting and to campaigns for so-called 'elimination' of animals, but can seriously affect conclusions in many other ways.

For example one of the keenest early observers of African mammals, and a pioneer in the development of the well known Kreuger National Park, after 30 years' observation, wrote in his book on African wildlife (Stevenson-Hamilton, J., 1947) that the breeding season of the common duiker was nearly throughout the year but with a peak mainly between November and March. The only month in which he had never observed a recently born duiker was July; many years later two observers collected 1552 skulls of duiker from a continuous campaign to eliminate wild animals to reduce the spread of sleeping sickness. Examining the proportion of very young animals confirmed one conclusion of Stevenson-Hamilton, that a peak period for calving occurred in December and January. At the same time a second peak occurred June and July in the very month Stevenson-Hamilton concluded there were least if any births. Although he was active in the field July is in the wet season when cover is most dense. Because he could not see the calves he assumed they were not there.

Wariness is often described in terms of flushing distance; that is how close you can get to an animal before it runs away. Along a road newly opened to tourists in Wankie Game Reserve I recorded warthog running away from cars at distances of over 100 yd (91 m). Three months later it was from the same car possible to photograph warthog as close as 15–20 ft (4.5–6 m). Eland, on the other hand, were in this area wary even at the end of the tourist season.

Observing animals in their natural environment is a basic way of learning. It is a good way and will probably always be with us as a tool for learning. However, failing to appreciate the significance of vegetative cover and animal wariness operating in combination can lead even the best of field observers to entirely or partly erroneous conclusions: conclusions that are important as they lead to practical decisions of large mammal management (see also Section 8140).

3426C *Cross-checking Census Methods*

There is a not too uncommon practice for biologists to use more than one census method to cross-check the population estimates or indices one against another to see to what extent the results are similar. If the two estimates give approximately the same estimate of population or index-order of comparison, then more faith is placed on the estimates or indices.

It is of course always good in principle to cross-check against one's own results. The chief danger in the approach lies in the difficulty in selecting *appropriate* cross-checks—that is, true cross-checks—which do not simply result in working out another technique which incorporates the same set of biases or assumptions inherent in the first technique.

A common example of census cross-checks which are not necessarily convincing, no matter how close the results, are the use of walking counts, road counts or other counts which depend on seeing animals, to check one against another. Obviously if the threshold of visibility were affecting any one of these techniques it would be affecting them all, and the fact that one got consistent results might be misleading. Cross-checking a technique based on observing animals with one based on counts of droppings on appropriately sited transects would be more appropriate.

3500   OCCURRENCE—ENVIRONMENT

The field observations that wildlife officers make of animals are greatly enhanced when the observations are related to the type of habitat in which the animals occur. This is because both its distribution and the continued abundance of an animal species is directly dependent on the extent that habitat requirements of that species are met. It is consequently essential, in any large mammal investigation, to be as aware of the habitat as of the animals living within and depending upon it.

Any general description of an environment may include such information as climate; topography; hydrology; soils; vegetation; man and his activities, as for example, managed forests or agricultural or pastoral lands. As W. Dasmann (1971) has noted for deer, food, cover and water are the main broad elements that habitat must provide to enable large herbivores to survive. Vegetation is usually considered the most important component of a description of animal environment. It not only supplies food but also provides shelter from the weather and screening from predators or man. Vegetation is of prime importance also because it is a reflection of other components of the ecosystem. If the vegetation of an area is well known, it is often possible to infer the nature of the climate, the soils and the hydrology, and it may be possible also to infer how man has in the past utilized the land. For these reasons it is often important to stress the vegetation component of the ecosystem in describing environments, but the other components should not be neglected. All, in one way or another, influence the presence, abundance, distribution and movements of game animals.

Several simple techniques for describing the occurrence of different environments or of different specific habitats are described below, but first consider two background considerations relating to animals and their habitat.

First, it is desirable to have some measure or assessment of environment or habitat even at a simple descriptive level and even if the study is directed at what you may consider as the animal population alone. The latter is, after all, a man-made idea of little value because, whether we like it or not, populations exist in certain environments and prefer certain habitats and as the environments and habitats differ so do the statistics relating to animal populations differ.

The second consideration relates to the specific objectives of your study or management activity and concerns the need to take information from both animal and habitat (environment) in such a way that the information may be closely compared and related. This will greatly facilitate discovery of cause–effect interactions between animal and environment. A simple example of a scheme for relating animal and habitat data is to make both kinds of observations along the same transect; for example, counts of droppings as well as measurements of grass, forbes, trees, shrubs, rock, etc. (see Section 8341 for further discussion of multipurpose transects). Another way to relate animal to habitat observations is to group observations of animals made in the vicinity of a vegetation sample together and associate them with that sample (transect or plot) or series of samples, while observations of animals in a different habitat would be associated with another appropriate sample or sample series.

3510 Checklist of Vegetation Types

Habitats are commonly described in terms of vegetation types. It is often useful to include mention of vegetation types in your field notes describing some aspect of animal behaviour, or to locate in terms of habitat some special information you have gathered about animals.

Rather than classify and describe all the vegetation types in an area, it is recommended that the field worker use a checklist of types already described for a given country or region. He should observe the major plant associations present and then try to place his observations into one or other of the types listed. This does not preclude the necessity of giving more detailed descriptions of the types observed if required.

Checklists not only vary with regions but also with authors. An example of a useful yet simple checklist, prepared by R. E. Atwell with nomenclature for vegetation types used in Kafue National Park, Zambia, is given below:

(a) *Dry deciduous forest:* closed stand with several strata; most trees of upper strata deciduous; understory of shrubs evergreen or deciduous; grass generally discontinuous or absent.
(b) *Thicket:* shrubby vegetation, evergreen or deciduous, more or less impenetrable, often in clumps; grass discontinuous or absent. A scatter of emergent trees may be present.
(c) *Riparian forest:* a narrow belt of closed forest along the banks of rivers. Found on nearly all permanent and many semi-permanent streams throughout the tropics.

(d) *Woodland:* open forest; tree stratum deciduous of small or medium-sized trees with the crowns more or less touching, the canopy remaining light; grass present but sometimes sparse.

(e) *Savannah woodland:* a grassland carrying a light open woodland with trees and shrubs; much mopane falls into this category.

(f) *Tree savannah:* a grassland carrying a few scattered trees and shrubs.

(g) *Grass savannah:* a grassland with trees and shrubs more or less absent.

(h) *Aquatic grassland:* open grassland, seasonally inundated, and remaining moist throughout the year, such as *Vossia* and *Echinochloa* swamps.

(i) *Herb swamp:* swamp containing *Papyrus* or other sedges.

Checklists of vegetation types can of course be much more detailed than that just given. M. F. Lamprey, of the Serengeti Wildlife Research Station, for instance, has prepared a checklist for Tanzania which includes only five main vegetation types but these are subdivided into twenty-eight sub-types. This classification is listed below:

A. *Montane forest and grassland*
   1. *Upland rain and mist forest* (characterized by hanging lichens)
   2. *Upland Miyombo* (dominated by *Brachystegia*)
   3. *Upland Acacia* (dominated by *A. lahai* and *A. rehmanniana*)
   4. *Highland grassland* (associated with montane forest)
   5. *Subalpine* (tree heaths, sedges and grasses and peaty soil)
   6. *Alpine* (sparse vegetation, alpine herbs, lichens, mosses)

B. *Miyombo woodland*
   1. *Pure Miyombo* (dominated by *Brachystegia* and *Pseudoberlinia*)
   2. *Associated vegetation types* (Miyombo with a *Combretum* layer)
   3. *Valley grassland* (associated with Miyombo woodland)
   4. *Regenerating bush in Miyombo*
   5. *Riverine forest within Miyombo*
   6. *Swamp in Miyombo*

C. *Thornbush (Acacia, Commiphora, Balanites)*
   1. *Thicket and thick bush* (more than 90 per cent bush coverage)
   2. *Medium bush* (Bush coverage of 30–90 per cent)
   3. *Open bush* (trunks more than 10 yards apart)
   4. *Grassland* (Serengeti plains, for instance, associated with thornbush)
   5. *Associated vegetation types* (minor associations, like *Terminalia*)
   6. *Valley grasslands within thornbush*
   7. *Regenerating bush in thornbush*
   8. *Riverine forest within thornbush*
   9. *Swamps within thornbush*

D. *Lowland forest*
   1. *Groundwater or coastal forest*

E. *Specialized vegetation types*
   1. *Mangrove*
   2. *Coastal thicket* (densely branching shrubs)
   3. *Saltbush thicket* (in saline conditions)
   4. *Itigi thicket*
   5. *Papyrus swamps*
   6. *West African vegetation*

These two checklists are only given as examples; similar checklists have been
prepared for many regions and should be consulted and utilized as guidelines
or modified for your own park or region. Remember, vegetation checklists
deal only with terms, with the words you decide to use in describing the
occurrence of vegetation. The vegetation still has to be described or measured
in a way appropriate to your objectives. Below are a few general and specific
ways of describing occurrence of the available habitats. As with other
techniques, before adopting, modifying, combining or inventing new techniques they must be assessed in terms of how appropriate they are to
accomplish your own particular objectives.

3520   General Subjective Description

Before any objective description of a study area, or a park or reserve, is
made, it is recommended that an overall perspective be obtained by means of
a quick general or superficial survey. This may be accomplished by a little
work with existing records and by viewing at least the major kinds of habitat
available. The latter can best be done on foot or by road but there are many
instances where an aerial flight or aerial photographs will be the best way to
form an early over-all view of the area. Sometimes traverses are made in each
different season of the year and in each major habitat to take into account
seasonal variations. In these ways a general subjective description of the area
may be made. This description should include general statements on:

1. *Climate.* Each area experiences during the year certain combinations and
   successions of high and low temperatures, precipitation, humidity,
   sunshine, wind. These make up the climate for a locality and constitute
   one prevailing control of wildlife distribution and movements.
2. *Topography.* This factor is especially important in extreme conditions,
   such as mountainous terrain where altitude may play a role, north- and
   south-facing slopes may considerably differ, or in very flat country poor
   drainage may create immense differences if the surface area is covered
   with water during the dry or the rainy season. Topography may be the
   determining factor in several kinds of animal movements, ranging from
   a pattern of daily movement to seasonal migrations.
3. *Vegetation types.* A general estimate of the total area covered by each
   vegetation type (using a checklist) should be possible: grassland, savannah

woodland, riverine forest, swamps, etc. For some species this estimate is not very meaningful unless it also expresses the degree of mixture of the vegetation types. This is referred to as *interspersion*. Large blocks of a single type of vegetation are not usually very good year-round habitat for most large herbivores although there are exceptions. The meeting place of two different vegetation types is usually referred to as *edge*, and many species are *edge species* (see Figure 13). It follows that the quality of an area is often proportional to the length of edge, or in other words to the degree of interspersion. Hence, if needed, one should be able to give an indication of the extent to which vegetation types are interspersed.

4. *Water.* Water is essential for all animal life. Some species require free water every day, others less frequently, and others no free water at all, for they are adapted to obtain sufficient water from the vegetation they eat. The distribution and movements of animals may consequently be influenced a great deal by the distribution of water within the area (rivers, lakes, ponds, swamps, springs, seepages) and this is an essential part of the habitat description. Not less important is the degree of permanency of the various water bodies, especially the conditions prevailing during the most critical season of the year, usually the end of the dry season in semi-arid areas or, in temperate regions, the winter season.

5. *Food and cover.* At this stage of the description, a general appraisal of the area may be given as to the availability and quality of food (for browsers, grazers and carnivores) and cover (both shelter from the elements and screening effect), and as to the distribution of the feeding areas in relation to the cover (interspersion). Each species has specialized requirements and preferences for certain food items as well as cover for resting, breeding and escape. Even a subjective description of these habitat components may be valuable.

6. *Human activities.* The presence or absence of people in the area, their abundance and distribution, their past and present land-use activities (agriculture, grazing, forest clearing, fires, hunting) and the influence of these activities on habitats available should also be described.

The quality and completeness of the information gathered and outlined in this general subjective description of the environment may be very uneven according to the circumstances. In many cases lack of time will not permit detailed studies and so the subjective description should be as thorough as possible. It may be done with the management of one animal species in mind or for the entire ecosystem. Emphasis may shift from all major factors in the latter case to a few factors or species requirements in the former case, but the concept of the ecosystem should never be forgotten.

For those working constantly in a park or reserve, it is often useful to write a short paragraph each week describing the chief characteristics of the environment over the past week. Over a period of years these summaries will provide a good record of seasonal changes and may often provide evidence of changes in specific habitats associated with seasonal movements, migrations or breeding seasons.

3530 Objective Description

In contrast to subjective description of habitats which is qualitative by definition, an objective description is based on actual measurements. These may be presented numerically or pictorially and then may be used for planning habitat improvements, to evaluate changes at a later time, or for comparison with other areas. Perhaps the main difficulty encountered in these objective descriptions is to decide on how many details should be included. It is easy to get so involved in figures, gadgets, instrumentation and complicated sampling designs that the field worker forgets the main objectives of his study. As a guideline, adopt techniques as practical as possible and in line with the objectives spelled out at the beginning of the investigation.

1. *Climate.* Data on climate should be based on information gathered by weather situations in the area, or as close as possible to the area, for a period of many years. The most meaningful climatic data, in tropical regions, are the total yearly and the mean monthly rainfall, usually expressed by a histogram, month by month. Temperature is rarely a limiting factor in the tropics but it might be useful to give mean monthly temperatures as well as minimum and maximum temperatures, at least in mountainous regions. Other climatic factors—such as relative humidity, cloudiness, prevailing winds—may be valuable in certain cases but even graphing temperature and rainfall is good evidence of the pattern of seasonal change.
2. *Habitat map.* A habitat map is normally essential to wildlife management. Such a map can be a great asset to planning if properly maintained. It will reveal the size, distribution and pattern of interspersion of habitats for various animal species. It may show the need, the desirability and the location of proposed wildlife habitat improvements. It can serve as a standard with which future surveys may be compared in order to determine what habitat changes have occurred as a result of either management or lack of management.

Existing maps of an area for forest or farm management are usually not suitable for wildlife management purposes because of the omission of smaller habitats, understories, non-economic features and other characteristics of great importance to wildlife. In developing a habitat map the field worker becomes well acquainted with his study area. When he finishes he will have not only a map but also much valuable data on the topography, soils, vegetation and water. In short he will know what exists in the area, how much, where, and the pattern and distribution of the various elements of the habitat.

A wide range of techniques are available for preparing habitat maps. The question is one of selecting an appropriate method to use, just as for animals we had to select an appropriate method for determining the occurrence of animals, their distribution and density. At one extreme a study describing for all plant species their distribution and density could be a long and tedious job if applied to a large national park or game reserve. When completed one would have a fund of detailed botanical information but not necessarily a better understanding of the habitats available. Likewise forest surveys, which have as their purpose the estimation of the number of board feet available to convert to sawn logs, are not normally a useful basis for making decisions requiring a perspective of animal habitats available.

3531 *Habitat Classification*

First determine the kind of terms with which you are most concerned, from the broadest most general major environments through a range of increasingly precise descriptions of more specific habitats within the broader categories. In the grossest terms, considering a cross-section of Africa, for example, one would consider the shores, the coastal plains, the first mountain ranges, the high or low plateaux, the interior mountain ranges, the deserts, the lakes and the major river systems. When in doubt a fairly safe way to start preparing a habitat map for a wildlife park, reserve or hunting area is to use some generally accepted checklist of broad vegetation types as described above in Section 3510.

There are no hard and fast rules regarding the selection of appropriate terminology for describing available habitats, or for selecting procedures for incorporating information on habitats into your study; however, the following four general guidelines may help.

1. Where large permanent parks or reserves are involved the numbers of habitat categories should not be too great. The terminology used in describing habitats is especially important for such large areas, for ideally they should be readily acceptable and capable of being used by different disciplines.
2. When smaller study areas are involved, if possible retain the same general terminology used for the larger areas but elaborate as required for your particular study purposes.
3. Of course if the habitat requirements are well known for the species to be studied, classification is no problem for the area can be mapped in terms of the several known habitat requirements. In situations where habitat requirements are unknown, or in a park or reserve with several species of herbivore then the range of habitats available should be noted as described above (e.g. Section 3510).
4. Where close comparison between animal and environmental aspects of an ecosystem is desired, as in discovering some environment–animal (cause–effect) relationship, then an important criterion for selection may

be the appropriateness of the method in producing data to be used in making close statistical comparisons between observations of animals and those relating to habitat. For example, data from both habitat and animal may be taken along the same transect or a record of habitat in which the animals are observed may be included as an integral part of a field form recording observations of animals in the field.

Examples of specific procedures are given below, the most detailed being listed first.

3532   *Alexander Method* (Alexander, M. M., 1959)

This is a procedure easy to apply to different areas, comparatively inexpensive and still capable of furnishing detailed information of considerable accuracy. It does not require highly trained personnel, and the variation between the work of several field men is not considered serious. It is more appropriate for detailed studies in fairly small areas, rather than in constructing habitat maps for large wildlife parks and reserves.

3532A   *Field Procedure*

The first step is to obtain an accurate boundary survey of the working unit. One should be made if none is available. The boundary should be mapped in a convenient scale, such as 100 m (110 yd) to an inch but not more than 200 m (220 yd) to an inch. If the area is large, it should be divided into convenient working units by lines along conspicuous natural features. Field work is made easier by cutting the base map into sections about 20–25 cm (8–10 in.) in size so that they will fit on a clipboard. One sheet may contain several working units.

Establishment of the unit boundaries is greatly facilitated by using aerial photographs. The common 5 by 9 in. picture with normal overlap between pictures and flight lines are suitable. By using an opaque projector, the picture can be projected to fit the base map. The roads, wood margins, ponds and rivers can be drawn in quite accurately. These become easily recognized unit boundaries. Stereo pairs of aerial photographs may be placed in a clipboard and taken into the field with a hand stereo-viewer. Many points of confusion concerning topography, vegetation and the irregular shape of features such as clearings can be clarified by studying of stereo pairs.

Prepared habitat description forms (Figure 21) help to standardize and simplify the procedure. When the area is subdivided into units, each should be given a number or other designation. This should appear at the top of each description sheet prepared for that unit. It is followed by another number identifying the portion of the unit being described. These numbers are assigned consecutively as the types are described; habitat 21-6 would indicate the sixth habitat described in Unit 21.

Date *15 July 1979*   STUDY AREA NUMBER *3*   Observer(s) *K. Brown*

Basic habitat type: *21-6*

Descriptive statement:

General: *Mature open woodland*

Overstory: (estimated percent total coverage): *± 15%*
   Species, age or diameter of trees.
      Mainly white thorn (acacia albida), many over two feet in diameter.

Understory: (or shrubs) (estimated percent total coverage) *± 20%*
      small scattered acacias and combretum mostly below 3 m. high, no thickets or clumps observed.

Ground cover: (estimated percent total coverage) *± 10%*
      mainly Aristida with some Digitaria and Cynodon impression is mainly of bare ground, very little litter.

Remarks: *Nearly all perennial grasses with bases over 2cm in diameter sitting on pedestals of soil 1-2cms above surrounding (packed) soil. Bare circles caused by harvester termites conspicuous, sparse erosion pavement with stones 3-10 cm. in diameter. In addition to the wildlife concentrations this part of the Park has been grazed by cattle for the past 10 years. Early burning has occured annually for the same period.*

Figure 21 Example of a field form used in describing habitat types

   The field man, upon entering a unit, describes the habitat at that particular point, determines its extent by pacing and using a compass if necessary, sketches it on scale on the map, and inserts the number on the sheet and on the map. He can check his type boundary line by studying the aerial photographs. He then moves to the next habitat. The entire unit should be surveyed systematically. One method is to work around the outside before going into the interior. The numbers placed on the field map are location or reference numbers only and will not appear on the finished map.

   The field man should complete a habitat description form for each habitat. One system is as follows: he should determine the basic vegetation type and write a descriptive statement. The latter can be of great importance in coordinating the work of several field men. The vegetation is recorded under three main headings: overstory, understory and ground cover. The overstory tree species are listed, with estimated percentages of cover, and sometimes with average diameters, height and distance between trunks. This likewise may be done for the understory of short trees and shrubs if desired. Understory is so classified even where no overstory exists. The ground cover will consist of the herb layer. It is often impractical to classify all the species in the ground cover; therefore only the more abundant or important ones are listed. An estimate of the total percentage of cover for each of the three categories is also indicated. The 'Remarks' space on the form provides the field worker a place to add anything that he thinks is important or of interest. Each form should be dated and initialled.

3533 *Making the Map*

Each habitat type that has been described and mapped can be transferred from the field sheets to a master map of the same scale with the aid of a tracing table. The procedure from here on depends on need. A complete work map may be made with all of the location numbers on it for quick reference to the data sheets. The final map showing habitat classifications may be in colour, or in black and white. Usually there is need for a map that can be reproduced by one of the common methods. This requires use of symbols rather than colour and can be done easily if commercially prepared sheets of overlay symbols are available. As with colour, a high number of habitat types results in a more difficult job of delineation. About a dozen types can usually be handled without confusion. Type boundaries have to be made heavy enough to separate the symbols clearly. Once the map is completed, the area of each habitat type can then be determined either with a planimeter or through use of the *squares* technique (see 3534A).

3534 *Aerial Photographs of Entire Study Area*

This technique may be too rough and simple for many purposes but it is a quick way to make a start and provide a reliable base on which one can, with appropriate ground checks, elaborate and which one can refine as required.

First see how many general types of habitat you can easily identify from the photographs. This may, in temperate mountainous country, consist of bare rock, grassland, shrubs, forest and grassy flats along a river. In tropical semi-arid areas it may be various kinds of woodland savannah and grassland, perhaps with a band of riverine vegetation bordering a permanent river. It may consist of various stages of development of a single vegetation type, as for example if parts of the area were burned within the previous few months, other parts were burned 10 years and still others 15 years before. If a vegetation type looks different the safe procedure is for you to treat it as if it were different until your accumulating experience shows otherwise.

Next select examples of all habitat types on the photos and see what they look like on the ground. Pay particular attention to edges between habitat types to determine whether or not a border zone consistently exists, which in fact may constitute another habitat. Even at this early and very superficial stage in an investigation, differences in concentrations of trails, tracks or animal droppings (that is, any evidence of differences in animal occurrence) may help in deciding if a habitat is sufficiently different from the viewpoint of your study objectives to justify separate identification and mapping.

Next, prepare a map. If only one or two photographs are involved you may trace the habitat boundaries directly from the photographs with transparent

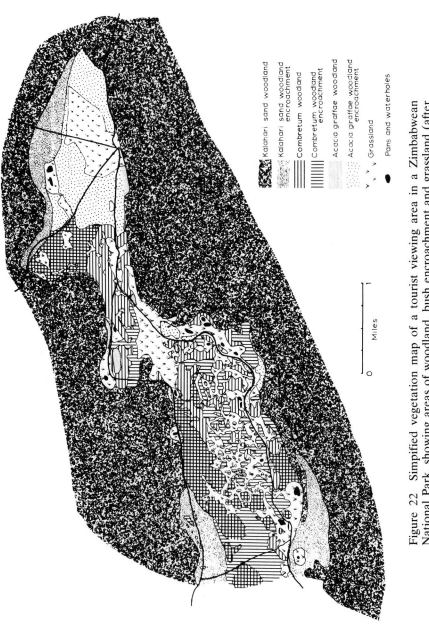

Figure 22 Simpified vegetation map of a tourist viewing area in a Zimbabwean National Park, showing areas of woodland, bush encroachment and grassland (after Riney, T., 1963a)

Kalahari sand woodland

Kalahari sand woodland encroachment

Combretum woodland

Combretum woodland encroachment

Acacia giraffae woodland

Acacia giraffae woodland encroachment

Grassland

Pans and waterholes

0   Miles   1

tracing paper; draw on the photograph with a grease pencil and trace from this; use an opaque projector and trace a larger map from the projected image, or use photography or any other means you can devise.

If the area is a large one requiring the use of many aerial photos to be assembled in a mosaic, consult foresters or land-use planners who have had experience with the preparation of maps in this way.

In mountainous country preparing habitat maps from aerial photographs has the disadvantage of not reflecting the true area because of the fore-shortening effect of the hillslopes. However, for management purposes such maps are normally a perfectly adequate means of reflecting the occurrence of various habitats, especially if you are interested in comparing areas whose topography is similar.

A habitat map may in itself be sufficient for your purposes. However, you may wish to determine the proportion of each of the various habitat elements in relation to each other or as a percentage of the total study area. Or you may wish to develop some objective index to interspersion. For both of these purposes observations may be made either on the aerial photograph or on the habitat map itself. If it is possible it is better to use the aerial photograph.

3534A *The Squares and Dot Methods of Estimating Percentage Occurrence*

The simplest method to determine the proportions of the various elements available is to use a squared or dotted transparent overlay. If squares are used, estimates are made for each square of the percentage area covered by the various habitats. The accumulated totals are then added to obtain a figure for the entire area. If dots are used, one systematically covers the entire area noting in which habitat each dot is located and recording each observation on an appropriate form. The dot method is the simplest to use.

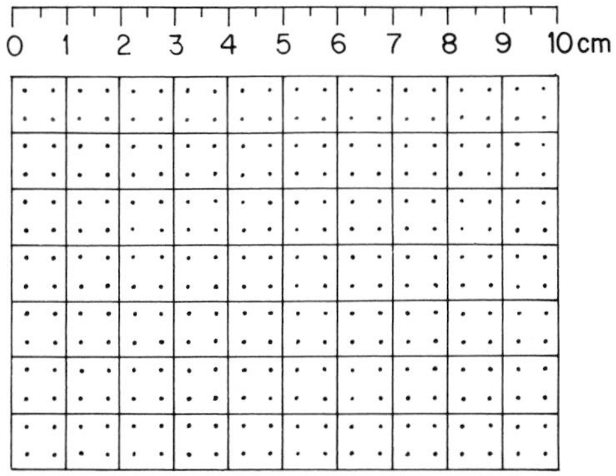

Figure 23   An example of a plastic overlay for use with aerial photographs. If squares are used, estimations of the percentage occurrence of various habitat elements within each square are made. If dots are used, the habitat element covered by each dot is recorded

A variation of this method is sometimes used where the photographic mosaics are themselves 'sampled'. For example, in a certain watershed one may wish arbitrarily to select large strips or squares from various parts of the study watershed. Squares or dots are then used for recording as noted above. Before accepting this kind of sample it is wise to check the sample results with an assessment of the entire area. If results are comparable for purposes of the study then the sample may be regarded as adequate and in future, in similar situations, samples alone could be taken. Figure 24 compares two such sets of data taken using a very crude classification of habitats into grassland, forest and bare ground or rock. In this example the six sample sections were large and an effort was made to select six representative areas, two each from lower, middle and upper parts of both study watersheds. The samples from the air photographs did not produce closely comparable results, but the difference between occurrence of forest and grassland (the main difference required to describe in this study) was in the same direction for each technique used. This area was characterized by steep and much dissected hillsides with small patches of forest interspersed with shrubs or trees. In such circumstances making a representative sample is always rather difficult. The time required to obtain the information was 6 and 3 hours for total air photo and

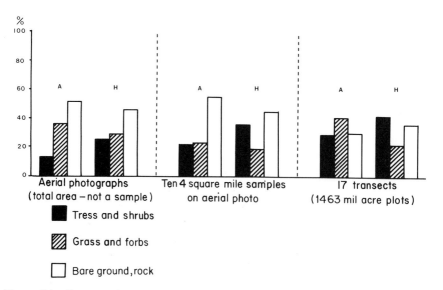

Figure 24 Comparative results of three different techniques for describing habitat. The best over-all perspective was by aerial photography, using the entire study area. In this example, although samples taken from the same aerial photographs were large, they did not reveal differences great enough for use in this particular study (see text). Transects, although underestimating the occurrence of bare rock, were nevertheless adequate for demonstrating differences in the proportion of forest and grassland in the two drainages and were best for integrating with other types of evidence taken along the same transects

air photo samples respectively. Obviously, in this case, an aerial assessment of the total area was preferred.

In areas with easily identified large blocks of comparatively uniform habitats, sampling becomes much simpler and results more easily duplicated using these same crude sampling techniques.

3535 Transects

When observations relating to habitat are incorporated into a line transect (see Section 8341) this can of course provide the most detailed quantitative information. In the above example, when more detailed information from transects was summarized into the general categories of forest, grassland and bare ground, the proportional occurrence of forest and grassland was comparable to the results taken from the complete aerial photographs. However, the transects underestimated the proportion of bare ground in the entire watershed as transects were located on the slopes of the hills and did not sample sheer cliffs, rocky bluffs and considerable areas of exposed rock on the tops of the ridges. On the other hand, in terms of grass habitats available to deer, the transects reflect a good perspective (see Figure 24).

The seventeen transects were run by two-man teams, taking 17 days (34 man-days). The Alexander method (see Section 3532) is not included in the comparison. Because of the large size and complexity of the study area it would be several times more time-consuming than the transects.

Transects provide an excellent way of objectively describing habitat elements available to animals. For example, measurements of the presence and the height of canopies formed by those shrubs or trees that cross points on a line point transect can produce a good estimate of the proportion of the ground covered by canopy and an estimate of the average height on different levels of canopy that may be present. Adding the percentage of ground covered by the basal diameter of perennial grasses to the average height of the grass produces an *index to the grass available*. Cutting and weighing and measuring the height of a sample of the perennial grasses in the study area will facilitate the conversion of the grass availability index to an estimate of weight of grass per unit if that is required. Both indices to cover and grass available may be incorporated into multipurpose transects, as described in Section 8341.

These are but a small sample of techniques for describing and comparing habitat elements in a study area. They have been selected for their simplicity and for the quickness with which the information may be obtained. For a range of other techniques consult *Wildlife Management Techniques* (Giles, R. H. Jr, 1969).

3536 Edge Effect and Interspersion

The term 'edge' is used to indicate the meeting place of two vegetation types. For large herbivores this may be important since one type will supply food and the other cover. When forest and grassland meet often a distinctly different band of edging vegetation occurs which may have special value as either food, shelter or both. 'Interspersion' refers to a mixture of vegetation that provides both food and cover. The more edges the more interspersion. Game managers have long recognized the importance of various kinds of edge effects and of interspersion of habitat types for a number of game species. Aldo Leopold, as early as 1933, considered that the density of game was directly proportional to the amount of edge for those species of low mobility that require more than one vegetation type.

Since this time we have of course learned much more of the movement behaviour and ecology of large herbivores which allows us to rephrase the A. Leopold concept as follows: *For a given species of herbivore, the more interspersed the habitat elements the smaller the individual home ranges and the greater the potential for the land to carry higher populations of that species.*

Since most species of herbivores have rather fixed home ranges it is clear that the extent to which major elements of the habitat are interspersed may, for many species, be even more important for the animals' welfare than are the total areas available of the various habitat types.

3536A   *How to Describe or Measure Interspersion*

There is no standard technique and perhaps this is a good thing as it facilitates using whatever method is most convenient and consistent with your study or management objectives. In the past the normal way of dealing with interspersion was simply to mention that it (or edge effect) was important and it occurred, or that the degree of interspersion was high or low. But very little has been done in the way of measuring interspersion in such a way that different areas may be objectively compared. A description of differences in the degree of interspersion between two areas is usually made by comparing aerial photographs of the two areas, providing the differences are great enough to be easily recognized.

If aerial photos or habitat maps are available it is a simple matter to produce numerical indices relating to interspersion by following one of several methods:

1. The number of different pieces of different habitat elements within a given area may be counted and recorded.
2. Using the squares overlay technique mentioned above (Section 3534A) for each square one records the presence of edges between two or more habitat elements. This will result in a total number of edges recorded in this way and the proportion of the total number which involves each habitat combination, as for example, forest–stream, forest grassland, forest–shrub.
3. Another way of expressing interspersion is to express the character of interspersion as a ratio between the numbers of pieces of habitat

(numerator) to the percentage of the total area occupied by the sum of those pieces (denominator).

4. If you use line transects which systematically cross the study area you may simply include a separate column in the transect field form for recording the points along the transect where one habitat type joins another. The best perspective of interspersion will be based on either actual photographs (e.g. 1–3 above) or on a detailed habitat map if available.

Although such arbitrarily obtained numbers work perfectly well as indices in comparing two or more of your study areas, if you wish to make your results more readily understandable to other workers you may wish to consider expressing your 'edge' contacts as so many per mile or kilometre and your numbers of separate pieces as so many per square kilometre or square mile.

Regardless of how it is done, if habitat elements are assessed the degree of interspersion should also be assessed even with a rough index. It will pay big dividends in understanding many species of herbivores.

3537 Water

The availability of water will normally be indicated on the habitat map of the study area. For many species, particularly if there are no marked seasonal differences in the availability of water, this is enough. However, if the species studied cannot survive for considerable periods without water, then a special effort may be required to map the available water during the most critical dry season.

3538 Shelter

Classifying and mapping the vegetation types will normally provide the basis for describing shelter or cover when this is necessary. However, in situations where vegetation is low or sparse, and in situations where a winter may be severe, evidence of sheltering effect depending on topography and prevailing winds may be needed.

Foresters have experimented both with the use of numbers of small flags planted in the ground and recording the differences in the time of their deterioration in the prevailing winds (Lines, R. and Howell, R. S., 1963) and others have recorded the angle of the skyline, taken at each of the sixteen principal compass points on selected areas and tried to combine these two approaches (Howell, R. and Neustein, S. A., 1965). These trials work well in some areas; not in others. This is why the technique is not described in detail. However, obtaining some good index to exposure would often be useful, particularly in management-oriented studies of over-wintering animals, and someone will one day develop a good way of objectively doing this. In the meantime it may be useful to keep in mind the potential importance of topographic shelter in areas with little vegetative cover and assess its importance through carefully recorded field observations in the season of greatest potential exposure (see also Section 3610).

3600   UTILIZATION—ANIMAL

Evidence on utilization by animals can be phrased in different ways depending on the relevant facts needed for particular study or management purposes. Thus, it may be appropriate to regard 'animal utilization' as evidence of use of available habitats, or as evidence of use of essential or minimum required habitat components. In planning research on utilization one should try to match the level of detail with the evidence available from, or designed to be taken from, the environment at the occurrence or response levels. Thus, if the habitats available are described in general terms as shrubs, trees, perennial grass, etc., the evidence of utilization should be gathered using the same terms and classification.

If more detail is required care should likewise be taken to obtain the same level of preciseness of environmental or habitat generalizations as is consistent with any of the other levels of observation. This is consistent with the approach elaborated in Section 2023 and will save much time and effort.

3610   Shelter

Shelter may well be an important element in the habitat of every mammal. It certainly deserves more attention than it generally receives, for lack of shelter can exert a major controlling or regulating influence on a population. Shelter is here used in several ways: as protection from the elements, as screening from predators or as a refuge from biting insects. Some ways of obtaining evidence of an animal's use of shelter are discussed below.

3611 *Direct Observations*

Simple direct observations may be made by observing and recording an animal's use of shelter in the field each time an animal is observed. Shelter commonly involves vegetation, and in observing one records the kind of vegetation, its height, size of trees, presence of understory, thicket or other relevant details relating to growth form. However, shelter does not necessarily involve vegetation. I have seen deer sheltering from wind, rain or snow on the lee side of hills, in road cuts, in steep-sided erosion gullies, in peat diggings, in abandoned forest sheds, under overhanging cliffs, in caves and abandoned mine shafts (see also Section 3538).

Observations on shelter should first be made at the time of the year when shelter is known to be important; for example, for deer in temperate regions, during the winter period. If a more specific question is considered—for example, involving the role of shelter in influencing calf mortality—observations would in that case naturally be made during the calving period and for the following few months. If the critical seasons are unknown, periodic recorded observations throughout the year will help in defining them.

To help speed the collection of information and to facilitate its interpretation, or when more than one observer is involved, observations may be recorded on forms of your own devising or modified or adopted from that of another worker, such as shown in Figure 25. Such forms may form the basis for weekly or bi-monthly summaries which should reveal not only seasonal changes in the use of cover but differences in use between sexes and age groups when this occurs.

Observations should be made during a number of contrasting weather conditions to answer questions such as 'Where are animals to be found in the heat of the day under extremely hot conditions?'

An animal's use of shelter under conditions of heavy rain, strong cold winds or during thunder showers is easy to observe and is often rewarding, as the response of each species differs and can be difficult to predict. For example, while in an African game reserve I wondered what use several common species made of shelter during heavy thunder showers. Eventually I found opportunity to observe animals in such a storm in an area characterized by grassy clearings interspersed with scattered Acacia and edged with bands of closed canopy forest. The storm was spectacular, with rapid lightning strikes, ear-shattering claps of thunder and torrential rain. At the onset of the storm elephant, buffalo, kudu and sable moved quietly out of the clearings, disappearing in the forested areas. A troop of baboon went into a frenzy of diffuse behaviour, almost constantly screaming while racing from one Acacia to another, climbing trees and jumping out of branches then racing to another tree and hugging the tree trunk, hugging each other and generally being quite beside themselves with panic. There was in fact a slow edging of the group towards the forest edge but this was not immediately obvious for at any moment individuals would be streaking about between trees in all directions.

Red Deer   Form Recording use of Shelter   J. Innes

DATE 1964	TIME	ANIMALS	LOCATION	WEATHER	DESCRIPTION OF COVER	USE MADE OF
7 Dec	07.30	3 hinds one yearling	Beech forest E at Loch Monk $\frac{1}{3}$ way to top	Snowing, following heavy snow in night.	Mature beech forest, see Transect No. 10 Photo taken.	Bedded on lee side of trunks over 3 ft in dia.
8 Dec	14.00	1 large stag	Hill N. of L. Monk $\frac{1}{2}$ way to top.	Heavy rain cold wind 1 ft snow on ground.	Dense beech thicket. No transect. No photo.	Bedded in best protected area.
10 Dec	11.00	Group of 12 feeding hinds and calves.	Tussock grassland above timberline on ridge W. of L. Monk.	Clear but with increasing wind.	Forest edge Transect No. 15 Photo taken.	When gusts of cold wind became strong enough the group moved into the protection of the forest understory thickets. I was down-wind no other disturbance noted.

Figure 25 An example of a form for recording the use of shelter

At the same time a group of warthog stood firmly in a clearing with their backs together and heads facing out in different directions as musk oxen do when danger threatens in the arctic. Reedbuck lay crouched on the ground in shallow depressions in the open, with their necks stretched forward and their heads nearly touching the ground.

Another approach is to observe individuals or small groups of animals continuously. If the animals can remain undisturbed and their progress and activities can be plotted on a map this is an excellent way to observe the use of shelter as an integral part of their daily activities. By accumulating such information you can gradually form a perspective of the use of shelter by your study species. Such observations can be used by themselves as evidence of use and even a few such observations may be useful in either planning, cross-checking or interpreting other more indirect evidence of utilization of shelter.

3612   *Indirect Evidence*

Many species of browsing and grazing animals have rather small and fixed home ranges and observing and recording the major trail patterns within these ranges can provide useful supplementary information on the use of cover. Normally well-established trails are found between feeding areas, watering areas and areas used for shelter. When these elements are close together the sizes of the individual ranges are normally small. When the elements are spaced out the home ranges become larger. Differences in size and shape of home ranges thus depend on the location of particularly these three habitat elements (Riney, 1950).

Many species, particularly of deer, consistently use the same bedding areas. As mentioned above these are normally linked with the feeding or watering areas by a trail system of some kind. When bedding or resting areas are in forest or shrubby areas it is difficult or impossible to observe the selection and use of such areas by undisturbed animals and indirect methods are necessary. In areas of very sparse populations in forest areas it is sometimes possible to follow individual trails from the feeding or watering areas to areas where shelter is required or preferred for resting or bedding. However, this is often difficult. In most situations the use of a good, quiet, tracking dog on leash is invaluable. Locate a fresh trail leaving a feeding area, then quietly move along until you contact the animal, or its recently vacated bed.

Some species concentrate in numbers in response to extreme weather conditions and develop special daily or seasonal patterns of utilization. Some seek shelter from the sun during the hottest part of the day, others concentrate away from cold winds, or deep snow. Feeding areas may shift seasonally because of more favoured shelter in certain areas. These possibilities could be extended to make a long list; it is up to the investigator to discover evidence for what does actually happen with the species concerned in a particular study area.

Special mention must be made of the use of cover as screening or protection from insects or predators, including man. The various ways of obtaining information on use of screening vegetation in these instances is similar to the methods suggested above, but special care must be taken in interpreting the results if evidence is based on observations of animals, for numbers of such observations are almost always deceptively low. This is due largely to the operation of various thresholds of visibility as described in Section 3426B.

For example, in several parts of southern Africa, habitat associated with the greater kudu almost always involves areas with very good cover, often thickets in the Acacia and Combretum savannah woodland. Kudu are known to feed on nearby farmland or grassland but such feeding is mainly nocturnal. In certain national parks, however, kudu are commonly observed feeding during the day in open savannah woodland. The contrasting use of cover and difference in the timing of feeding and watering is due mainly to disturbance by man. For all large mammals the use of cover differs before, during and after clearly defined hunting seasons. In interpreting evidence obtained on the use of cover, various disturbing factors should thus be taken into account. This quite naturally often involves comparison with (obtaining similar evidence from) other areas which have differing combinations of disturbing elements.

3620   Food and Feeding Behaviour

3621   *General*

Studies of food habits form a large part of the activities of game departments and of wildlife research throughout the world. A thorough knowledge of the food habits of species can contribute significantly to broadening the fund of knowledge used as a basis for making management decisions. However, this is not always the case. In my opinion the main difficulty to avoid is that food habits studies are often irrelevant and the following approach is oriented with this in mind.

Specifically what is meant by food habits studies depends largely on what kind of evidence is required, although such studies generally deal with what and how much animals eat. Relevant evidence may consist of the species composition of the diet or seasonal or monthly variations in its composition. It may be important to know the parts of the food species being utilized, for example, fruit, seeds, leaves, twigs, roots, bark, bones, flesh, skin, viscera or entrails. The daily amount of food consumed is sometimes necessary to know, especially when estimating the numbers of animals a given area can carry, but not all estimates of carrying capacity require this information.

Food, in the broad sense, means more than plant or animal matter. The intake of water and minerals, e.g. from either natural or artificial saltlicks, is also part of the 'food habits' and these aspects should not be overlooked by the investigator. In some areas the unavailability of water or minerals may constitute important limiting factors on occurrence of abundance of animals.

Chemical analysis of the various foods composing the diet are often useful in interpreting the significance of items from a nutritional standpoint but this is a more detailed type of food habits study than is appropriate to this handbook (see Giles, R. H. *et al.*, 1969).

Sometimes included in food habits studies are studies of feeding behaviour, which treat more with how the animals get their food, and through such studies evidence is produced of the proportion of an animal's time spent in feeding and where and when (how often and how long) feeding occurs. In defining problems of competition between one or more species, studies of feeding behaviour is often more relevant than determining which food the species have in common.

The techniques reviewed below give examples of how these types of evidence may be obtained.

3621A *Principal Foods and Preferred Foods*

Before reviewing techinques for obtaining data two categories of food must be mentioned; most important, key, or principal foods, as distinct from preferred or most palatable foods. It is important to distinguish between these two categories of food.

The principal foods of an animal population are those which it eats in greatest quantities. These form the largest percentages of food items in the animal's diet. A rumen may contain twenty easily identified species of plant while the main bulk of the food consists of only five or six species, often less. These few most-used species are the principal foods. When measuring animal use on plants, the principal foods are generally known as the 'key plant species' and will be referred to again in Sections 3731 and 3914.

Another caution is that no matter how good the information on principal food preference, as here used, refers to an animal's choice of available foods, following W. Dasmann (1971) and Stoddart and Smith (1943). Food preference is thus the extent to which a food is consumed in relation to its availability. An index to food preference must therefore combine evidence taken from both the 'plant-occurrence' and 'animal-utilization' levels of an ecosystem and involves the following simple calculation:

$$\text{Food preference index} = \frac{\text{Percentage of food species present in diet}}{\text{Percentage of food species available in environment}}.$$

Thus, the higher the resulting number the more preferred, or palatable, the plant species.

The greatest difficulty in using such an index lies in the fact that the percentage of a food species generally present in a given environment is not necessarily that available to the animal concerned, although it may be so. A knowledge of the home ranges of a few individuals should put one in the position to estimate the extent that this may pose a problem in interpretation (see Section 3221).

Another caution is that no matter how good the information on principal foods and preferences in one area, care must be used in applying the findings to another area (even a nearby area in a different successional stage) because of the different spectrum of plants available.

3621B *Adequate Sampling Levels*

Another important aspect of food habits studies is the assessment of the point at which an adequate level of sampling has been reached by using techniques described below. Too much sampling is a waste of effort and money. One simple and commonly used approach is as follows: the sample is subdivided into a number of equal subsamples; the cumulative information from the subsamples is then plotted on a graph as shown in Section 3426A (see Figure 19); the amount of change created by adding subsamples is then studied. A point will be reached at which the amount of change obtained by adding more subsamples will be very small: this usually is an indication that the proper level of sampling has been reached.

Two considerations may be useful: (1) initially it may be necessary to carry sampling beyond the desired level in order to assess the adequacy of the sample; (2) the amount of sampling will ordinarily vary in accordance with the numerical importance of a particular species in the diet; the species most represented in the diet requires considerably less sampling than that least represented. For most management purposes if the principal foods of a population are sampled this is usually good enough.

3622 *Stomach Analysis*

Stomach analysis is used mainly in areas where sport hunting, game cropping or extensive game control operations are carried out as part of the management scheme of the area. The shooting of animals for the sole purpose of studying their food habits is not usually considered as warranted unless the species is particularly abundant, or unless the animals are also needed for other studies.

3622A *Rapid Rumen Analysis in the Field*

The speed of this technique results from reducing and simplifying the decisions to be made, thus making possible the accumulation of evidence of utilization, even when little time is available or when only untrained field men are available for recording. The technique was developed in New Zealand for use by hunters under the supervision of a biologist. Early trials tested the extent results could be consistent between a given observer making repeated assessments of different samples taken from the same rumen and between different observers in assessing a range of rumen samples. From these trials it was concluded that classifying the rumen contents into one out of four or five categories was unsatisfactory unless a very large sample of animals was taken. However, a three-way classification proved easily repeatable and results were highly consistent between different observers even with a few (under twenty) samples. Each sex and age group is of course assessed separately. The technique consists of:

1. Opening the rumen.
2. Mixing the contents with a knife or stick.
3. Making one of three decisions by answering the question—do the rumen contents consist mainly of:
   (a) shrubs or trees (woody material);
   (b) grass (herbaceous material); or
   (c) is woody and grassy material so mixed it is difficult to decide?
4. Enter the decision on a form with separate columns for specimen number, date, location, male, female, calf, yearling, adult, mainly shrubs, grass, or mixed.
5. Under a final column it may be useful to identify the leaves of main species which are still complete and undigested. Each investigator or team should have readily available a collection of leaves from the main plant species known to be used by the animal studied. If a leaf cannot be identified in the field it should be put in a small envelope, labelled with the specimen number and collector's name and later identified and recorded.

Results in terms of broad category of food taken are presented as the proportion of the total sample occurring in each category or as a ratio between different categories or between one category and the total observations. Even this crude technique will provide evidence of seasonal changes in diet and of differences in use when comparing areas whose habitats markedly differ. Although crude it is remarkably sensitive to seasonal change and to comparing gross differences in stomach contents between areas.

The chi-square method, described in any book on statistics, is normally used to test if the apparent differences are statistically valid.

The list of main identified species encountered (the final column on the form suggested above) will give slightly more detail of the diet and indicate seasonal changes in proportional composition. This information can also provide a useful guide to the selection of the key plant species.

3622B *More Detailed Rumen Analysis in the Laboratory*

One of the clear situations in which a detailed analysis of plant species eaten is required is where several species of herbivores are feeding in the same habitat type—for example, on the same area of grassland. This more detailed information may be obtained from animals by stomach or rumen analysis, by analysing faecal material or by direct observation.

For detailed stomach analysis of large mammals it is generally sufficient to collect 1 litre of the stomach contents. With smaller animals the whole stomach is preserved. In ruminants the sample should be collected from the rumen, or paunch, because the plant material there is less broken down and therefore more recognizable. Before collecting the sample, it is important to thoroughly mix the contents in order to achieve a representative sample. If possible, the sample should be placed in a fine-mesh ordinary kitchen sieve and flushed with water before being preserved. This will remove the fine unidentifiable particles and gastric juices, and thus facilitate preservation. Polyethylene bags, provided that they are watertight, are more convenient than glass jars for storing the samples under field conditions. The recommended preservative is a 8–10 per cent solution of commercial formalin and for proper preservation it should represent at least seven times the volume of the sample to be preserved. The samples should be rinsed again before analysis, or better, soaked in a deformalizing solution made of one litre of water, 55 g of sodium bisulphite and 35 g of sodium sulphite. The analysis of the samples can be carried out with either wet or dry specimens.

The first step in the analysis is to segregate the material according to various categories. These, of course, depend on the objectives of the study. For example, it may be sufficient to determine the proportion of vegetable v. animal matter in the diet of an omnivorous animal, or the proportion of woody vegetation or grass in the diet of a herbivore. Segregation, under such circumstances, is a relatively simple matter. Most often, however, the diet is to be broken down to the species level, otherwise such a detailed technique is inappropriate. This requires a good knowledge of the plants of the study area. A reference collection of material likely to be encountered in the diet should be at hand and frequently consulted. Fragments belonging to the same species are placed together. Unidentified fragments are kept for later identification. Proper identification often requires that the specimens be examined under the microscope and compared against slides from a reference collection. Once all fragments have been properly identified, the respective quantities are measured. Measurements can either be effected in terms of weight or volume. The relative frequency of occurrence of fragments can also be used to describe the diet. Volume is most often measured by water-displacement. Weights are usually measured on air-dry specimens. In any case results should state whether the weight expressed refers to the dry-weight or to moist weight.

Stomach content analysis is a practical way of learning about the food habits of animals where large samples of animals are collected for other purposes and when sufficient time, equipment and trained personnel are available to do the job. This more detailed evidence of utilization may be used in the same way as evidence obtained by the rapid rumen analysis.

Caution should be exercised in interpreting results obtained from such studies. It has been observed with animals submitted to a known diet and then slaughtered for stomach analysis, that findings can be distorted as a result of differing rates of digestion of the various foods eaten. It is good practice to supplement the findings with field observations in order to help in interpreting the data and, if necessary, to apply corrections.

3623 *Droppings Analysis*

Most mammals pass droppings of a specific shape and size. Because of their distinct appearance, their identification should become relatively easy to an attentive observer who has spent time in the field (see Section 3423D). Because fragments of nearly all foods pass through the digestive system without being digested, knowledge of the food consumed by an animal can be gained from study of its droppings. Although the study of droppings can shed light on the food habits of most animals, it is especially applicable to carnivores. Hairs and feathers are for the most part ingested , but not digested. The structure and colour pattern of hairs, and/or feathers, vary with the different prey species and can easily be identified under a low-power microscope. The guard hairs (coarse body hairs) are the most easily identifiable of the hairs. Keys for the identification of guard hairs exist for the mammals of man localities; if such keys are not available, a reference collection can easily be prepared by the investigator himself.

Both with carnivore and herbivore studies, interpretation is often complicated by the fact that indigestible fragments are more likely to occur in some food species than in others, thus distorting the results. Moreover, fragment identification is complex and often requires the attention of a specialist.

Samples of droppings of carnivores can be preserved in a dried condition. However, it is best to preserve them in a solution of formalin for two reasons:

1. Organisms which consume faecal matter may be present in the samples when they are collected and, if they are to be stored over a lengthy period, the samples or parts thereof may be lost;
2. The droppings of carnivores almost always contain eggs of parasites. Often, such eggs are infective to man and can constitute a serious health hazard.

Investigators should never handle the droppings of animals with their bare hand. Gloves, forceps or small sticks should be used and disinfected thoroughly or thrown away after use. After droppings have been stored in formalin, they are safe to handle.

A method of analysing plant material in faecal droppings that has proved useful in Africa where eight different species of ungulates were grazing on the same pasture has been described by Stewart, D. R. M. and Stewart, J. (1970) as follows:

1. A selected pasture was observed for 6 days to determine which species consistently fed there during this period.
2. On the seventh day ten fresh piles of droppings from each species were located and 70 g (fresh weight) of faeces were collected from each pile and stored in formalin acetic alcohol. An attempt was thus made to sample ten different individuals of each species, although it is possible that occasion-

ally more than one pile of droppings had been deposited by the same animal.

3. After the collection of the faeces a record of the vegetation in the area covered by the animals during the preceding 6 days was made. Grasses and herbs were each given an abundance symbol and notes were made about their degree of maturity, height and condition. The list thus made should therefore contain all the species which might be found in the faecal samples. In addition to allowing a comparison between food available and food selected (food preference as described above) such a list of plant 'possibles' greatly reduces the time required to identify fragments in the faeces.

4. Analysis was done by first agitating the storage tube, then using a shallow scoop of appropriate size to remove sufficient material to provide a suitable density of fragments under a 3.8×1.9 cm coverslip. If too much material is placed on the slide, fragments overlap making identification difficult. The slides were marked with parallel lines whose distance apart was slightly less than the diameter of the microscope field of view at the magnification of ×100. The slide was traversed systematically along alternate pairs of lines. Every fragment of plant material which could be identified using a key, reference slides, descriptions or microphotographs of such material and which fell wholly or partly between the lines was recorded. Alternate traverses were omitted to avoid duplicate recordings of fragments overlapping the lines. Fragments which were large enough to bear sufficient characters for identification but which nevertheless could not be identified, were listed as 'unidentified' and included in the total of 100 fragments recorded for each sample.

Woody material in the diet was assessed on a subjective basis by means of an abundance scale of five categories ranging from abundant to nil.

It will be appreciated that the success of this technique depends on the existence of a reference collection against which one can compare the plant fragments found in the droppings. This in itself is a time-consuming and detailed task but if it is to be done it would be well to consult another paper by D. R. M. Stewart (1967) for more detailed instructions. Another prerequisite to the use of this particular technique is that the herbivore should be one that eats grasses and herbs and little or no woody vegetation. Using one objective method of assessment for herbaceous material and another subjective rating method for woody vegetation in the same stomach is difficult to interpret if the data are combined because grasses are more quickly digested than shrubs.

The technique is mainly of use in determining differences in utilization of grass in different areas and with different species.

The results provide evidence of the frequency with which various species of plant were found in the droppings of several species. All species of grasses were not easy to identify by this method, but it was a simple matter to separate grasses from woody vegetation. Thus if the technique were used to separate food into broad easily recognized categories it is much simpler to apply. The method as described provides an idea of species of plant consistently eaten by most of the animals present as well as species sometimes eaten but often ignored. Preferences between animal species for certain plants were revealed and differences in the range of species (few or many) were shown.

One of the reasons for mentioning this technique in detail is to emphasize again the need to anticipate the kind of results you can expect to achieve using a given technique and then to select, modify or devise methods of achieving your objective with a minimum waste of time, money and effort. For herbivores collecting samples of stomach contents should be considered only if other simpler methods are not possible. For carnivores it will probably always remain a useful standard method of assessing food habits.

3624 *Mouth Contents*

If a relatively large number of individual herbivores are to be shot for some reason, limited information can be gained through study of the food found in the mouths of the specimens collected. This method has the advantage that foods are not yet broken down by digestion. The disadvantages of this method result from the fact that because the sample represents only a mouthful, a large number of samples may be required to obtain an accurate picture of the diet. However, even a limited amount of such samples, used in conjunction with stomach or dropping analysis, can provide supplementary and possibly corrective information.

3625 *Direct Observations*

This method is probably the most widely utilized method with the large herbivores. Herbivorous animals are usually watched through binoculars or a spotting scope while they are feeding. Records are made of the plant species taken and of the frequency with which they are eaten. Observations are recorded either in number of bites or mouthfuls taken from each plant species, or in the amount of time spent feeding on each species or simply occasion by occasion when it is possible to make a positive identification. Choose one, but do not mix these different kinds of observations. In studies which require the constant watching of animals, it is advisable to use a tape recorder in order to record the information, unless an assistant is at hand to write the tally. Otherwise, important behaviour may be overlooked while the investigator is busy writing information. If timing is involved, the use of a good stop-watch is strongly recommended, for errors are less likely to occur with their use than with other watches.

Observation periods should, if possible, extend over both day-time and night-time as animals often use different parts of their range, and different foods, at different times in the 24-hour period.

As with other methods, direct observation has shortcomings. When herbivores are feeding in dense stands of mixed vegetation, it becomes very difficult to identify the plants that are taken. Under such circumstances the investigator is advised to examine the plants in that particular site after the animal leaves, and record the amount of fresh signs of feeding on the various species present. In this connection the use of tame animals can be invaluable for observation at close range makes it possible for nearly everything eaten to be identified.

With the larger carnivores, the direct observation method is easy to apply and likely to yield good results. Food habits can be studied either by watching the animals continuously over relatively long periods, or by recording the species of prey on which the carnivores were seen feeding, during random tours or set circuits of the study area. As pointed out earlier, it is important that sampling covers various periods of the night-time as well as various periods of the day. Night-time observations are usually carried out during full-moon periods, when the use of headlights is not necessary. Using a good pair of binoculars, the animals can easily be watched from a distance, without being disturbed by the observer. It would remain to be demonstrated, however, that feeding behaviour of animals during full-moon periods is representative of that during darker nights. Special image-intensifying optical equipment is available which permits excellent observation at night. At present it is too expensive for normal use but in the future, if prices are considerably lowered, image-intensifying equipment should be considered. In the meantime a great improvement on normal vision is possible by using good-quality binoculars of low (two to four power) magnification.

One advantage of direct observation is that it is easy and natural to combine such observations with those of feeding behaviour. When this approach is used the examination of a few stomach or dropping contents becomes a useful cross-check against purely observational evidence.

3626 *Methods for Determining Daily Intake*

Sometimes studies of food habits include a calculation of a daily intake of food per adult animal. This is useful in calculating the carrying capacity of an area when measurements are also taken of the amount of edible dry weight produced and removed. Two ways of estimating the daily intake of food of a herbivore are described. One may obtain information under field conditions on the frequency of meals per day, and the amount of food ingested per meal. Alternatively, with animals confined to pens one can measure the weight of food consumed. The latter method is more accurate. Under field conditions it is much more difficult to assess the food intake of herbivores than of carnivores. With herbivores, it is only possible to get an approximation of the intake, although the use of tame animals will make the estimates more realistic. Even with tame animals care must be exercised in applying observations made on young animals to adults, or on females to males.

Several field methods of assessing the intake of herbivores exist. One approach that often yields fairly good results, and is simple to apply, is to determine:

1. The average total time that the species being studied spends feeding, by watching sample animals over a series of 24-hour periods. If this is impractical a daily pattern of time spent feeding can be built up by more observations over shorter periods. Observations must be made at different times of the day and continue until a pattern of feeding times becomes clear;
2. The number of bites or mouthfuls taken per unit of time;
3. The average quantity of grass, or browse, per mouthful, which is done by weighing an equivalent quantity.

By use of these data, the investigator can make a general estimate of the intake. It should be noted, however, that since the most important part of the equation is the weight of a mouthful, the investigator must exercise considerable care in obtaining an estimate of this value.

Many game parks maintain orphanages on their grounds. Animals kept in orphanages can be used with advantage to complement food habits studies, particularly for daily intake determination and for preference ratings. Such animals, and animals raised as pets and which tolerate very close observation when away from small enclosures and under field conditions, are obviously ideal for such studies.

With carnivorous animals, and especially the larger ones, daily intake determinations can be more accurately determined by a relatively easy procedure. The frequency of meals can be determined as outlined above for the herbivores. The intake per meal can be measured in two ways. When fresh kills not yet eaten by the predators are discovered, it is relatively easy to chase the predators away and to weigh the prey, or take measurements that will permit a reliable estimate of the weight (Talbot, L. M. and McCulloch, J. S. G., 1965); see also Section 3841E. The carnivores chased away will usually lie down to wait, a few hundred yards away, and will return to their kill after the investigator has left. After the animals have completed their meal the remains are collected and weighed. If it is permissible to shoot animals in the study area, the investigator can shoot prey, weight them, offer them to hungry carnivores and then pursue the experiment as outlined above. Normally the carcass should be kept under observation from a distance as other species of predator or scavenger may arrive and complicate the observation.

3627 *Records from Predator Kills*

For many species in many areas predation is still poorly understood. In order to gain a better knowledge of predation, a considerable amount of information is needed over a wide range of subjects. Information on predation should not only include identification of the prey species but a record of the sex and age groups and physical condition of the prey as well as the circumstances of its killing. The parks and game reserves of several regions provide excellent opportunities to collect information on predation because of their staff of guides and rangers who tour regularly throughout these areas. These employees, as part of their routine work, can record information pertaining to predator kills encountered during their patrols. Such information has been gathered in some African game parks over many years and invaluable information has been amassed. The following is a list of observations which may be recorded from predator kills on forms following your own design.

1. Information pertaining to the quality of the prey.
   (a) *Sex*—this should be recorded in the field.
   (b) *Age*—some assessment of age should be made or material collected and labelled for assessment at a later data (for example, see Section 3832).
   (c) *Physical condition*—some index to physical condition should be taken. A range of possible techniques for doing this, with their uses and limitations, is found in Section 3800. In addition the prey should be examined for infirmities, broken limbs, or injuries suffered prior to being hunted and which may have favoured its killing. The presence and numbers of parasites should be noted and unidentified organisms should be collected for later identification. If the prey is a female, it should be noted whether she is pregnant or lactating.
2. Information pertaining to the circumstances of the killing.
   (a) *Date and time of kill*—if known or by estimation.
   (b) *Weather and light conditions*—at time of kill.
   (c) *Location of the kill*—in relation to topographical and vegetational features; the screening effect of existing vegetation, etc.
   (d) *Method of killing*—suffocation, tearing apart after the prey had been hamstrung, neck-breaking, or other.
3. Information pertaining to the predator.
   (a) *Species.*
   (b) *Number of predators on kill.*
   (c) *Sex of predators on kill.*
   (d) *Age classes*—Adult, sub-adult or cub.
   (e) *Apparent physical condition* of predators (see also Section 3800).

3628   *Guidelines to Consider Before Starting Food Habits Studies*

The use of information on food habits alone as a basis for management can lead to trouble. There is probably no other kind of wildlife research activity that, over the past 30 years, has led to more wasted time, effort and money and that has produced more misleading conclusions and half-truths. Concern over the amount of misdirected energy applied to food habits studies was one of the main reasons for developing the approach to classifying techniques used in this handbook. When facts about food habits are in hand there is still a need to evaluate their significance in relation to other habitat requirements for, considered in isolation, such facts simply provide good evidence of what part of an environment an individual or sample of animals has utilized as food. Food habits are not synonymous with habitat requirements, neither for large wild mammals nor for humans.

Most of the pitfalls associated with food habits studies could be avoided by considering the following five guidelines before starting a study.

1. To what extent are food habits studies essential to meeting your research or management objectives? What kind of information on food will meet your requirements best? In your working plans avoid circular arguments like the following: '*Object* To determine the food habits of the Javan rusa. *Value* A detailed knowledge of food habits is pre-requisite to the management of the Javan rusa.' This may be true, but if it is then one should spell out specifically how this knowledge would be used in managing the deer. One may wish, for example, to compare two areas which differ in kind and quantity of vegetation available, by testing to see if differences in responses exist between populations living in both areas. Translating in terms of the classified techniques, as described in Section 3100, would involve obtaining evidence from both the environment at the *occurrence level* and the animal at the *response level*, and would not necessarily involve evidence taken from the animal at the *utilization level* (that is, food habits) at all. However, if an understanding of the mechanism by which these responses were brought about was desired, then some kind of food habits studies would be essential (see Guideline 5 below).

2. Avoid over-generalizing. A change of soil type on a nearby area will normally result in a different vegetation and consequently different food habits. Before applying the results of a study carried out elsewhere, one must make certain that the components of the habitat occur in about the same proportions and patterns as exist in your study area. To do this use some simple, easily applied cross-check (see also Sections 3100, 8170, and 8321). The ultimate largest perspective of food habits of a species would require covering the entire spectrum of conditions under which the species lives, and during all seasons. However, this is normally a practical impossibility. At present the simplest way of obtaining such a perspective is to compare results of different studies made in different habitats and from different countries. Only the broadest of generalizations will be possible because different kinds of facts will have been gathered by different observers for different purposes and in different ways. It will often be impossible to interpret the results because the report available will not have described exactly how the information was obtained.

3. Avoid too much detail. Simplify. Food habits studies lend themselves to consuming time and to the use of gadgets. Obtaining detailed information on numbers and percentages of plant species composing the food is possible, is sometimes necessary and some of the techniques for doing so are

mentioned above. If you have to have this kind of detail to answer your questions, then get it. But it is surprising how often a simplified breakdown of food for herbivores into a few rough classes, such as woody vegetation, grass, roots, bark, etc., can be just as significant, for a basic understanding of animal–habitat interrelations and for more immediately practical purposes of management, as are more complex techniques.

In this connection it may be useful to be reminded that fourteen species of ungulates have been introduced into New Zealand starting in the middle of the nineteenth century. Following their introduction most of them became well established; some have become numerous enough to be regarded as problem animals. There is no question but that all habitat requirements for these animals have been adequate, or more than adequate, for some of the herds of deer produced world record heads of antlers. The species came from many parts of the world—for example, a partial list of countries includes Java, Ceylon, the Himalayas, Great Britain, Switzerland, Germany and North America. With this, and food, in mind it is relevant to observe that for most of these species, living as they mostly do away from agricultural lands, none of the kinds of plant they were accustomed to in their native lands are found in their New Zealand diet. So perhaps the species of plant is not as important as we have been led by a vast volume of literature to believe. However, I have been impressed to observe that the kinds of plants are similar, that is each species seeks out seasonally a balance of woody vegetation, grass, herbs, minerals, etc., similar to that used by the species in its homeland. These are simple facts which cannot be ignored. It may be useful to keep them in mind before deciding finally on the precise nature of your food habits investigations.

Another way of simplifying, particularly in the early stages, of investigating a species, is to give priority to that period of the year which is known to be the most critical to the animal concerned, since the foods available during this period may limit the carrying capacity of the range. If this period is not known it can be determined by systematic recording of changes in the physical condition of your species as the seasons change (see Section 3840).

4. In selecting the animals used as samples of a population, effort should be made to assure a fair representation of sexes and ages. The animals easiest to observe, or most easily sampled, do not always make a representative sample. Food habits often differ according to the sex and with the age of animals, and consequently animals of both sexes and of various ages should be sampled. The sex and age class of the animal under observation should always appear on the observation sheet in order that the information can be analysed separately to determine if such differences in food habits do occur.

    Sections of the population most important to sample, if not all can be sampled, are adult females and young of the year.

5. A useful general guideline to keep in mind is this: if either the adult sexes or any age groups are geographically separated (i.e. live in different

habitats) during any season of the year, one can consider the possibility that differences in the availability (occurrence) of the various essential habitat elements may be reflected in differences in utilization by the animals. Differences in use of essential requirements in turn cause different responses.

Since this is essentially a summary of one simple chain of cause–effect relations by which the sex and age structure of a population may be moulded by the environment, it merits elaboration with an illustration. Consider, for example, two New Zealand valleys occupied by deer with rather small and fixed home ranges. In summer and early autumn when the deer are improving in physical condition before the onset of winter the sexes are geographically separated, the males occupying the upper parts of the mountain slopes and feeding mainly above the upper forest edge in grassland. The females dwell in the lower forest reaches and obtain their summer grazing from clearings inside the forest and from areas near the streams. In this example, the areas occupied by females were essentially the same in both valleys. The areas occupied by males were however, different as one valley held a considerably higher proportion of grassy areas above the forest than was found in the other valley. Proportions of grass in the stomachs of adult females (utilization) were about the same in both valleys as was an index to the fat reserves (response) taken from samples shot in both valleys. In adult males, however, a higher proportion of grass appeared in the stomach contents of animals in the valley with the most grass available and fat reserves were also significantly higher in samples taken from this valley. Under these circumstances it is only natural that mortality was higher among males who entered the autumn breeding season and the winter season in poorest condition and that consequently the proportion of males to females differed between the two valleys.

3629 *Water*

Some idea of the significance of water as an essential habitat requirement can be learned by direct observations or, indirectly, through a study and recording of tracks and trails associated with available seasonal or permanent watering places. In some areas, where drought conditions occasionally or periodically prevail, indirect evidence may be useful in assessing the importance of available water as an essential habitat requirement. In extreme cases some species can be seen concentrating around waterholes in the process of drying up and eventually, when the water is gone entirely, their carcasses are observed in the vicinity of the dried-up watering points. In the same period other species show no sign of visiting waterholes and indeed apparently exist for periods of 3 and 4 months without drinking water. Some species, like the addax of the Sahara, can exist entirely without drinking water. Other species, like black rhinoceros, must have water in dry periods in some areas, while in other areas supplied with certain plant species with high water content, the rhino can fulfil its water requirements with water from the plants alone.

3700   UTILIZATION—ENVIRONMENT

In wildlife investigations evidence of utilization of the environment is obtained mainly either by measuring or estimating the percentage of annual growth removed by animals or through the use of some index to the intensity of utilization, as by classifying the key species into use classes. In general direct measurements are taken of length of growth, or weight or number of grazing ratings or other qualitative assessments. There are many such techniques for extracting this evidence and some of these, appropriate for use either on woody vegetation or on perennial grasses and forbs, are described below.

As with other types of evidence mentioned in this handbook there is little point in getting it unless it is necessary, either to fill some gap in understanding or to provide further basis for making some decision in management. The following examples illustrate three ways of using such evidence.

For example, when evidence of plant utilization is combined with evidence of the occurrence of these plants, this gives an index to the food preferences of the animal, thus assisting in identifying or more clearly defining preferred species and key areas of use (see also Sections 3621A and 3730).

When evidence of plant utilization is taken from the key plant species, during the animals' most critical season, and combined with evidence of the trends shown by these species, this allows standards of allowable use to be formed. For example, for many key browse plants on wintering ranges, be-

138

tween 50 and 60 per cent of the annual growth can be removed without detriment to the plant species. In other words the plant will continue to grow, reproduce and remain in about the same level of abundance with this amount of use continuing on a permanent basis.

When the concern is with animal damage, in forests for example, the standard would be set on another basis, that is in terms of the level of utilization that may be tolerated within the context of the major management objectives of the forest. Measures of utilization would thus be made on those species of tree of most concern to the forester, not necessarily the species most preferred by animals. The standard of use is thus based on the forester's tolerance limit or on a criterion set with some other form of land-use in mind rather than on limitations in a plant's ability to tolerate animal use as in the previous example.

3710   Removal of Annual Production of Vegetation

Some plant-eating animals are classed as browsers because they feed principally on the growth of shrubs and trees; others are classed as grazers because they commonly feed on grasses and *forbs* (broad-leaved herbs). Actually most such animals eat both classes of forage at some time during the year. When the grasses and herbs dry and lose food value, many kinds of grazing animals seek to improve their diets by eating browse. Likewise when browse has low food value during the late dormant season, the appearance of fresh green grasses and forbs will attract browsing animals, because at that time of year these offer a better food. Very often the investigator must determine how important is each class of forage to the well-being and survival of the animal about which he is concerned. For instance, it may turn out that a shortage of browse may have an adverse effect on grazing animals because they are not able to maintain good health and condition on herbaceous plants alone during the dry period or the winter period. In such a case, it would not matter how abundant and vigorous is the stand of grasses and forbs in summer or during the rainy season, the critical need will be either to provide more browse or else reduce the animal population to a level at which the present stand of browse will offer sufficient food to the remaining animals.

Plants are similar to animals in that they must have food in order to live. Most of the higher orders of plants manufacture their foods through chemical reactions that take place when the sun shines on the green parts, namely the leaves and fresh twigs. The food so manufactured is stored in roots and also (for shrubs and trees) in the tips (or growing parts) of the branches. If the animals remove too much of the leafage and green twigs before the plant has stored enough food in the roots to maintain itself, the plant will suffer. It will start to die back to reach a balance between its demand for food and the stored supply. If severe cropping by animals goes on long enough to exhaust the food reserves of the plant, it will wither and die. Likewise, if too many of the growing parts towards the end of the twigs are cropped by animals, the

140

next year's growth will be scanty and there will be little or no viable seed produced. In most plants and animals, the food needs of the individual will take precedence over the need for reproduction. Hence a plant that is browsed or grazed so heavily that it is unable to maintain full health and vigour, will begin to slack off on reproduction and use most of all of the deficient food supply to maintain its own structure.

In light of the above, it becomes necessary for the investigator to determine how much cropping a stand of plants can stand without loss of health and vigour. This amount is called the *allowable forage crop*. The part of the growth that must be left to maintain the health and vigour of the stand, and the tilth and fertility of the soil which supports the stand, may be called the *maintenance reserve.*

Investigators have revealed that most shrubs and small trees can maintain themselves in a productive condition if no more than 40–70 per cent of their annual growth is removed, dependent on the species, the site, and the time of year. Ordinarily, trees and shrubs can withstand heavier browsing during the dormant season than when growth is most active. Trees and shrubs growing on dry sites are usually less tolerant to cropping than those growing on areas where soil moisture is more plentiful. But many plants are stimulated to produce more than normal growth under moderate browsing. When the degree of browsing results in decadent trees and shrubs of poor vigour, it must be considered destructive (see also Section 3914D).

Perennial grasses and forbs also vary widely in their maintenance needs. Those growing on moist or sub-irrigated sites may often be grazed down evenly without harm to a 2–4 in. stubble by the end of the season of use, so long as use is gradual and not so heavy or concentrated as to prevent manufacture and storage of food reserves. Perennial hebaceous plants growing on dry sites (particularly where there is a long rainless season) are much less tolerant. On such sites there may be need for the vegetation to be left at the end of the season with a patchy (rather than an evenly grazed) appearance and with stubble height averaging 6–8 in. Most stands of dry land perennial herbaceous vegetation will deteriorate if there is uniform heavy cropping.

The simplest time to measure utilization is before the start of the new growing season. Then observations can include the amount of plant material produced in the previous season as well as the amount removed. For example, for those temperate ungulates that depend on woody vegetation to see them through the winter, observations will be made on the key browse species before the new growth starts in spring. For those ungulates, in semi-arid regions, that depend on perennial grass to see them through the dry season, observations will be made at the end of the dry season before the new spring growth takes place.

3720   Utilization Surveys

Some investigators have made utilization surveys an important feature in their management systems; but annual fluctuations in forage production coupled with fluctuations in animal numbers make annual forage utilization surveys an insecure base upon which to manage wildlife populations. It is wisest to use both range condition and trend measurements as a base for management recommendations (Sections 3914E and 8160). Such management decisions are based on the effect of animal use on the condition of the entire range, rather than simply upon the amount of the total forage that is consumed each year. There are, however, occasions where utilization information is needed for one reason or another.

Graziers everywhere are interested and concerned with the use of the basic resource on which their animals depend. Often this has simply involved looking at the area occupied by animals and judging how much the vegetation has been used, that is unwritten observations followed by an assessment of the situation based on experience. This 'method' can of course be very reliable if done by an experienced observer. But the present degraded or degrading condition of marginal lands occupied by domestic animals in all the developing regions of the world provides evidence enough that some refinements in procedure are needed. The following techniques, by including some form of measurements in the observation process, allow the eventual judgments to be on a sounder, and in terms of protecting the plant resource, a safer, basis.

The evidence is normally based on measurements of height or weight, on counts of grazed twigs or stalks, or on certain qualitative measures. Such observations involve samples, either over a considerable area, or within selected smaller areas of special significance to the population.

Several methods recommended for use by wildlife range investigators are described below. However, before using any of these utilization survey techniques, refer to Sections 3914E and 8160 and read carefully the sections below dealing with key forage species, key areas, range sampling and exclosures. It is important to check these other sections when looking for alternative methods.

3721   Trees and Shrubs

3721A   Tagged Twigs

The tagged twig method works best where seasonal growth of browse tends to be more or less in a straight line. With this method, one to several twig clusters on each bush or tree in the sample are marked with short pieces of brightly coloured insulated wire or plastic tape. A measurement is then made of the length of annual growth on each twig in the marked cluster.

A first measurement is made of the current production after the plant has made full growth in late summer or early fall in order to determine the amount of twigage available for browsing. Usually, only leaders that measure 1–2 in. or more in length are included in the sample. Shorter lateral spurs may be ignored. A second measurement is taken just before the start of the next growing season. This is made to determine the amount of twigage removed by browsing up to the end of the growth year.

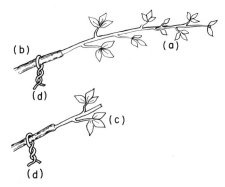

Figure 26 The tagged twig method of assessing utilization by measuring twigs before and after browsing: (a) shows annual growth before (or without) browsing, (b) last year's growth, (c) annual growth remaining after browsing and (d) coloured wire tag

If browsing takes place during the growing period it will be necessary to record measurements for uncropped leaders separately from those of cropped leaders during the first measurement. If browsing during the summer is so heavy that most leaders are cropped at the time of measurement, there will be a need to protect some plants from browsing so that annual production may be determined. Different plants should be chosen for protection each year for a shrub or tree that has been protected for several years will produce differently from one that is browsed.

Good sampling will require distribution of marked twig clusters at different height levels on the shrubs or trees. Usually, each twig or leader in the twig cluster is measured to the nearest $\frac{1}{2}$ in. (or 1 cm), or in some cases to the nearest $\frac{1}{4}$ in. (or $\frac{1}{2}$ cm) in order to determine the total linear growth.

It has been found best to mark fresh twig clusters each year, rather than remeasure the same clusters year after year. Under the latter practice, the pattern of growth will become more complex and harder to measure accurately with each successive season. For the same reason, simple rather than complex twig clusters should be chosen for measurement where possible. Often, toward the end of the growth year, it becomes difficult to distinguish the old growth from the current growth. Where twig patterns are simple and well-defined, the possibilities of error from this source are much reduced.

3721B *Marked Bush*

The marked bush method is a modification of the tagged twig method. With this approach, only bushes or trees—not individual twig clusters—are marked. On each of the marked bushes, twig clusters are grasped at random, and the current growth on the five twigs nearest the thumb is measured. A total of between thirty and fifty twigs are measured on each bush. No effort is made to remeasure the same twigs when rechecks are made. The number of twigs measured is considered large enough to give a representative sample.

Figure 27   The marked bush method of assessing utilization. On each marked bush twig clusters are grasped at random and the current growth on the five twigs nearest the thumb is measured

All twig measurement methods give best results when browsing occurs after the growth of the year is complete.

Yet another alternative method is to take measurements only once a year, just before a new growing season. Twigs that have not been browsed are measured to obtain a length from which the average length of the browsed twigs can then be subtracted to obtain an index to utilization. In cases where all new growth has been utilized some twigs will have to be protected as mentioned above to obtain a figure relating to annual production.

While data from twig measurement methods are usually presented as average percentage utilization by plant, by plot, or by range, there is a great advantage in expressing the findings in terms of inches of growth, as can be appreciated by considering Table 5. It will be seen in this table that a fixed demand for food, i.e. 2 in., can result in a 33 per cent utilization one year and a 67 per cent utilization in another year, solely as a result of fluctuations in annual browse production. Where the data are presented solely in terms of percentage utilization, increases or decreases of animal numbers may be inferred that are not warranted by the field data.

3721C   *Twig Counts*

The twig count method is a simple operation. Eight or ten twig clusters are chosen at random on each bush. A sample count is made of the browsed and

Table 5   Comparison of browse growth and utilization

Year	No. of twigs measured	Average leader length (in.)	Amount consumed (in.)	Percentage utilization
1959	500	6.0	2.0	33
1960	500	3.0	2.0	67

of the unbrowsed twigs *within reach of the animals*. The percentage of browsed twigs may be used directly as an index of use, or it can be converted into percentage utilization of linear growth, or of weight, by means of correlation curves or tables.

3721D   *Age Classes, Browse Lines and Hedge Forms*

Observations of age classes, browse lines and hedge forms can be used as evidence both of animal utilization and of plant response. Discussion of these relevant techniques is deferred to Sections 3914C and 3914D.

3721E   *All Methods*

In all methods described above, it appears best to eliminate the 'mostly unavailable' form class of browse from the sample (see Section 3714). Such plants offer only small amounts of available forage and are often only lightly cropped even where the general range use is heavy. Also, when the forage is mostly unavailable to mature animals it will be out of reach to the young. The inclusion of such plants may obscure the vital story of what is happening to the range. The younger and available browse plants make up the class of forage which especially should be protected from over-browsing for it is these plants upon which the future forage supply will depend.

When first deciding on a technique it may prove useful to try the last-mentioned simple twig count alongside of one of the more elaborate techniques to see to what extent both methods will produce approximately the same results.

3721F   *Production Index*

It has been stressed already that the supply of forage on a range may change greatly from year to year as a result of changes in annual rainfall or in favourable growing weather. An index of forage production is sometimes helpful in the evaluation of range trends. Since a poor year or a good year for one browse species is apt to be the same for all browse species, measurements of current twig growth on one, or two, important species will be enough for index purposes, for use in determining trends from one year to another.

Sample measurements should be taken from a minimum of ten, and preferably twenty or more, shrubs of each species. The average length of leader or unbrowsed twig growth for the range may be used directly as a growth or production index. The index gives a rough measure of the supply of food becoming available each year. These measurements should be in the same areas in which measures of utilization will be made.

3722   *Grasses*

The use of transects, small plots and photographs is now discussed as basic simple approaches to obtaining either evidence of the percentage utilization or in index to the amount of grassy vegetation removed by animals. Transects of various kinds are perhaps the simplest to design, use and often to interpret. By contrast plots take more time, and must be carefully located. While photographs can be used as a basis for obtaining evidence of utilization, they are normally best used to illustrate differences demonstrated by using other techniques.

3722A *Transects*

When evidence of grass use along a transect is desired this is usually done by measuring either height (an index to the weight) or weight of grass. Before selecting a technique the section on multi-purpose transects should be read (Section 8341).

3722A-1 *Toe point*[1] (dryland areas of perennial grass)

1. Make a paced transect of 100 points on each sample area and check whether or not the plant closest to the marking point is grazed or ungrazed.
2. Record the number of grazed plants out of the 100 plants checked.
3. Select enough transect locations to give a representative sample of the key area (Section 3732).
4. Calculate the average percentage of grazed plants as an index of current cropping.

In using the toe point method for utilization measurements in dry country often one finds very little perennial grass existing at the end of the dry season. This means transects must be very long to allow them to include enough grasses to provide a measure of utilization. Under these circumstances an alternative procedure is to follow a transect course in the usual way but instead of waiting to 'hit' the base of a grass tussock, simply measure the plant nearest the toe point. If such a decision is made it should of course be included in your description of method used.

On dryland bunchgrass ranges in western America it has been found that under heavy cropping 75 per cent or more of the plants will be grazed, and under extreme cropping 95 per cent or more. In many areas in Africa, the Middle East and Australia most palatable grasses are grazed level with the ground in dry seasons of dry years.

3722A-2 *Toe point* (wetland, seasonally inundated and higher rainfall areas)

Follow the same procedure as for dryland range, except check whether or not the average stubble height within a 4 in. (about 10 cm) diameter wire loop

---

[1]As Section 3722A-2 describes, the toe point is but one form of marking point. These sections are equally applicable to other forms of marking such as wire loops of various sizes, or a wheel with one mark on its edge.

held immediately in front of left (or right) foot is shorter than 4 in. (rather than grazed). Then calculate the average percentage of plants grazed shorter than 4 in. as an index of cropping. Both the size of the loop and the critical stubble height can be varied to fit the situation at hand. If the grasses occur in broad clumps, the diameter of the loop may be enlarged. If condition and trend surveys reveal the grass stand is deteriorating under grazing to a 4 in. stubble, the critical height can be increased as needed to halt the decline.

When using toe-point or wire loop methods in both dry and wet areas either the presence or absence of grazing, or the height of grass, or both may be recorded and used as an index. Paced transects are usually to be preferred for most purposes where an index to utilization is required because of savings in time and expense.

If the weight of grass consumed is required a simple method is to select the tussock nearest a transect toe point. At the end of the growing season remove half of the plant by cutting as close to the ground as possible, dry and weigh. Number and label both cut and uncut halves and clip the remaining half of the plant after the grazing season and before new growth commences. Subtract the second weight from the first to determine the percentage removal of perennial grass by weight. Take a large enough sample to provide a consistent sample result for the area.

Because of the effect of grazing and clipping in stimulating plant growth, estimates may be much lower than the amount of vegetation actually removed by animals. While this is an obvious drawback weight removal measures will nevertheless always be related to the amount of use and can thus be used as an index to utilization for comparative purposes, between years or areas.

Weight and height of grass are also closely correlated and if it is desired to estimate weight removed from height measurements the correlation should be determined for each key species of perennial grass (see also Section 3535).

3722B   *More Detailed Analysis*

There are of course much more elaborate techniques for measuring utilization. These usually involve belt transects or permanently marked square plots, marked out in grids and in which each plant is mapped and repeatedly observed. Such detail is not normally required in field investigations.

3722C   *Qualitative Measures*

Qualitative measures usually involve some form of rating system. Rating systems are often used, because they are usually less time-consuming than measurements; but the lack of measurement does not imply that such techniques are inferior. Where criteria for classifying observations can be clearly established and understood by participating observers results can be duplicated and just as much reliance placed on the proportion of various classes represented as on evidence based on measurements. For example, the 'primary forage plant method' described by M. H. Deming (1939) as discussed in Stoddart and Smith (1943) is

> based upon the principle of recording at the close of a grazing season, specific information about each of the main forage plants which carry the principal load of grazing use on a range area, describing briefly the other factors which influence the use made, and considering all these to reach a conclusion which assigns the area to one of nine described classes of degree of use.

3722D *Use Classes* (example)

Utilization estimates are made at various points over a study area and placed on a map. This is used as an index to uniformity of grazing, showing points where corrective management methods may be necessary.

Plants that are obviously the dominant ones from a forage-production standpoint (usually about six) are considered (see Section 3621A). Attention is given to plant abundance and vigour, soil-erosion conditions, topography, watering facilities, season of use, rodent activity, fires, and any other factors that might influence grazing use. Considering all these factors, the investigator decides into which of the following comparative use classes to place the area:

1. *Unused* — No use by ungulates.
2. *Slight* — Practically undisturbed.
3. *Light* — Only best plants grazed.
4. *Moderate* — Most of the range covered. Little or no use of poor plants.
5. *Proper* — Entirely covered. Primary forage plants correctly grazed.
6. *Close* — Completely covered, with some repetition of grazing. Some use of low-value plants.
7. *Severe* — Hedged appearance and trampling damage. Primary forage plants almost completely used. Low-value plants carrying grazing load.
8. *Extreme* — Range appears stripped of vegetation. Primary forage plants definitely injured. Low-value plants closely grazed.
9. *Destructive* — Much loss of primary species by death. Only remnants of good plants survive. Range in a critical condition (Stoddart and Smith, 1943).

The value claimed for the method is that it eliminates the problem experienced by an observer estimating a definite percentage of utilization. Instead he merely places the use within a certain class and suggests that utilization percentages be assigned arbitrarily to these nine classes.

The difficulty with this kind of rating may be appreciated by recognizing that it may be possible to identify downgrading trends in terms of increasing amounts of bare ground anywhere between classes 5 and 7 as listed above. Alternatively, trend assessments may indicate an improving condition present in recent years even in the last three classes (see Section 3912). Thus if one wishes to spot trouble at as early a stage as possible, a monitoring of trends would seem more sensitive; it is thus more appropriate than the application of such systems of utilization ratings.

Techniques used to describe occurrence, if sensitive enough and if used periodically, can also provide evidence of utilization. And techniques which essentially measure response automatically provide evidence of past use. A good example is the classification of shrubs or young trees into hedge form classes, a method described in Section 3914D.

3722E *Photographs*

Photographs can be useful in illustrating utilization on individual plants. One simple system is to photograph marked plants before and after the grazing season (or a known period of grazing) against a white background board marked out in squares of appropriate size (5–10 cm). Photographs may also be useful to show the appearance of different use classes within certain habitats.

3723 *Forbs*

The methods for measuring utilization described in Section 3722— *Grasses*—may be used also on rangelands supporting stands of forbs.

3730 Sampling and the 'Key Species' and 'Key Area' Concepts

The systematic selection of species and areas for sampling is commonly used in both wildlife and range management. Plants and areas thus selected for use as indices of utilization are known as 'key species' and 'key areas'.

3731 *Key Species*

Key species are the most commonly used species. In practice it would be difficult for wildlife or stock managers to base their estimates of utilization on all the plants consumed by the animal. Only a few plant species will furnish the bulk of the food at any given season and it is these species that concern us most. They are commonly referred to as 'key species' or 'utilization indicator species'. When considering food as one of the essential habitat requirements for large mammals it is the key species that most appropriately provide the evidence in terms of either availability (occurrence), utilization, or response (e.g. trend).

For some ungulates the key species will always be grasses; for others, shrubs and for some a mixture of the two. For many species in both temperate and tropical regions feeding in one season may involve mainly grasses; in another season, shrubs.

If any season is easily identified as the most critical period of the year for the well-being of the animal species studied, then the key species used during that critical season provide a most important habitat element for measuring, for monitoring, for understanding.

3732 *Key Areas*

The character and location of areas selected for sampling to obtain evidence of utilization will, as always, depend on management or research objectives. Since utilization by wild animals is not normally uniform throughout the areas they occupy, special areas called 'key areas' are usually chosen to obtain evidence of utilization. Three different approaches to sampling are summarized below in terms of the kinds of areas selected as key areas, and alternative definitions are proposed for the last two approaches.

1. As used in this book, key areas refer simply to areas of concentration of feeding activities. Such areas show much heavier use on the key forage species and are the most vulnerable as regards over-use of the vegetation. Standards for allowable use set for such areas may range from being close to the maximum a plant species can withstand, or minimal to allow a species to recover.
2. Another approach identifies 'key areas' as index areas other than those heaviest used. Thus if there were over-used areas, moderately used and lightly used areas, the 'key areas' would be those moderately used, recognizing that this would mean under-use on some and over-use on other areas. The latter are sometimes referred to as 'sacrifice areas'. It is suggested that 'average-use areas' would be a more appropriate term. It should be clear that under some circumstances the same rate of stocking might well be determined by using either definition (1) or (2).
3. Still other workers refer the term 'key area' more to response than to utilization. Thus 'key areas' would be the most vulnerable areas. If the most vulnerable areas are managed properly, so they reason, then the entire watershed will remain healthy although it is recognized that this may well mean that most of the watershed will be under-utilized. 'Critical conservation areas' would seem a more appropriate term and this term is used in the following sections.

These most vulnerable areas may be exactly the same as the areas of heaviest feeding concentrations, but they are not necessarily so. For different species they may, for example, involve bedding areas, areas of concentration to shelter from weather, over-intensive trail systems or areas of concentrated use through territorial behaviour.

3740   Enclosures

'Enclosure' is a term used for a fenced area enclosing vegetation for one reason or another. Fences can be designed to keep all animals out, or to allow entrance of some animals and exclude others. They are commonly used in research stations where a given number of animals can be enclosed along with the vegetation for precise periods. With enclosures various combinations of stocking, burning, reseeding, etc. can be tested and measured under controlled conditions.

The question of whether or not to use enclosures has on occasion become controversial. This is understandable when one observes an expensive enclosure erected by research officers without a clear idea of what it will contribute to their understanding; or when enclosures are erected and maintained for 10 years, for example, only to find that the questions asked of the enclosure 10 years ago are now no longer relevant, having been already answered by other simpler, quicker methods. This does not mean to imply that enclosures have no used in research; far from it. However, before using enclosures for research purposes review carefully the working plan (Section 2030) and see if the evidence needed could not be obtained in some other quicker, simpler way than by erecting enclosures.

Enclosures can, in certain areas, be an excellent means of *demonstrating* what will happen to vegetation if animals are excluded. Even as methods of demonstration, however, they must still be appropriate for the time, animal, the habitat and the people for whom one is demonstrating. For example, I once heard elders of a well-known African tribe making sport of a visiting pastoral expert behind his back. He had constructed an enclosure and when asked by an elder 'Why?', had replied, 'To see if the grass would grow if we keep the cattle away'. The tribe refused to take him seriously thereafter, explaining: 'Surely anyone who knows anything about grass and cows knows that if you keep the cattle away the grass will grow. Why did he do it?'

3800 RESPONSE—ANIMAL

Reviewed in the sections on occurrence (3200–3500), were methods used to describe the presence, abundance and distribution of animals and plants in a study area. In the sections on utilization (3600–3700), a few methods were outlined on how to learn which parts of the habitat the animals use for food, shelter, water and how to measure this use. The present section is concerned with measuring the response of animal populations to changes in their environment. Response of the environment to utilization by animals, or other causes is covered in Section 3900.

3810 Principles

First, a few basic definitions and generalizations. A population consists of a number of individuals of one species living in one region. In that region there

152

may be many populations; each of these being composed of individuals belonging to a species different from those of other populations. These populations of different species, together, form a 'community' consisting of the sum total of all the populations in the region. One may have animal and plant communities. The biotic community (combining both plant and animal communities) taken together with the inert, physical environment (rock, soil, climate) is referred to as the ecosystem.

In describing response within the context of an ecosystem three obvious levels at which we can direct our observations are the individual, the population and the community. Individual animals form the basic units of a population and it is convenient to regard populations of various species of plants and animals as the basic units of the community. Since management of wildlife is usually in terms of a population, techniques are commonly selected to emphasize evidence at this level. The population, in fact, has properties different from those of the individuals that compose it: the individual is born and dies, but the population has a birth rate and a mortality rate. Once an individual is killed, it is gone forever, but a population can be harvested and still flourish. The population has a density (number of individuals per unit of area) and a distribution, which an individual cannot have; so obviously, when considering the conservation, destruction or management of wild animals, we must refer to wildlife populations and not to individuals. The following techniques are designed to assess some kind of population response.

Recognizing that our basic units of management are populations does not imply ignoring evidence from individuals, for it is by sampling and building up cumulative observations of individuals that a perspective of the population emerges. This is true regardless of the kinds of samples with which we are concerned. For example, sex and age ratios are obtained by looking at enough individual animals from the population to satisfy us that our ratios are representative of that population.

Wildlife, being a product of the land, will reflect in its abundance and quality the conditions prevailing in its habitat. A good fertile soil covered with a rich and nutritious vegetation will support a healthier and more abundant wildlife population than will a poor soil with impoverished vegetation, other factors being equal. If habitat conditions change in an area, either by upgrading or downgrading, the animal populations will respond more or less rapidly by a change in their densities, birth rates, mortality rates, or general physical condition. The responses of animal populations to habitat conditions can be measured by several means. These can be grouped into three categories: density of animals, structure of populations, and physical condition of individuals.

3820 Evidence Based on Density

One could argue that the two principal kinds of evidence sought at the animal-occurrence level—that is, distribution and density—could also be interpreted as evidence of response because if the environment were not suitable the animals would not be there. This is perfectly true and such under-standing, even at this superficial level, may assist planning for management.

Density, expressed either as individuals per unit area, or as an index to numbers (e.g. number of faecal droppings per 100 stations), is often a valu-able and necessary piece of information for management, for it is part of an inventory of what is present as discussed under Animal—Occurrence (Sec-tion 3400). If measured during the same season in consecutive years, changes in density-related figures will indicate increases or decreases in the popula-tion. However, care must be used in interpreting density changes. Since an area may be under-populated (following over-hunting, poaching, recent invasion or introduction) or over-populated in relation to the carrying capacity of the habitat, it becomes obvious that density in itself, while being a valid indication of response to some influence operating within the ecosystem, does not neces-sarily reflect the existing condition of the habitat. The same holds true for biomass (weight of animal life supported by a unit area) which has been often wrongly used by biologists as a measure of the capacity of the land to support animals. Density and biomass supply helpful information on what exists at a particular time (present status, standing crop, occurrence) but do not neces-sarily allow one to pass judgment on the suitability of the present environ-ment, even less on environmental trends and least of all on the environmental potential for production of wildlife.

In other words, density or biomass, taken as such and considered alone, may be very misleading unless it is viewed in relation to the status and trend of the habitats concerned. A high density of animals, for instance, may be exciting to find but it will have completely different significance in a situation where the vegetation status is stable or the trend improving as compared to a situation where the trend in habitat is definitely downward (see Section 3900).

If differing densities are used as evidence of response in comparing differ-ent areas, then, as a precondition, one should have good evidence to show that each population has been in some kind of stable adjustment with its environment for the previous several years. When two such areas are com-pared then differences in density (taken at the same season) may be inter-preted as responses associated with more or less suitable environments.

The same generalization and caution may apply to other measures of response such as age structure, young to adult female ratios or some index to physical condition, for each may be equally misleading when considered in isolation.

3830   Evidence Based on Population Structure

One property of an animal population is its structure, that is the proportions or ratios between the sexes and ages of individuals that compose it. A popula-tion is not fixed and static; there are always births and deaths which result in seasonal fluctuations in the size of the population. If, year after year, the birth rate is greater than the death rate, the population will increase, and vice-versa. The structure may also be affected by the movements of immigrating or emigrating individuals. It has repeatedly been verified by biologists in many regions that the birth and death rates of a population will change in response to changing conditions of environment.

It is not possible within the scope of this handbook to discuss the complicated field of population dynamics; techniques have been limited to those which may help in establishing birth and death rates and population structure.

The information can take the form of percentages of the population in various groups or in ratios between different classified groups. Data can be taken either from live or from dead animals. In either case a sample of sufficient size has to be made to permit duplication of results either by the same or different observers. The procedure outlined above, in Section 3426A, works equally well for these kinds of observations.

3831   *Value of Animals of Known Age*

Animals of known age are often invaluable as cross-checks against other kinds of observational evidence used to estimate ages. They can be most useful in helping a field observer 'get his eye in' on yearling and in some cases second-year animals. They help in determining the line beyond which you will personally not go in terms of how many categories you use in classifying animals in the field.

Ages may be known in several ways: by keeping recently born animals, and raising them as pets, by marking and releasing newly born animals, by identifying and subsequently observing young individuals through natural markings or abnormalities, by cooperating with your nearest zoo and taking advantage of animals born in captivity. When an animal of known age dies it is important to obtain as much information on body measurements, tooth formulae, or other criteria (see below) used for ageing. It takes time to build up a collection of material of known age but it is well worth while for purposes of reference, or for in-service training courses within conservation departments. For species that breed twice a year throughout the year it seems essential to build up a collection of material taken from animals of known age.

For species with a single distinct breeding season, a simplified classification of animals into three, sometimes four, approximately known age groups can usually be made accurately enough for purposes of management. Even in the areas most favourable to observing recently born animals, the exact data of birth will normally not be known. In such cases the general approach used is to become acquainted with approximately aged animals. The approach is based on the fact that most of the calves will be dropped within a fairly short period of 2 weeks, although some calves may be born up to 3 weeks either side of the main calving period. These figures will naturally differ between species.

An observer living in the field and seeing the animals each day will have no hesitation in classifying the youngest animals seen 3 months after the calving period as this year's crop. Even when realizing that he may be a month out in his assessment he can still confidently separate this new contribution to the population from older animals. As he continues to contact the young growing animals in the field they will still be identifiable as this year's crop for several months. Under these conditions the question is, at the time of the next calving crop, 1 year later, can at least three categories of animal be easily identified as: the present year's crop; the previous year's survivors (now the yearlings); the older, adult animals.

Even this simple three-way classification is not always possible. For example, even when kept in pens, at 6 months of age some small ungulates, like the common duiker (*Sylvicapra grimmia*), cannot be distinguished from adults. In other species yearlings can be identified but only with difficulty and only by experienced observers. For still other species yearlings are fairly readily separated from other groups by most observers. As a generalization, the larger the adult animal the easier it is to identify yearlings.

Whenever a new animal is studied it is first necessary to determine which categories of classification are possible. Then set clear criteria for all cooperating observers to use in classifying animals. Train inexperienced observers until they can produce classifications consistent with those of experienced observers when both are observing the same animals. Then proceed with the classifications. Anyone who has not classified your particular species should be regarded as inexperienced until demonstrated otherwise.

3832   *Visual Classification of Living Animals*

In classifying a population the attempt is made to determine the proportions of different recognizable sex and age groups. Whatever system is used it is very important to record in detail the criteria used in classifying, for slight difference in technique sometimes make considerable differences in the end result. It is better to devote an entire morning or day to classification than to classify animals incidental to other types of work. The following instructions have been used with a variety of African species and wth a number of species of deer in temperate regions.

1. Aim at classifying entire groups. Do not classify if the group is in such a situation where young cannot be seen, or if males cannot be distinguished from adult females. It is better to spend an hour or more being certain of a group of ten than to enter 100 classifications done superficially.
2. Classify into age groups only in so far as you can be absolutely certain of every animal in the group. For many species this will mean classifying only calves, yearlings and male and female adults.
3. Enter the classification figures immediately after making the observation. For many species it will be necessary for two people to cooperate in classification, one person with binoculars and calling; the other recording.

A small portable tape recorder is invaluable if classifications are done alone.

4. Do not repeatedly classify the same group in the same week. However, there may be good reason to classify in the same area after a few weeks have elapsed between observations, to record changes in group composition, calf mortality, etc.

The most appropriate time of year to classify a species will vary with the species and with the special limitations to observation that apply to particular areas. For several species of deer and African ungulates, with clearly defined breeding seasons, it has been found that classification at a period when the young are 3–4 months of age is an appropriate time for getting calf to cow ratios, and for many larger species it is then still possible to recognize and to classify yearlings as well. For species whose breeding seasons extend over a period of 6 months or more, select a period or periods you feel most appropriate for classifying and then, to obtain comparable figures in the following year, use the same period. If the best season or month is not known one may for the first year wish to classify at monthly or quarterly intervals, then select one or more of these periods after considering the results of the trial classification periods.

Sex ratios and age ratios are discussed separately below, although such classifications are usually made at the same time.

Sex ratios

The significance of the sex ratio varies with the mating habits of the species of animals involved. In monogamous species, an equal sex ratio will favour maximum production of young, but in polygamous or promiscuous species, a distorted sex ratio in favour of females will result in a higher production if no other factors controlling productivity were operating. For example, a herd of 100 breeding polygamous animals of even sex ratio with production potential of 100 calves per 100 cows, can produce no more than 50 young; but if the sex ratio was changed to 20 males : 80 females, the herd could produce 80 young. Such distortion in sex ratio is advantageous up to the point where there are not enough males to service all females. However, there is much more to understanding measurements for management than observing sex ratios, as the rest of this section should demonstrate.

Sex ratios can be obtained either by observing live animals in the field or by examination of kills. Classification may be easy with certain species but much less reliable for others. In all cases, care must be taken as to when and where the sex classification is made, otherwise the conclusions reached may be erroneous. The main limitations result from segregation of sexes—a common phenomenon. Males and females react differently to each other and to the various parts of the environment at different times of the year. Even if sexes stay together, one sex may be easier to observe than the other. The sex ratio of kills is as subject to limitations as is observation of live animals due to segregation, difference in wariness between males and females, hunter selectivity, or hunting method used.

It is obvious that there can be discrepancies between the composition of either a live or killed population sample and the actual composition of a herd of free-ranging animals in a given area. Field observation and kill classification techniques are reliable only when they have been adapted and modified to fit local circumstances. To be meaningful, the classification must be made at about the same time each year, and also at the same time in relation to seasonal phenomenon (calving, weaning, breeding) of the species involved. It is also suggested that classification be done each time in the same habitat type and, if possible, by the same or similarly trained personnel, if trends are to be detected.

Age ratios

Knowledge of the age ratios of animals within a population enables the investigator to determine the structure of a population and of the kill. Age ratios are often more important to know than is the size of the population. Age classification divides the population into age groups or classes, such as young of the year, yearlings, breeding females and other individuals. It measures production of the year and survival—fundamental information in population dynamics.

In most cases, when age ratios are obtained by field classification of live animals, the identifiable groups will be limited to young of the year, yearlings and adults; for some species even yearlings are omitted because of difficulties of identification. Even with only two or three age groups, however, the results can be meaningful when considered along with evidence of the physical condition of the females and of the recent trends in those habitat elements on which the female population depends. Examples of criteria based on size and shape are shown for African elephant and buffalo in Appendix 2.

As with sex ratios, the determination of the age structure of a population of animals may be easy or difficult, according to species and to habitat (visibility). The limitations are similar; namely: herd segregation, differential wariness, visibility, seasonal timing. As noted above, it is important always to record in detail the criteria used in classifying animals; slight differences in technique sometimes make considerable differences in results. An example of a classification form is shown in Figure 28.

Whenever the opportunity arises to examine dead animals, age should be determined by one of the methods described next. These methods often yield data which will allow more sophisticated analysis of the population structure.

A word of caution must be entered concerning the interpretation of sex and age ratios—particularly the latter. Interpretation is not always simple, or straightforward, and comparing two different areas is more difficult than comparing successive seasons or years in the same area.

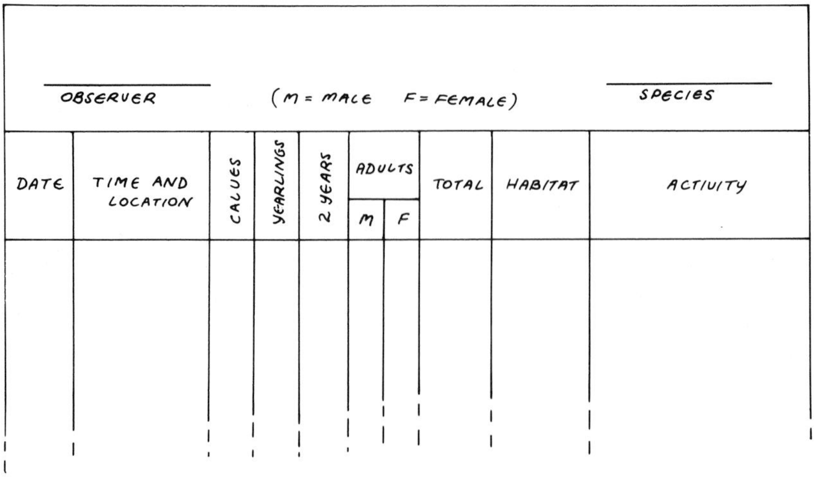

OBSERVER (M = MALE F = FEMALE) SPECIES

DATE	TIME AND LOCATION	CALVES	YEARLINGS	2 YEARS	ADULTS		TOTAL	HABITAT	ACTIVITY
					M	F			

Figure 28 Example of an observation form for classifying a population into sex and age groups

In comparing different years the main consideration is to ensure that the index is obtained in exactly the same way, at the same season and, preferably, by the same observers. In comparing different areas an important additional precondition should be met before these figures can be interpreted: good evidence should exist to demonstrate that the populations in the areas in question were in some kind of stable relationship with the environment for the previous few years. Evidence of such stability must be obtained from measurements or objective observations of the environment (see Section 3900). Comparisons might also be made if it could be demonstrated that the two areas concerned were in the same phase of an eruptive oscillation; however, at present the latter kind of comparison may be difficult because of the current lack of precision in identifying the point along an eruptive curve at which the study populations should be placed.

Once these conditions are met then one can consider making comparisons between the two areas concerned, based on sex and age ratios as described above. However, if the conditions cannot be met then the possibility exists that the population is in one phase or another of an eruptive oscillation; either increasing or decreasing.

A difficulty likewise applies if one uses only age and sex ratios to indicate a healthy balance between population and environment. The difficulty in interpretation stems mainly from the fact that a population whose individuals are in good physical condition, whose adult females are reproducing well and whose mortality is low (in other words a healthy population) is not necessarily in a balanced or healthy relationship with its environment. High productivity and low mortality are, perfectly normally, associated with populations in either early, or rather advanced, stages of an eruptive oscillation when the animals' environment is demonstrably downgrading (see Section 7100 for a further discussion of the eruptive oscillation).

The proportions of animals classified into even these few categories can, when used with care, provide evidence of an animal population's response to its environment. If changes in the environment take place which modify the essential elements of species habitat then changes in the proportions of animals classified will eventually follow. If, in comparing two demonstrably stable areas, one is found to be more suitable than the other for the production or survival of a study species then this will be reflected in the sex and age ratios. When the above-mentioned conditions are met, age and sex classifications can thus provide an objective means by which different environments or different years can be compared. The ratios will also reflect differences in weather conditions between years, and the effect of hunting seasons in modifying adult sex ratios.

Photographic techniques

It may be useful to include reference to two interesting photographic techniques, one for elephant and one for warthog.

A method for assessing age structures in elephant populations has been refined by H. Croze (1972) who also reviews similar previously used methods. The method is quick, inexpensive, and provides useful usable information on the age structure of elephant populations. The method consists of photographing elephant breeding herd groups vertically, measuring the relative lengths to form a mean breeding herd growth curve.

Equipment suggested is: a motorized 35 mm camera equipped with an extra large exposure back, loaded with fine-grain film and a 55 mm lens. It is

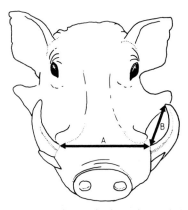

Figure 29   The measurements used to calculate the tusk length: snout width ratio. The snout width (A) is the distance between lip extremities, measured between the tusks. The tusk length (B) is the shortest distance from the base of the upper tusk, where it protrudes from the lip, to the tip (drawn from a photograph by Bradley, 1972)

suggested that the camera be mounted vertically through a hole in the belly of a light aircraft.

In using photographs to age warthog, Bradley (1972) photographed warthog facing the camera. From the photographs a tusk length : snout width ratio (Figure 29) was calculated as a percentage and plotted against the dates on which the photographs were taken. Regression lines were calculated for those individuals photographed several times and other less photographed animals were assigned to age classes as determined by the regression line into which their tusk ratio fell on the date photographed. Using all points within an age class a single regression line was calculated and plotted and a 95 per cent confidence interval established.

The technique is mentioned here because Bradley suggests the principle could be applied to any species in which the individuals are recognizable; the animals are tame enough to photograph; and two body characteristics, visible in the photographs, show a proportional relationship that continues to change with age.

3840   Evidence Based on Dead Animals

Before describing techniques for obtaining evidence of population response taken from dead animals it is important to list a major limitation. It is normally not advisable to use sex and age classifications based on dead animals as representative of a population because this is not normally the case. The problem of selecting animals killed in the proportion in which they occur in the population is not surprising for the differential wariness between different classes of animals, the different habitats occupied, different thresholds of visibility that operate in the different habitats, selection by the shooter; all these exemplify factors which are important in making it difficult to obtain a representative sample of the existing population.

Having said this is it still possible to obtain useful evidence of response from certain classes of dead animals, and to classify this evidence, if not the animals, in various ways. Indeed, most of the techniques described in this section depend on and involve dead animals. The uses and limitations of some of these techniques are now discussed.

3841   *Production of Young*

Workers associated with animal control operations, or otherwise having access to freshly killed specimens, should be aware of the value of collecting data on birth rates. This is particularly important because of the difficulty in getting information on early stages of productivity from wild populations. Birth rate (natality) data are expressed differently for different groups of animals, but with big game species they are commonly expressed as the number of young produced per year per 100 breeding females (50 calves per 100 adult females, for example).

Ideally, to understand productivity more precisely, we should have some evidence of the percentage of males that is capable of reproducing during the mating season. For females, ideally one would like some measure of the percentage of adult females that conceived, the percentage losses of embryos before birth, losses at birth and losses during the first few weeks following birth. Where possible, and when opportunity presents itself, biologists may occasionally be able to obtain this kind of detailed information. However, practically and for purposes of obtaining data on response relevant and useful to management, observations may involve embryo–cow ratios, wet udder percentages and/or calf–hind ratios when calves are 3–4 months of age. (The latter observations consist normally of visual observations taken from wild populations as mentioned above.)

3842 *Embryo–Cow Ratios*

With wild populations it is usually difficult to locate newly born young. Also, many losses may take place in the very early days or weeks of the life of animals. An early stage of production of young may be determined by counting the number of embryos in the uteri of females collected as a sample from the population. Where large numbers of seasonally breeding animals are being killed, the number of embryos present per 100 cows will provide data for later comparison with the numbers of young observed at the age of 3 or 4 months. This will show the early mortality rate of calves and is often useful in reflecting response to varying habitat conditions.

In cases where breeding is spread out during most of the year, embryo counts per female will still give an indication of production and frequency of breeding if observations are made throughout the year.

3843 *Counts of Corpora Lutea*

To determine potential production and to obtain a basis for comparing subsequent losses, information may be obtained on productivity at the earliest source practically possible to observe and record, the corpora lutea.

Following the discharge of an ovum, a white or greyish body called the corpus luteum is formed in the ovary and gradually disappears after the ovary is discharged. If the ovum is fertilized and pregnancy results the corpus luteum grows to a recognizably larger size and remains so throughout pregnancy. Even if an embryo is lost, the corpora lutea of pregnancy can still be recognized as such several months later.

It is thus possible to obtain evidence of losses between conception and birth by counting the number of corpora lutea in the ovaries and comparing these counts with those of embryos near full term. The differences between such counts give an indication not only of the proportion of adult females that have conceived young but of the losses due either to resorption or abortion.

However, while it is easy to count whitish bodies on ovaries it is not so easy to interpret; to distinguish between the corpora lutea of menstruation and the corpora lutea of pregnancy, or between recently formed bodies and those remaining from previous pregnancies. The technique, in short, is not for inexperienced personnel. It is necessary, even for a fully qualified biologist first to consult a specialist, preferably with experience of your study animal, in interpreting observations of corpora lutea. But it is interesting for field men to know that such observations are possible. With careful instruction by an expert the technique can be mastered and used but it is not an appropriate technique to elaborate here.

3844   *Wet Udder Ratios*

Just before dropping a calf, a female starts producing milk in her udder and she continues to do so until she loses her calf or until weaning time. When a calf is lost the female's milk soon ceases. When a freshly killed female is examined, the presence of milk should be verified by gently pressing the udder and teat between the fingers. If milk is readily expressed, a cow is ready to drop her young or she is feeding it; if milk is absent, she has no young or it has been weaned or has died.

If, *in seasonally breeding species*, a fair sample of females are shot each month, starting before the calving period and finishing after the weaning period, examination of udders will yield data that can be used to show a ratio between wet and dry udders. This ratio can be expressed as the percentage of wet udders in samples taken and grouped into weekly or monthly periods. If plotted, the resulting curve will slope upward as more and more females drop their calves until a plateau is reached with the end of the calving season, when 80 or 90 per cent of cows have wet udders.

The shape of the curve in the few weeks following a calving peak will largely depend on the mortality of young calves. If the proportion of wet udders drops sharply, it will indicate that a high number of calves have died soon after birth; if the proportion remains fairly stable, little or no mortality will have occurred over this period. As the weaning period approaches, the proportion of wet udders obviously will gradually decline, but a sharp drop following the peak is an indication of a sudden loss of young calves (Figure 30).

Such information, therefore, is useful not only for determining the period of birth but also for comparisons of early mortality rates of young calves between two or more areas, or between years.

As a matter of interest the drop in the proportion of wet udders shown in Figure 30 for Area (a) between February and March, was considered due to predation. Area (a) was the only one of the three areas with predators; packs of feral dogs, who fed largely on red deer calves during this period. After weaning, Areas (b) and (c) has similar young to female ratios, while Area (a) remained distinctly lower. As the figure shows, classifications made 3 months after a peak of the breeding season were still lower. One year later in Area

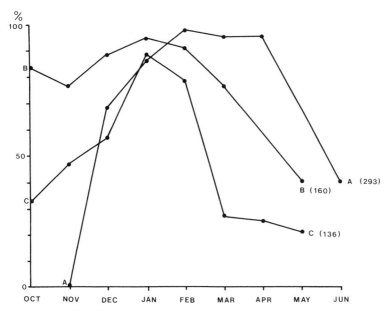

Figure 30   Monthly differences in the percentage of wet udders in adult female red
deer, sampled in three New Zealand areas in the 1952–53 calving period: A, Marl-
borough–Nelson area; B, Canterbury area; C, Central North Island area. Changes in
the percentage of wet udders observed were used to more clearly define the calving
season and as indirect evidence of early mortality. (Sample numbers are in paren-
theses)

(a), yearling to adult female counts were about the same as the April wet
udder counts of April in Figure 30. In Areas (b) and (c), similar yearling
counts were but half the wet udder counts of the previous April. In other
words, the main effect of predation in Area (a) was that the first year mortality took place sooner after birth than in areas with no predation. The percentages of yearlings in these three areas were not significantly different.

For Areas (a) and (b), some observatons were made on two successive
years and for each area the proportions of females with wet udders were not
significantly different between those two years.

It must be stressed that the wet : dry udder ratio technique is most appropriately used with *seasonally breeding species*. The most serious difficulty of
course is to get a sufficient number of observations. Where it becomes necessary to hold down costs by securing help from untrained personnel, difficulties
may arise if personnel do not understand or care about the need for careful
observations and records. Opportunities vary from species to species and
from place to place, but field workers, especially those associated with control
operations or with utilization schemes, may occasionally find the technique
worth considering.

3845   *Breeding Age*

A proportion of the individuals in a population are able to breed, and the population birth rate is produced by this segment. Other individuals are either too young (immature) or too old (senescent) or breed. It is often useful and sometimes essential to know which age groups make up the breeding segment. To get this information, record the age or age group each time an opportunity arises to examine a breeding animal. (Techniques for age determination are described below—Sections 3850–3854.) In females, the occurrence of breeding is relatively easy to ascertain by the presence of embryos or of corpora lutea, by scars left on the uterus after each pregnancy, by the presence of milk or of accompanying young. It is more difficult with males, where the weight of the testes may be meaningful, or the presence of sperm in the genital glands. An experienced biologist may be required to determine the breeding age of males. In any case it is the proportion of the female population that breeds that is normally important for management; and the knowledge of the age of puberty is more important than the age of senescence, for old animals usually form only a small proportion of the population, insignificant for management purposes.

When data on breeding and age have been gathered and compiled it will become possible to conclude which age groups make up the breeding segment of the population. This information is valuable not only in calculating productivity; minimum breeding age can also be evidence of response to habitat conditions. In North America, for instance, yearling cow moose will breed in good habitat but not in poor habitat.

When studies on a species are first started, the age at which females first breed is usually unknown. From the literature it is comparatively simple to find the average number of young produced and the gestation period for many species of large mammal (e.g. see Asdell, 1946), but the age of puberty is not usually included in such references and must be determined locally.

The information is not easy to obtain, for it requires animals of known age and it must accumulate in any way possible and convenient: from zoo, pets, or experimental animals. The timing of the onset of puberty can be still another variable responding to differences in environment quality. For this reason the breeding ages that are usually most significant for study or management purposes are those taken in the field from wild populations. This can be done by re-examination of animals marked as calves, or by autopsy of species where a reliable technique for ageing is available.

Finally, a word of caution about transferring knowledge of minimum breeding age from one closely related species to another. The onset of puberty may be the same, between two species and especially two species of about the same size. However, such similarity cannot be assumed and it should be cross-checked with animals of known age from the species and areas concerned.

3850   Age Determination

Although many techniques for ageing exist only a few will be discussed below. The precision of the techniques will vary with the age of the animal and with the general kind of observations made.

In general the younger the animal the greater the possibility of ageing more precisely; the older the animal, the less precise the ageing.

Precision depends also on the kind of evidence on which criteria are set. For example, when a sequence of erupting teeth is used, sufficiently large collections either of skulls of young animals or of their lower jaws can provide good evidence of the sequence of the erupting permanent teeth, following which eruption classes can then be assigned. However, animals of known age provide the only satisfactory evidence for deciding which of the eruption classes are most useful in separating age groups. Less precise are data taken from animals in the hand, either dead or captured alive. Least precise are of course the observations based on free-ranging wild animals which are classified and grouped in various ways as described above.

Techniques based on material either from animals in the hand or through field observations are usually less precise than the literature describing the techniques imply. However if one accepts that for most management and research purposes an approximate age grouping is adequate then most of the techniques described below have some value.

3851   Age Determination from Tooth Eruption

Most mammals have during a lifetime two sets of teeth: milk teeth, which are replaced, and permanent teeth. The eruption and replacement of these teeth in most large herbivores is commonly used as the basis for ageing young animals.

The normally extended breeding seasons, individual variation in the timing of tooth replacement and the variation due to different environmental conditions make it difficult to use tooth replacement as a criterion for exact ageing of most ungulates. However, tooth replacement still remains an excellent basis for assigning at least some age classes. Using tooth formulae it is almost always possible to identify several sub-adult age classes and to be certain of an adult tooth formula when it occurs. When these formulae are recorded for appropriate samples of animals from the same population, year after year, it provides an adequate basis for a kind of understanding of age structure useful for many management and research purposes.

Recognition of teeth and tooth formulae

The 'milk' or deciduous teeth appear around the time of birth, or soon afterwards, and are replaced by the permanent teeth, of which, in most ungulates, there are ten in each half of the lower and six in the upper jaw, arranged in

166

two groups. For the lower jaw, from front to back, there are four front teeth, then a space, then six cheek teeth.

The first three front teeth are the incisors (I); the fourth tooth, usually looking like an incisor, is actually a modified canine tooth (C). The first three or four of the six adjacent cheek teeth are premolars (P) and the last three are molars (M). In the upper jaw, for many ungulates only the upper cheek teeth (premolars and molars) are present; however, in some species canines are present and, in some species, four premolars.

Milk teeth consist only of incisors, canines and premolars, thus only these teeth can be replaced by permanent teeth. Molars are always permanent, even in young animals.

Milk teeth will be readily distinguishable from permanent teeth once a collection of material of known age is available for comparison. As eruption of permanent teeth progresses while milk teeth are still present, the latter will have a darker stain and show more wear than recently erupted permanent

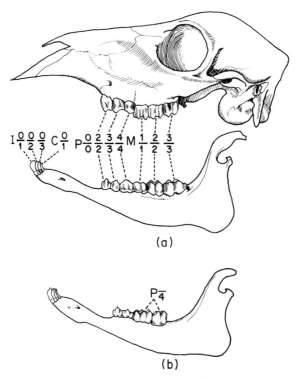

$$I\frac{0}{1}\frac{0}{2}\frac{0}{3} C\frac{0}{1} P\frac{0}{0}\frac{2}{2}\frac{3}{3}\frac{4}{4} M\frac{1}{1}\frac{2}{2}\frac{3}{3}$$

(a)

$P\overline{4}$

(b)

Figure 31   Differences between adult teeth (a) and milk teeth (b) of an ungulate (common duiker, *Sylvicapra grimmia*) and the conventional way of writing a tooth formula. Note that the last lower milk premolar is actually larger than the corresponding P $\overline{4}$. (a) represents a fully adult female with light wear on the cheek teeth, while (b) indicates eruption class 1, as described in the text (from Riney, T. and Child, G., 1962)

teeth. Furthermore, milk teeth are normally smaller in size and less robust in shape. One notable exception, the third milk cheek tooth, that is the last premolar (P4), is particularly easy to distinguish from its permanent replacement. This tooth is usually larger than the permanent P4 and has three distinct ridges as compared with but one or two on the permanent P4 (see Figure 31).

In most species the deciduous teeth are replaced progressively from front to back, but here too there are exceptions and the investigator is advised to familiarize himself with the usual order in which it takes place in the species in which he is interested. This can usually be done from selected specimens with teeth in various stages of development, with lower jaws taken from known-aged animals, etc. (see Appendix 1 for several examples).

The labelling of the first premolar P2 instead of P1 has an historical explanation. This is because, while in the process of evolution, most ungulates have lost the first premolar, it can still appear as an abnormality in several species and is in fact normal for still other species. Similarly, the canine still persists in the upper jaw of certain species. In the lower jaw, where the canine normally occurs, it has become modified to look like an incisor.

Figure 31 shows the position of the teeth in the upper and lower jaws of a common African ungulate, the common duiker (*Sylvicapra grimmia*). Teeth formulae normally describe only half the number of teeth in an animal as the teeth are the same number and kind on both sides of the head. Since in this example there are three incisors, one canine, three premolars and three molars, the general or summarizing way of writing a tooth formula for this species is $I\frac{0}{3}$, $C\frac{0}{1}$, $P\frac{3}{3}$, $M\frac{3}{3}$. A more precise formula that names and numbers each tooth would be written thus:

$$I\frac{0\ 0\ 0}{1\ 2\ 3}\ C\frac{0}{1}\ P\frac{0\ 2\ 3\ 4}{0\ 2\ 3\ 4}\ M\frac{1\ 2\ 3}{1\ 2\ 3}.$$

Recording of tooth formulae

Once the names and numbers of the teeth are known it is a simple matter to record the number and position of each milk (m), erupting (2), or permanent (p) tooth for any ungulate by entering the appropriate letter on a prepared form.

To speed the processing of observations, overprinted hand-punch cards can be used; a simpler procedure than trying to prepare master sheets including a tooth formula data. Figure 32 shows a card thus used to produce the information shown in Figure 35. Following this procedure, a separate card was used for each skull. On the cards milk, permanent or erupting teeth were so indicated under 'tooth formula'. Also recorded were date, sex, and location as well as information relevant to another study.

168

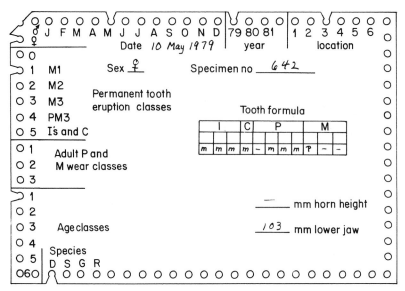

Figure 32   An example of a punch-card used for entering data on each skull. Letters at the bottom of the card indicate species, for example, D (duiker), S (steenbuck), G (grysbuck) and R (reedbuck) (after Riney, T. and Child, G., 1962)

Converting tooth formulae into age classes

One fairly safe procedure for developing ageing criteria from tooth formulae is first to convert the formulae into eruption classes based on the sequence of tooth replacement; then convert these eruption classes into age classes. The latter ideally would involve large samples of animals of known age, but this is of course rarely possible. More commonly the evidence consists of samples of skulls and measurements taken on known dates and compared with formulae and growth measurements taken from comparatively few animals of known age.

The example mentioned above in Figure 31 provides a convenient illustration of the criteria used in assigning *eruption classes*. In this example, skulls without sign of the first molar were placed in class 0. Erupting first, second and third molars represented the first, second and third eruption classes respectively. If any premolars were erupting the skull was placed in class 4, and skulls with permanent dentition were called adult and placed in one of three adult classes based on the amount of wear on the teeth. If wear on the premolars was so slight that the cusps appeared clean and sharp the wear was called 'light'. If wear was down to less than 3 mm above the gum line, wear was called 'heavy'. All other adult skulls were called 'medium'.

Since the order of tooth replacement is not the same for all species eruption classes will differ with different species. For example, in the mule deer (see Figure 33) permanent incisors are usually fully erupted before the appearance of the last premolar (P4).

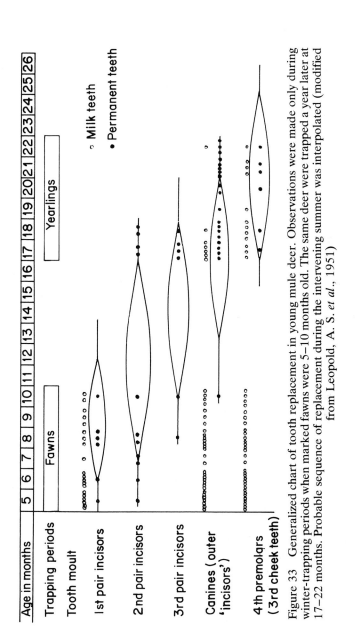

Figure 33 Generalized chart of tooth replacement in young mule deer. Observations were made only during winter-trapping periods when marked fawns were 5–10 months old. The same deer were trapped a year later at 17–22 months. Probable sequence of replacement during the intervening summer was interpolated (modified from Leopold, A. S. *et al.*, 1951)

In general the smaller the animal the more one tends to use groups of teeth in assigning eruption classes. The larger the animal the slower the growth curve to adult size, and emphasis is placed more on the eruption of certain individual teeth as Figure 33 shows for male deer. The longer the time taken to grow up, however, the greater the opportunity for variations in eruption time to occur, due either to environmental causes or to individual variation.

Figure 33 is worth closer inspection, for it illustrates variation in tooth eruption dates, a major limitation of this method, even under conditions most favourable to its use. In this mule deer area the period of calving was short, occurring mainly in the first half of July, but extending from early June to the latter part of August. A few fawns were trapped and tagged when only a few days old. Ninety-eight of the observations summarized in Figure 33 were made on trapped and individually marked deer 5–10 months later. One year later fifty-two further observations were made of these same tagged individuals. Thus, although the annual calving period occupied no more than 3 months, Figure 33 suggests that the variation in the appearance of the permanent teeth was about 1 year. Although this evidence makes clear the general sequence of eruption it is equally clear that it would be unwise to use each erupting tooth as a criterion for ageing. With only this evidence in hand only two age classes (those two not overlapping) could be used with confidence:

1. The first pair of incisors were replaced between 5 and 10 months, but perhaps as late as 10 or 12 months.
2. The last premolar was replaced between 17 and 22 months, perhaps a little earlier and a bit later for deer were trapped only during the winter months.

In winter, then, fawns could be separated from yearlings by the occurrence of these two permanent teeth alone.

Evidence has been presented elsewhere to show that different levels of protein in winter can effect the timing of tooth eruption (Steenkamp, J. D. G., 1970). With this in mind and referring to the Figure 33 study area, where deer sampled in these observations came from two habitats with markedly different feeding conditions, the variation in eruption times may have been greater than if observations had been made on animals from a more uniform habitat.

Least favourable conditions for assigning an approximate age to tooth eruption classes are found in those species which breed all year round. For these species the normal, and often the only, method is to gradually accumulate a collection of skulls of known-age animals. These animals will normally have been pets, animals marked and released shortly after birth, or animals of known age in zoos.

As noted above, determining the sequence of tooth eruption, describing eruption classes and assigning usually rather broad age classes are not particularly difficult, providing enough skeletal material and animals of known age are available. Because of the latter provision, however, there are comparatively few animals for which this has been done. Appendix 1 includes some extra summarized examples taken from widely differing types of large mammal. Anyone interested in further exploration of age determination of wildlife should consult the excellent bibliography compiled by R. M. Madsen (1967).

3852   *Ageing based on Tooth Wear*

Evidence of comparative age of erupted teeth can be based on either colour or tooth wear or a combination of both.

The colour of newly erupted permanent teeth is light and as time goes on the tooth gradually takes on a darker stain. This is useful in recognizing newly erupted or most recently erupted teeth.

In permanent teeth the height of the tooth crown above the gum line is obviously related to wear and the wear in turn takes time. These truisms have been used by many workers in developing ageing criteria for adult ungulates based essentially on the amount of wear on the permanent teeth, especially the premolars and molars.

There is no doubt that several, perhaps even most, of these criteria for ageing animals work well in the study area where the technique was developed. But variations in wear of teeth due to differences in soil types, habitats and level of nutrition are so great that it would be misleading to present them in this handbook. Criteria for ageing based on tooth wear cannot be set in a way generally useful throughout the entire range of a species because differences in the rate of tooth wear between different areas are too great. If used at all, care should be taken to develop ageing criteria based on known-age animals taken from the study area concerned. Even within a given study area, if soil types and/or kinds of food vary, the rate of tooth wear can also be expected to vary.

When tooth wear is used in ageing it is usually recorded either as general wear classes or more specifically using measurements. The commonest general classes are light, medium and heavy wear. When measurements are used they are usually of the height of a selected tooth above the gum expressed in millimetres.

3853   *Other Selected Ageing Techniques for Large Mammals*

Growth curves

Any measurable part of the body can be used in determining a growth curve. It is well to remember that the rate of growth is not the same throughout the body as at any given period some parts of a sub-adult mammal will be growing faster than others and the part you select may not reflect total growth as accurately as you may wish. Zoologists have constructed growth curves in many ways for many animals, but if the growth curve is to be used as a basis for, or as an aid to, ageing, measurements that show the least variation in adult samples will be more useful.

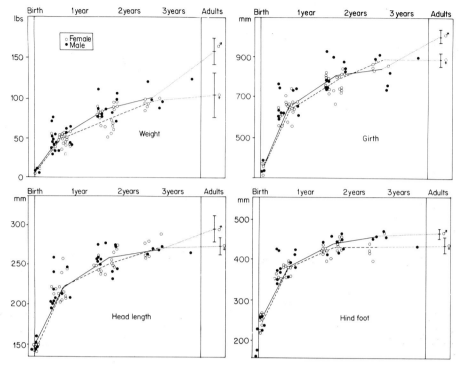

Figure 34   Growth curves for weight, girth, length of head and length of hind foot, based on measurements of marked and recaptured mule deer. Vertical bars represent standard deviations among adults (from Leopold, A. S. *et al.*, 1951)

Thus the length of the tail, which is notoriously variable, would not be used. Weight also shows too much variability to be used as a guideline or in cross-checking age in large mammals, as Figure 34 suggests. In this California study the cluster of measurements (range of variation) taken for each sex at various ages was slightly smaller for hind foot than for other measurements. It will be appreciated that even the length of the hind foot would be useful as a basis for ageing only in the first few months of a deer's life.

Figure 35 illustrates another, more appropriate, use of the growth curve in ageing: as a cross-check against other evidence. In this example advantage was taken of a campaign to eliminate animals, and 1552 skulls of the common duiker (*Sylvicapra grimmia*) were collected from hunters' camps each month for a period of 14 months. Tooth formulae were recorded, as was the length of the lower jaw. In this sample the length of the lower jaw showed less than $\frac{1}{2}$ in. difference between animals of 6 months of age and adults. In other words growth to adult size was achieved well before the permanent teeth erupted and the growth curve was mainly used to provide a framework of reference against which one could compare eruption classes with a very few known-age animals.

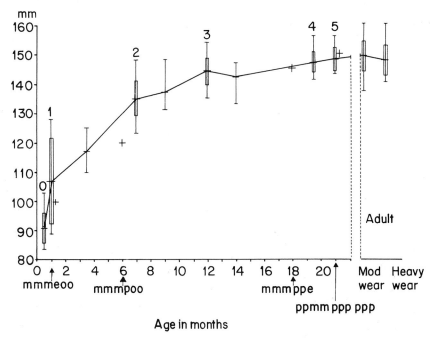

Figure 35   The growth of the mandible of female duiker as it relates to tooth eruption classes. Crosses indicate status of known-age animals. Rectangles indicate standard deviation from the mean. The unnumbered averages and extremes represent intermediate series, between the eruption classes as defined in the text (after Riney, T. and Child, G., 1962)

Growth curves were thus used in combination with tooth formulae of known-age animals to extend the use of tooth formulae beyond the first two eruption classes in this example. In some larger species tooth eruption may be complete before adult sizes are obtained.

Age determination from tooth rings

The determination of the ages of mammals from annual rings formed in the outer (cementum) layer of teeth is a technique requiring considerable time, equipment and experience in interpreting the carefully prepared tooth sections. It is most useful in certain middle- and high-latitude species where there are distinct changes in climate or physiology, or where there are distinct wet and dry seasons during which the browsers or grazers experience marked changes in feeding conditions.

Annual tooth rings, when obtainable, can be applied as a cross-check against other methods of ageing and can extend this classification to older groups. Although the technique is useful for carrying ageing into early adulthood, J. D. G. Steenkamp (1970) warns of the need to develop criteria in the

174

same area of the study population because of differences in seasonal changes and in quality of food. The older the animal the less reliable is the interpretation of the rings.

This interesting technique was originally applied to ageing seals (Scheffer, 1950) and later to ageing elephant seal where the yearly cycle included two periods of complete fasting which were reflected in the ring counts (Laws, 1952). Since then it has been tried with varying degrees of success on many species of large mammal. Spinage (1973) in reviewing age determination by means of teeth in African mammals, suggests that this may be especially promising in ageing carnivores in Africa. It has been successfully used on black bears (Stoneberg, R. P. and Jonkel, C. J., 1966), grizzly bears (Craighead, J. J. et al., 1970) and coyote (Linhart, S. B. and Knowlton, F. F., 1967) in North America.

Two main procedures are used in counting growth rings. A microscopic technique involves softening the tooth, embedding it in wax, preparing thin

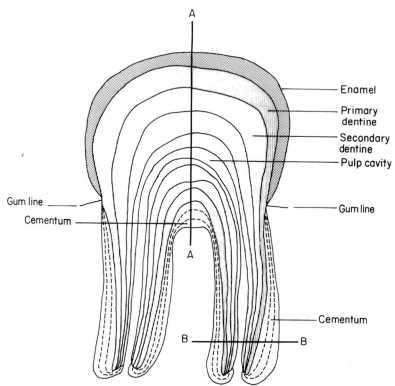

Figure 36   Location of cementum and dentine on a molar tooth and two common cuts used for exposing a section for polishing (A–A and B–B). Counts can usually be made in young animals on either secondary dentine or cementum but rings in dentine can be more difficult to interpret in older animals (modified from drawing supplied by W. A. Low and B. Mitchell)

sections and staining it until the rings in the cementum become visible, then examining under a 10 to 15× microscope. A simpler procedure is to cut through the earliest erupting permanent tooth with a jeweller's saw at the level of the cementum and then polish on a fine carborundum surface. The rings can then be counted under 10–15× magnification by using reflected light. High magnification can obscure the pattern.

W. A. Low and B. Mitchell have kindly provided a diagram showing the location of the cement pad on a molar and two photographs showing annual rings from teeth of a 4- and 6-year-old red deer in Scotland (Figures 36 and 37).

In interpreting the annual rings, Low and Mitchell stress the importance of initially using teeth from known-age animals. They also note the need to determine the age at which the first ring is formed in the permanent teeth and the time at which the tooth erupts, so that the number of layers in the teeth can be converted to years. In red deer, for example, the number of opaque cementum layers in the first molar is equivalent to the age. However, in mule deer, 1.5 years elapse before the first permanent incisor erupts and the first annual ring is not visible until the spring when the animal is 2 years old. Thus, 1 year is added to the number of annual rings to obtain the age of mule deer.

W. A. Low and B. Mitchell also note the difficulty in interpreting the presence of various other lines or rings which may not be annual rings and the possibility of confusion, especially when first using the technique. An observer must be thoroughly experienced before much reliance can be placed on the counts. This, coupled with the variability in the number of rings due to different areas or different levels of nutrition as noted by Steenkamp (1970), should make biologists in developing countries think seriously before adopting this technique.

Although cementum rings are usable on many species, for most large mammals the simpler, cheaper and reasonably accurate technique of age determination by tooth replacement is sufficiently precise to distinguish younger age classes for most purposes of research and management.

Age determination from horn rings

Some species of hollow-horned ungulates lay down a growth ring on the horn during the winter or dry season. These rings should not be confused with the ornamental rings that most ungulate horns possess and which can be laid down at any time of the year. Only the growth rings can be used for ageing.

The first problem in applying this method is to determine whether the species under study deposits growth rings. A growth ring differs from an ornamental ring in exhibiting a superficial break or overlap of horny material on the outside of the horn sheath. Several horns should be examined carefully. If rings that appear different from ornamental rings are present, and if they get progressively closer towards the base, they may be growth rings.

The next step is to check the conclusion by counting the 'growth' rings of a

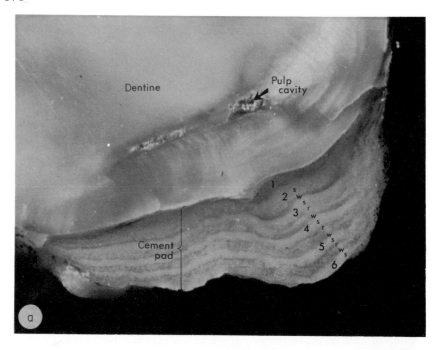

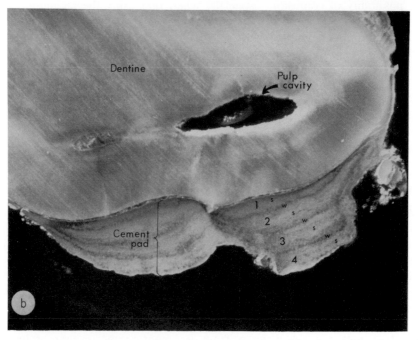

Figure 37   Photographs of sections of molars taken from (a) a 4- and (b) a 6-year-old red deer. Sections follow line A–A as shown in Figure 36 (photographs by W. A. Low and B. Mitchell)

specimen of known age. Should the count be equal to, or one or two less than, the age in years, the rings are very likely growth rings. Should the number be higher than age in years, the rings are probably ornamental and will have little relationship to age.

In Himalayan thar, for example, the first growth ring is normally laid down at about 20 months of age and a further growth ring is laid down every 12 months thereafter (Caughley, C., 1965). Some species may lay down the first ring at 8 months of age. An inspection of horns of several yearling and 2-year-old thar of known age is necessary to establish which case applies.

When a species lives in scrub country or is kept in an enclosure with wire fences, horns often become worn to the extent that the first growth ring is rubbed off. Unless this is known and allowed for, many determined ages will be a year less than the true ages.

In summary, if this technique is to be investigated for ageing:

1. Look at several adult horns for evidence of growth rings.

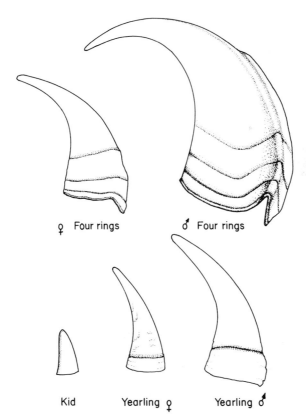

Figure 38 Growth rings on the right horns of Himalayan thar adults, inferred yearlings and a kid (modified after Caughley, G., 1965)

2. If they appear to be present, check the number of rings on several more animals of known age.
3. Should this examination show that the rings counted are probably growth rings, examine a number of horns from known-age yearlings and known-age 2-year-olds to find the age at which the first ring is laid down.
4. Determine whether the first ring is usually worn off the horns of adults, and if so, adjust subsequent determined ages for older animals accordingly.

For the usual case where the first growth ring is laid down in the second winter of life, the age of an animal examined after a calving period is 1 year more than the number of rings. Figure 38 shows the horns of Himalayan thar aged 5 years, and the horns of yearlings and a kid. The apparent ring on each of the yearlings' horns is not a growth ring. It flakes off in the second year of life.

Age determination from weight of eye lens

In most animals, including man, the weight of the lens of the eye increases as the animal advances in age. This forms the basis for the use of the lens weight to determine age distribution in a population of wild animals. The technique is less accurate than those based on counting annual rings in teeth or horns, but may be sufficient to classify species into two or three age categories and is most suitable for rabbit-sized and short-lived animals. In Africa, for instance, it could be of great help to determine age categories (e.g. immature, yearlings, adults) in populations of economically important animals like springhares, dassies, the giant rat or other large rats that can be secured in large numbers and bought on the markets.

Specimens should be taken, labelled (noting at least the sex, date and location), fixed in a 10 per cent formalin solution and packed according to instructions from the laboratory or biologist to which they are to be sent for examination. In the laboratory the lens is extracted, blotted or dried to a standard procedure and weighed on a sensitive scale. If more details are desired, consult R. D. Lord (1959).

The technique has been used successfully on a number of large mammals, including elephant (Lawes, R. M., 1967), warthog (Child, G. et al., 1965), kudu (Simpson, C. D. and Elder, W. H., 1968), pronghorn antelope (Kolenosky, G. B. and Miller, R. S., 1962), Columbian black-tailed deer (Longhurst, W. M., 1964) and white-tailed deer (Lord, R. D., 1962). See R. M. Madsen (1967) for further references.

The biggest difficulty in its use for large mammals is the time and equipment needed and the difficulty of obtaining adequate samples.

Age determination from skeletal features

Some skeletal features are also useful for age determination of certain mammals, but since they are not commonly employed for large herbivorous species, only brief mention is made of them here.

1. The degree of fusion of the skull sutures may be used to distinguish young from adults in certain species.

   As is the case with man, other large mammals have more or less soft spots on the tops of their heads when very young. If you check a series of skulls of animals of different ages you may find, for example, that in the youngest skulls the bones that will come together at the very top of the head are not yet completely joined. Later these bones become firmly joined throughout their points of contact. They are not joined together in a smooth line but in a kind of zig-zag line and as animals get older the zig-zags become more pronounced. In some mammals a ridge appears along the line of zig-zags, and in some species, as widely separated in character and geographical distribution as giraffe and sambar, a deposit of bone forms on top of the joined bones until it entirely hides the zig-zag join and the top of the skull is covered with an increasingly thick layer of solid more or less smooth bone. This kind of observation, as with several other kinds of observation based on skeletal features, is often useful to a field man simply to give him a little more information than he would have if he had been unaware of this process. For example, if he came across the top of a skull and could not find any other part of a carcass, he would at least know that the specimen was from a very young, a young, a fully mature or a very old animal.

2. The length of the long bones, up to maturity, has also been a useful criterion.

3. The long bones of mammals grow from the tips. Each long bone has, during its growth, a cartilaginous zone at each end covered with a bony cap called the epiphysis. As the animal matures, the cartilage is progressively ossified (the process by which cartilage turns into bone) until, in the mature animal, the bony cap and the bone shaft are solidly fused together. So the presence of the cartilaginous zone, or a line representing it or its absence, are sometimes used as criteria for age determination.

4. In certain mammals, notably in the bear, dog and weasel families, the males have a bone (called the baculum) in the penis. The size of that bone has been used as an indicator of age.

Generally the above methods prove of limited use to a field worker.

The use of antlers in ageing

A variation of the use of skeletal features involves the use of size and shape and number of points of antlers in male deer. Although only a few biologists have been concerned (e.g. Huxley, J. S., 1931) antlers are widely used by sportsmen in many regions to estimate age. Biologists are understandably wary of using antlers as a basis for assigning age because of the tremendous individual variations in wild populations. Two sources of variation occur, genetic and environmental, and the latter alone should rule out the use of antlers as a general widely applicable criterion of ageing. I have seen red deer in wild populations in New Zealand produce six tines in their first antler set, while in other areas mature deer not less than 4 or 5 years of age have borne similar sets. These are admittedly observations of extremes, in these cases due to extreme differences in environment rather than to genetic variation. But as a basis for sampling the age structure of a population, antlers are obviously unsuitable unless criteria based on known-aged deer are developed for a particular study area for use only within a consistently uniform environment.

3854 *Summary Discussion of Criteria for Ageing*

The obvious most relevant source of evidence relating to age comes from animals of known age, regardless of which criterion or technique is being observed and evaluated. Information from as many such individuals as possible should be located and recorded.

All techniques are not equally applicable to all large mammals, although for each mammal several different techniques for ageing could be used if necessary. In selecting a technique to obtain from a population some measure of response that is based on proportions of different age groups one must obviously consider certain characteristics of the technique available. First there is the need to produce information in such a way that different areas and years can in fact truly be compared; and almost always there are the limitations of time and money and appropriately trained men to do the work.

Normally these requirements are best met by those techniques depending on classifying various observations made in the field. Techniques depending on dead animals, laboratories and special expertise may be absolutely essential for particular research problems. But to base each year's management decisions on this kind of information can result in unnecessarily heavy drains on a department's recurrent resources.

Particularly if one works, or is to work, in a country which is developing its wildlife resource, attention should be drawn to Appendix 2 which shows outline drawings of African elephant and buffalo. More of these simplified approaches to ageing need to be developed, and their uses and limitations explored.

3860 **Evidence based on Physical Condition**

The physical condition of an animal is a sensitive and easily measured response to the condition of its habitat. It follows that the physical condition of an adequate sample of individuals will provide a measure of a population's response to a particular habitat at current population densities. Determination of population response by this means has several advantages. It usually provides a more direct measure of the response than other commonly used indices, such as those based on the relative sex and age structure of the population or on its reproductive rate. Changes in population structure or of the reproductive rate are often a secondary manifestation of the physical condition of members of a population and are less clearly defined. Further, physical condition responds quickly to environmental changes and so provides a good index to a population's response to these changes after a minimal time-lag, thus facilitating a more precise understanding of cause–effect interrelations between animal and environment. For example, aside from some assessment of condition, one of the most commonly used measures of response are calf–cow ratios, taken as soon after birth as possible. This implies a period of a few days to 2–3 months after the period of calving, depending on the species studied. Since the gestation periods of the species which principally concern us occupy periods of between 6 and 12 months, this means a possible time-lag between cause and response of between 7 and 15 months. Comparable times for larger animals such as elephant, hippo and rhino would of course be considerably longer.

Most animals, but especially those living under natural conditions, are subject to fluctuating levels of stresses of various kinds. These may affect certain segments of the population more than others or they may affect the whole population differently in different years. The effect of stress often needs to be taken into account when interpreting the observed level of physical condition. Factors which may cause stress include:

(a) seasonal and longer-term climatic changes, vagaries in the weather, removal of forests, charcoal burning, calamities such as floods or fires or changes in competition either within or between species, all of which may affect the availability of food or water, or shade and escape cover; and
(b) changing physiological demands at different stages in the life cycle, such as the demands of growth among young animals, or those caused by pregnancy, lactation or rutting among adults.

These factors can be ignored for practical purposes when comparing the relative physical response in a given animal species on two or more areas, provided comparisons are made between similar segments of the populations, and assessments of physical condition of animals from different areas are made within the same period of time. However, if there is a question of whether or not a population is too numerous on a year-round basis in a given area, the stress factor must be considered. If animals that should be under added stress because of seasonal physiological demands are found to be in reasonably good condition at the most critical time of the year, then the population is probably not too numerous. This is particularly true if the observations are made in a year which is not favourable for the species for example, severe drought or unusually cold winter. On the other hand, if animals which should not be experiencing undue stress are found to be in poor condition, it is possible that the habitat resources may be inadequate. If the population remains in poor physical condition during less severe periods of the year, then it is probably too numerous for the habitat, either as the result of its own increase, or because of a decline in the habitat resources.

Emphasis above has been placed on the use of measurements of physical condition as an aid to determining interreactions between a population or populations and the environment. This seems important enough to elaborate by listing several specific ways in which indices to physical condition have been applied:

1. As a means of elucidating the ecology of deer populations by revealing annual patterns of seasonal changes in physical condition.
2. Comparisons of differences in levels of population responses in different environments is a tool useful for understanding mechanisms by which wild ruminant populations are adapted or are adapting to environments. An implication of such a demonstration for management is in providing data, in combination with sex and age ratios, for estimating rate of recovery in severely reduced populations. This in turn may help understand the effectiveness of control schemes, harvesting populations at various levels, etc.
3. Physical condition can not only be related to, but may be indicative of, various stages in eruptive oscillations (see Section 7100, also Caughley, G., 1970).
4. Condition indices have been used to demonstrate that deer and several African ungulates tend to remain in a given area and to give an indication of the extent to which populations are localized. This has obvious significance in emphasizing the need for more intensive rather than more generalized management practices.

In short, indices to physical condition provide the most sensitive measure of animal response available. The uses of such indices are only beginning to be explored.

3861 *Fat Reserves*

A measurement of fat reserves provides the best index to physical condition in a wide range of mammals. During the earliest growth in farm animals, bone is the tissue which develops fastest, later muscle grows faster, and finally fat becomes the fastest growing tissue. Hence, fat makes up a significant part of the body in mature well-fed animals. Wild animals generally live on a lower plane of nutrition than well-fed domestic stock, but may still carry sufficient fat to render its measurement of value in determining physical condition.

It has been suggested that large fat reserves are seldom developed in tropical large game animals because they never have to face critical winters, similar to those experienced by deer, for example, in ranges with winter snow. Sufficient work has been done in several parts of Africa with an adequate variety of species to indicate that this is not true. A possible explanation for this misconception is that most autopsy material tends to originate from areas where over-population is the motive behind culling large numbers of animals. Under these circumstances it can be expected that the bulk of the population will have limited fat resources. It is here assumed that, wherever one has opportunity to observe populations of the same species under a variety of differing environmental conditions, ranging from good to bad, then it will be possible to devise some method of assessing changes and detecting differences in physical condition.

In most species of large mammal the main centres in the body in which fat is deposited are the subcutaneous connective tissue, the abdominal cavity, and the intermuscular connective tissue. The fat content of the bone marrow is also of particular practical importance under certain conditions. It has been found that the order in which fat is laid down or resorbed from given centres in the body is consistent in several distantly related taxonomic groups. In most large mammals the bone marrow is the first easily observed deposit to respond to improved conditions following very poor conditions. This is followed by abdominal fat, and finally subcutaneous fat is laid down. Loss of fat during deteriorating environmental conditions is in the reverse order.

For practical reasons it is usually sufficient to limit the measurement of fat reserves to those in three centres of the body, as this provides a scale that covers animals ranging in conditions from very poor to very good. Fat beneath the skin has been measured by means of the back fat index, the abdominal fat by means of the kidney fat index, while the fat content of the bone marrow is determined from a femur marrow index. In many ungulates the kidney fat provides the best index for a wide range of physical condition and is considered first. Changes in the back fat overlap changes at the upper end of the kidney fat index and are useful for judging the condition of individuals that are high on a theoretical scale extending from emaciation (and death) to animals in peak condition. Differences in the femur marrow index are useful at the opposite end of the overall scale and overlap the lower limits of changes in the kidney fat. Figure 39 shows the extent to which various field indices may reflect changes throughout the entire range of physical condition. For more details including trials of other kinds of indices see Riney, T., 1955a.

Kidney fat index

This index expresses the weight of a standard sample of fat from around the kidney as a percentage of the weight of the kidney itself in order to standardize readings from different-sized animals. The calculation is simple:

$$\frac{\text{Weight of kidney plus fat}}{\text{Weight of kidney with fat removed}} \times 100 = \text{kidney fat index.}$$

184

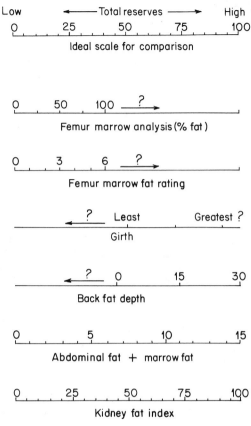

Figure 39   Diagrammatic comparison of the range of usefulness of various indices discussed in the text against an 'ideal' scale (after Riney, T., 1955a)

One kidney is usually anterior to the other and the consequent relationship to other organs of the body may influence the amount of fat around each of the kidneys. For this reason it is preferable to standardize which kidney is used for determination of the fat index. In studies of red deer, for example, I used the left kidney. The fat around the kidney is separated from other abdominal deposits, and the kidney and a standard sample of surrounding fat are removed with a sharp knife, using the standard cuts shown in Figure 40.

The kidney fat deposit is usually clearly separated from other abdominal fat deposits. Even in very fat animals there is usually some obvious discontinuity, such as an area of almost fat-free mesentery, between this and neighbouring deposits in the vicinity. If the whole organ is hidden by fat, the standard cuts are best achieved by slicing through the fat at 1–2 mm intervals until both margins of the kidney are reached. After weighing the kidney and fat, the fat is peeled off the organ from the side opposite the urethra and blood vessels. This is done quickly and easily by making a shallow cut along the outer curve

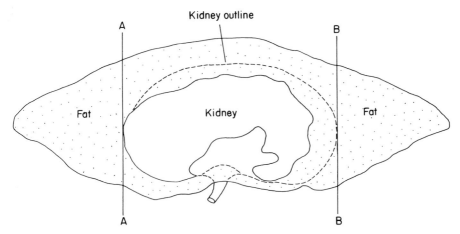

Figure 40   Position of cuts at each end of the kidney (A–A and B–B). The kidney, with its surrounding fat, is then weighed, the fat removed and the kidney reweighed. The weight of fat is then expressed as a percentage of the weight of the kidney to form the kidney index (modified from Riney, T., 1955a)

of the kidney and peeling off the thin kidney covering membrane to which the fat is attached. The kidney without fat is then reweighed.

The size of an animal will largely determine the accuracy with which such weighings should be attempted. With most large African game animals a spring balance sensitive to the nearest gram is satisfactory, while with the very large mammals a sensitivity to the nearest 5 or even 10 grams is adequate.

Observed values of this index from antelope have ranged from under 10 in several species to over 300 in a fat eland (*Taurotragus oryx*).

Back fat index

For animals in fair or poor condition there is usually no fat to measure and this reserve is, in deer and in several other ungulates, commonly one of the last to build up as an animal improves in condition. Thus back fat is more appropriately used as a means of comparing animal condition between two areas in the most favourable season of the year for the species, rather than as evidence taken from the same population in different seasons. When used as suggested, as a way of comparing two or more areas when the animals are in their best condition, back fat is somewhat more sensitive; that is differences will appear sooner and with a smaller collection of animals, than with the kidney fat index and under these special conditions is to be preferred.

This index is based on the depth of fat near the root of the tail. Record to the nearest millimetre after making a cut forward from the base of the tail at about 45° to the dorsal processes of a lumbar vertebra and measuring the greatest depth of fat along the cut. It is sometimes difficult to distinguish between remnants of fat and surrounding connective tissue and it is hard to measure thicknesses under 3 mm. Where depth of fat is under 1 mm it is preferable to ignore it altogether; if between 1 and 3 mm it is usually better to describe it as a trace. If the skin must not be cut, great care must be taken while skinning this part of the body so as not to disturb the underlying fat.

As an example of the kind of evidence obtained in this way, the greatest depth observed in a medium-sized antelope, the lechwe (*Kobus leche*), was 30 mm (1.2 in).

It is of interest to note that as an all-round index the back fat index was more appropriate than kidney fat for one small mammal, the marsupial possum (*Trichosurus vulpecula*) (Riney, T., 1955a).

Femur marrow index

A close relationship exists between the amount of fat in the femur marrow and its colour and texture. This is fortunate, since it is difficult and time-consuming to extract and analyse the fat content.

The marrow of the femur, or upper hind leg bone, is exposed as indicated in Figure 41, by cracking the bone midway along its length. In ungulates colour may vary from white, or white with fine red streaks, through straw colour to brown or red. Marrow in best condition (with the highest fat content) is light and becomes progressively darker when fat is absorbed as an animal declines in condition. Simultaneously, there is a change in the feel of the marrow. If a small piece is squeezed between the fingers the change in texture can be felt as fat content declines from a firm waxy feel, to one that is greasy. Finally it becomes like a watery jelly or even like thick blood.

Changes in texture and colour merge gradually and it is important to sample animals that died in good condition and compare with others that died in very poor condition before attempting to use these criteria in the field.

The condition of the marrow may be broken into the three classes, by colour and texture. As a rule, it is good practice to consider whether the marrow is either 'good' or 'poor' and if neither to record it as 'fair'.

Detailed comparisons of bone marrow classifications made by two or more observers should be used with caution, unless the observers have used the same criteria and checked each other's results on a trial basis. Table 6 suggests criteria for use in assigning one of three condition classes.

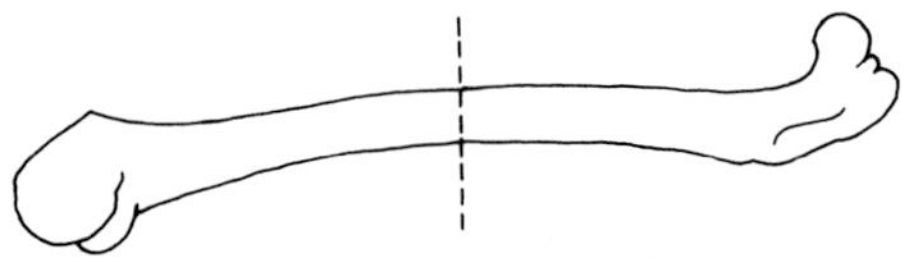

Figure 41   Marrow at or near the centre of the femur is used in assigning a 'condition rating' based on the colour and texture of the marrow

Table 6   Criteria for use in assigning bone marrow condition classes

	Good	Fair	Poor
Texture	Waxy or firm	Greasy	Watery or jelly-like
Colour	White or off-white, sometimes with red streaks	Intermediate	Brown or red

The marrow fat index is convenient to use on recently killed animals. However, in each species and area, tests must be made to determine more precisely how long after death the rating can still be used with as much confidence as if the animal were freshly killed.

Dr Graham Child has suggested (personal communication) that, in standardizing the way of assessing bone marrow, sufficient time should be allowed for immediate post-mortem changes to be complete before an assessment is made. For example, for certain large mammals, such as buffalo, the 'waxy firm feeling' of marrow classed as in 'good' condition may not be evident immediately after death. Child suggests that this may be remedied by removing the femur and exposing the bone to the air for about 20 minutes.

Uses

Fortunately, in temperate and sub-tropical regions, bone marrows, for most species are easily assessed several weeks after an animal dies, which allows these observations to be made on carcasses discovered in the field.

Marrow fat is a good index to condition in the lower third of the condition scale, that is when the animal died in poor condition and had been declining in condition; for example, in the latter part of the winter season in temperate regions, or the latter part of the dry season in tropical semi-arid regions.

Another less widely known occasion for use is shortly after a habitat or environmental change for the better, as in early spring following a severe winter, or after the first rains following a drought. Under these circumstances, response in the marrow can be so quick that an animal can be literally skin and bones, in very poor general condition, and yet the marrow observations may indicate fair or even good condition. Observations of marrow can thus be used to learn how quickly a population in poor condition is responding to a more favourable environment change.

Recording of fat reserves

Separate sets of records should be kept of the indices to physical condition of each species, unless such information is included on an autopsy form for each specimen. The condition form illustrated in Figure 42 may be usefully adapted for summarizing the information from such autopsy forms. It is important to note the date, locality, sex and if possible the age category of

Observer _____

Species _____

Specimen number	Date	Locality	Sex		Age (e.g.)			Adult	Kidney index	Condition				Remarks (Activity etc.)
			M	F	0–1	1–2	2+			Back fat (mm)	Marrow			
											G	F	P	

Figure 42 Example of form for recording indices to physical condition based on fat deposits. G = good; F = fair; P = poor

each specimen. It is often relevant to further classify females as pregnant or non-pregnant, or as with or without milk in the udder. The form includes space for recording back fat and marrow fat indices as well as the kidney fat index. Such a form may be most useful to biologists: in situations where condition indices have previously been little used; where it is considered desirable to test for oneself the correlations between these three indices; or to test the generalizations made above concerning the limitations of back and marrow fat indices.

In the absence of established ageing criteria, ages may be expressed in terms of tooth eruption or arbitrary permanent tooth wear classes (see Section 3852). The remarks column is valuable for recording observations such as incidence of parasitism or disease, or other possible relevant influences on condition. Note that a specimen number is necessary for cross-reference to other types of data relating to the same specimen when information is recorded in more than one place or form.

If it is possible to take advantage of carcasses available through commercial cropping, normal hunting season kills, or animal reduction, control or elimination campaigns, the above-mentioned indices may prove useful. Normally, however, such material from dead animals will be unavailable. For this reason it is worth considering to what extent differences in animal response, in terms of physical condition, can be recognized in living free-ranging animals.

3862 *Weight*

Weight is commonly selected by big-game workers as the criterion being either most indicative or convenient for assessing the condition of deer. For example, L. A. Davenport et al. (1944), working with Virginia deer; E. R. Doman and D. I. Rasmussen (1944), L. A Stoddart and D. I. Rasmussen (1945) and A. S. Leopold et al. (1951), working with mule deer; O. J. Murie (1951), with wapiti, and others have all interpreted loss in weight as a loss in physical condition.

Consistent increases between amounts of fat and carcass size have been described by J. Hammond (1942) as a characteristic pattern of growth for various tissues in farm animals. Demonstrations of consistent increases in fat and weight during growth have been made for mammals as widely separated as sheep (Clark, E. A. and McMeekan, C. P., 1952) and mice (Fenton, P. F. and Dowling, M. T., 1953), but consistent fat increases paralleling weight increases during growth of immature animals do not imply that weight is therefore a suitable index to condition for animals in all stages of physical condition.

Most of the above-mentioned studies obtained their information during the hunting season, when males were at or near peak condition. This seems justified where the objectives are to compare animals of a given sex or age group, in good to fair condition, in different years or from different areas. However, because of the close relationship between girth and weight, the same limitations apply as shown in Figure 39, weight being most useful as an index when animals are in the upper parts of a condition scale. It is less useful as an index to compare condition between sexes or age groups or between different areas at times when animals are in poor condition.

The value of total weight as an indicator of total fat content, in starvation, has been questioned by several workers, mainly because of the formation of metabolic fluids accompanying mobilization of fat reserves. According to S. Brody (1945): 'The use of weight as a criteria [sic] [of condition] is seriously questioned because of the homeostatic process involved in the trend toward maintenance of a standard weight in the healthy.' Lack of correlation of total weight with changes in the total amount of depot fat in the body has been demonstrated in various stages of starvation in several types of animals (*man*: Kireilis, R. W. and Cureton, T. K., 1947; Brozek, J. and Keys, A., 1950; Keys, A. *et al*., 1950; *mouse*: Hodge, H. C. *et al*., 1941; *rat*: Dible, J. H., 1932; *cattle*: Moulton, C. R., 1920; *pig*: Hilditch, T. P. and Pedelty, W. H., 140).

Precise information is lacking on the extent to which total fat can be correlated with total weight in ungulates in various stages of starvation or recovery. In view of the above-mentioned lack of correlation between total weight and total fat reserves, it may be well to re-examine the value of weight as a precise indicator of condition in large mammals, particularly when animals are in poor condition or where sex or age comparisons are desired.

3863   *Comparative Uses of Several Indices to Fat Reserves*

Figure 39 compares with an ideal scale the fineness of scale and range of several indices to fat reserves. Ideally a technique for assessing physical condition would be sensitive both to increases and to decreases in physical condition and to keep registering changes in condition at either end of the condition scale. I know of no technique that will do this, but for most ungulates I have worked with, the kidney fat index comes closest. Table 7 reviews the usefulness of these indices.

3864   *Judging Physical Condition in Live Ungulates*

Even a crude assessment of the physical condition of individuals in a population may be of value as an indication of the population's response to prevailing environmental conditions. Under most field conditions it is impossible to judge the condition of antelope, or any other wild animals, precisely and in such a way as to allow comparisons with evidence of condition taken from carcasses. However, populations commonly include individuals in either good or in poor condition so that, if sufficient care is taken, the proportion of animals from each recognizable sex and age group, that fall into three broad condition classes, can be determined. In practice careful field assessments as described below have proved as useful as the various condition indices based on dead animals.

Table 7 Comparison of uses and limitations of indices for assessing conditions of deer populations

Index	Uses	Limitations
Kidney fat index	(1) Measures condition in all seasons. (2) Enables different-sized animals to be compared on a uniform basis. (3) Enables measurement over a wide range (scale 0–300).	Showed less change than the back fat index when measuring differences between animals in good condition, but this index is considered the most satisfactory all purpose index tested.
Back fat index	(1) Measures condition between top condition animals (e.g. adult males in autumn; adult females in early winter).	(1) Measurement of condition restricted seasonally due to inability to measure the approximate lower third of the condition scale. (2) The scale has a narrow range (0–42). (3) Different-sized animals not placed on comparable basis.
Abdominal fat index	(1) To measure entire range of condition. (2) To rate different-sized animals on a comparable basis.	(1) Scale is narrow (0–9). (2) Rating is by visual estimate.
Femur marrow fat index	(1) To compare gross differences within or between populations in mid- or late winter. (2) To assess condition on deer dead several weeks or months. (3) To learn speed of improving condition following improving environmental conditions.	(1) Poor for measuring changes in upper parts of condition scale. It is therefore of little use in spring, summer, autumn, or early winter. (2) Poor for demonstrating differences in condition between different sexes. (3) Scale is narrow (0–6). (4) Rating is by visual estimate.
Combined index (abdominal fat plus femur marrow)	To measure condition in all seasons.	(1) Due to influence of marrow scale, this index is: (a) distorted seasonally; (b) of less value than abdominal fat index in illustrating sexual differences. (2) Scale is narrow (0–15). (3) Rating is by visual estimate.
Weight	Measures only inter-population or seasonal differences.	(1) Not consistently related to fat reserves in all stages of inanition, which implies limitations for seasonal comparisons.

Table 7   Continued

Index	Uses	Limitations
		(2) Different sizes not place on comparable basis, which implies limitations for inter-population condition comparisons. (3) A gross measurement difficult to take in the field.
Girth	Of little use in measuring condition in deer.	(1) More a measure of size than condition. (2) Of no use for inter-population comparisons of condition.
Antlers	Of doubtful value.	(1) In most species of deer applies to males only, thus eliminating intra-population comparisons. (2) No known direct connection with fat reserves.
Field assessments of condition	(1) Reflects condition in all seasons. (2) Sensitive to assessment of condition when expressed as a proportion of one condition class to all animals classified within a particular sex or age group. (3) Useful when study animals cannot be killed.	(1) Requires special effort for two or more observers to standardize techniques of observation. (2) Less suitable for animals with long hair.

As an animal's condition deteriorates and fat and eventually protein reserves are mobilized, the rump appears progressively more angular. Studies have shown that an objective assessment of physical condition in a sample of deer and some African antelope was possible, if individuals were assigned into one or three arbitrarily assigned condition classes, based largely on the appearance of the hind quarters, (Riney, T., 1960).

Determination of condition classes

Figure 43 illustrates the criteria for assigning deer and several other ungulates to condition classes designated as good (1), fair (2) and poor (3). As the fat reserves of the animal diminish the tail appears more angular (a), the outline of a point of the pelvic girdle can be seen (b), a distinct angle appears at the point indicated by (c), the lateral processes of the backbone vertebrae can be seen as a faint line (d), and the outline of the ribs are visible (e). If the angles

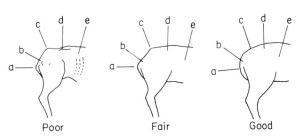

Poor   Fair   Good

Figure 43   The general appearance of the hindquarters of ungulates that are arbitrarily classified into good, fair or poor condition. As the fat reserves of the animal
diminish the rump appears more angular (a), the outline of a point of the pelvic girdle
can be seen (b), a distinct angle appears at the point indicated by (c), the lateral
processes of the lumbar vertebrae can be seen as a faint line (d), and the outlines of the
ribs are visible (e). If the angles at points (a) and (c) cannot be observed the animal is
classed as in good condition. If any two of the points indicated at (b), (d), or (e) can be
observed the animal is classed as in poor condition. 'Fair' represents an intermediate
category of individuals, not clearly classed as 'Good' or 'Poor'

at points a and c on the diagram cannot be observed, animals are classed as in
good condition (Class 1). If any one of the points indicated at b, d, or e can be
observed the animal is classed as in poor condition (Class 3). Class 2 represents the appearance of individuals who are not clearly in either Class 1 or
Class 3. In this class angles can be seen at points a, or c; points b, d and e
remain unrecognized.

As with other field techniques of this type certain arbitrary criteria have to
be followed, which vary to some extent with local conditions. These decisions
will depend largely on the species under consideration and the nature of the
vegetation and terrain where the observations are made. It is usually only
possible to make satisfactory assessments if the light is good, if the animals are
seen from the side and are either standing or walking. Binoculars or a telescope should be used unless the animals can be approached closely. The
maximum distance at which assessments are made will have to be determined
separately for each species and area studied. Condition classifications should
be made on undisturbed animals as on several species the hairs conspicuously
rise along the back and rump when the animal is disturbed.

Classify all animals in a group; otherwise a biased sample may result from
the greater ease with which animals in one of the condition classes can be
recognized. This is especially true if some of the animals are very thin, since
these tend to stand out more than those in better condition. If visual assessments are made on dead animals these records must be kept separate from
records based on living animals as they are unlikely to be comparable.

Assessing condition in this way is exacting and time-consuming and should
not be done by the same person at the same time sex and age classifications
are made. Furthermore, although condition ratings are naturally grouped by
sex and age, these observations should not be used as a basis for sex age
classifications. If both condition and herd composition classifications are

desired from the same animals at the same time, it is usually easier for the two classifications to be done by a team of two persons, each making one kind of classification. Even when only one kind of classification is attempted it is useful to have another person helping by writing down classifications as they are made. In this way the observer can be free to keep his attention focused through the telescope or binoculars and take full advantage of the particular observational opportunity. If such classifications are made by one observer working alone, then serious consideration should be given to obtaining a portable tape recorder whose microphone can be fastened near the mouth. Classification decisions can thus be entered on the tape, then later transcribed to a prepared form.

When first trying this condition rating there is a tendency, especially for those who know the animals well, to ignore arbitrary criteria and assign ratings on the basis of their considerable knowledge of the animal and the area. Thus one frequently hears the statement that: 'considering the poor season of the year the animals are in pretty good condition'. This kind of judgment must be strongly resisted and clear criteria set and adhered to if different areas, seasons or years are to be compared.

If new observers are being trained in classification, trying different approaches is an excellent way to gain an appreciation of the need to have clear criteria as a basis for making field classifications and of the importance of initially concentrating on one kind of observation at a time. With this in mind, trying to apply the approaches *not* recommended in the preceding paragraphs should be useful as an exercise in training.

Recording of physical condition

Figure 44 indicates a form which is suitable for recording the physical condition of free-living antelope. It is important to separate the records for the different sex and age classes as far as possible, to locate the observations; to be able to relate the observations to a particular habitat type. The evidence can be presented as the proportion of animals in each condition class or as a ratio between numbers assigned to a given class and the total animals classified at a given locality and season. On several occasions, in a single afternoon, I have demonstrated statistically significant differences in the condition of adult females in different habitats less than 5 miles apart and by as few as twenty careful classifications of adult females in each habitat.

Suitability for various species

Species on which this technique has been successfully used are red deer (*Cervus elaphus*), fallow deer (*Dama dama*), sambar (*Cervus unicolor*), feral goat (*Capra hircus*), chamois (*Rupicapra rupicapra*), and thar (*Hemitragus jemlaicus*) in New Zealand; and, in Africa, kudu (*Tragelaphus strepsiceros*), duiker (*Sylvicapra grimmia*), steenbuck (*Raphicerus campestris*), reedbuck

FIELD CONDITION FORM

Observer _____   Species _____

Date	Locality (as precise as possible)	Habitat	ADULT						CALF			YEARLING						Remarks
			Male			Female						Male			Female			
			G	F	P	G	F	P	G	F	P	G	F	P	G	F	P	

Figure 44   Example of form for recording visual assessments of physical condition

(*Redunca arundinum*), sable (*Hippotragus niger*), impala (*Aepyceros melampus*), eland (*Taurotragus oryx*), blue wildebeest (*Connochaetes taurinus*), Grant's gazelle (*Gazella granti*), Thomson's gazelle (*Gazella thomsoni*) and dorcas gazelle (*Gazella dorcas*).

Slight modifications must be made for several other African species whose body conformation differs enough from deer and antelope to preclude use of the same criteria, e.g., Cape buffalo (*Syncerus caffer*) and wild pigs. For animals of considerably different body conformation such as zebra, giraffe, rhinoceros, hippopotamus and elephant, satisfactory criteria have still to be devised. In devising such criteria the obvious first requirement is to observe animals at both extremes of physical condition. Then, the selection of suitable criteria is not usually difficult.

Uses and limitations of visual assessments

The assessment of condition of live animals is, for an individual, not as precise as assessments based on fat indices and taken from a dead animal. However, in terms of comparing the physical condition between a given sex or age segment of a population, this field technique seems to work as well as those based on autopsies. Although the scale for an individual is small, there being but three choices, the scale for a population segment or population is as large as the number of individuals assessed in each area, since the results are presented either as a ratio of animals in a certain condition to the total animals assessed, or as a percentage of the total.

Thus, the demonstrably better condition of animals in one area as compared with a poorer condition in another area at the same time of year presents the case for the differences in the suitability of contrasting habitats as eloquently as if the animals themselves had spoken instead of their protruding ribs. The uses of other condition assessment techniques mentioned above in Section 3860 apply equally well to visual assessments. Such visual assessments can be applied at any time of year provided comparisons are restricted only to animals of similar age, sex and breeding condition.

The chief drawback lies in the fact that two or more observers cannot be certain their findings are comparable, unless they have first observed together in the field in order to standardize the criteria upon which their decisions are based. However, with this qualification met the method is useful, especially in situations where it is impractical to obtain carcasses to assess condition.

3870 Mortality

Death is the most extreme measure of animal response and it has already been mentioned in various forms, for differential mortality between areas automatically reflects in differences in age structure, cow–calf ratios, sex ratios, wet udder counts, embryo counts and so on. Likewise big drops in population may be measured by some index to total numbers as described in Section 3423.

The simplest way of measuring annual mortality in a population which has a clearly defined calving season is to obtain an index to numbers immediately after the calving season and, nearly a year later, immediately before the next calving season. If an estimate of the population has been made this percentage loss through mortality can be converted to numbers. If the area has been harvested, subtracting the harvest will leave an estimate of natural mortality. This figure can be used in calculating harvesting rates as described in Section 7212B.

When mortality is particularly heavy and it is desired to obtain some measure of the number of carcasses per square kilometre, for example, field counts of carcasses can be made using sampling strip counts as described in Section 7212C. See also Robinette, W. L. *et al.*, 1956.

Another kind of evidence is to record the sex and ages of carcasses found, on a detailed map of the study area, to learn in which habitats or situations mortality was particularly heavy.

3880   Animal—Response—Summary

Response is what wildlife management is all about, regardless of what our objectives are. When we can observe or manipulate a change in either an animal population or its environment and then predict the consequent responses, we should be satisfied.

Examples of uses of evidence of response are given in various parts of Chapter 8000 which deals with problem analysis and with various problems of management. Evidence of response is not always required in defining management problems but is almost always required for understanding them. Since some of the techniques are easily and inexpensively obtained they provide useful cross-checks against ideas arising from examining other kinds of evidence. The indices to physical condition provide a quick way to asking relevant questions about the extent to which various habitat elements may be regarded as required, the combinations in which they become optimum, or the extent to which they are irrelevant.

3900 RESPONSE—ENVIRONMENT

3901 INTRODUCTION

The environment is composed of a complex of interdependent living and non-living elements. Some of these (parent rock, climate, weather) are less subject to modification than others (soil, plants, animals). If any of these elements change, the interrelationships operating within the environment also change. This section is especially concerned with measuring the response of habitats to changes in animal use.

For browsing and grazing animals vegetation normally supplies basic requirements of food and shelter, thus knowledge of how the vegetation responds to use by animals is often vital to management. Changes in vegetation are commonly reflected in changes in ground cover. As vegetation downgrades as a result of over-use by unconfined wild or feral animals it does so in a patchy, discontinuous pattern. This means in practice that even when the overall condition of an area is good, if the area is downgrading, at least some small patches of bare ground may be exposed. This is especially noticeable in steep mountainous areas and in semi-arid lands. As the habitat continues to downgrade more and more ground becomes exposed; the stage is thus set for rain or wind to sculpt the surface of the land into the final forms of degeneration: the shapes and patterns that characterize accelerated erosion. With these considerations in mind the two main aspects of environmental response here considered are (1) vegetation, and (2) ground cover and accelerated erosion.

As here used environmental response means, essentially, trend. The common question is: Is the element concerned increasing (improving) or decreasing (downgrading) or remaining approximately the same (stable)? Two quite different approaches are used in obtaining information on which decisions regarding trend are based.

First, evidence relating to the occurrence and availability of habitat elements (Section 3500), if taken in the same way over successive periods will of course show changes and lead to a recognition of trend. If in the same area, sampled in the same way at the same season, percentages of ground cover were 60 per cent in one year and 40 per cent 5 years later, the trend would be recognized as clearly downgrading.

The second kind of evidence may just as confidently allow the recognition of trend but does not quantify it in the same way and for this reason is referred to as apparent trend. Thus, if the same area were sampled in the year of the second sample mentioned above, ground cover would be still described as 40 per cent and as having decreased over the previous several years. This is done through the use of various indicators or indicative observations and by using a syndrome approach as described below (Section 8120). If a date of a change in land-use or management practice is known and other nearby unchanged areas are still available for cross-checking it may still be possible to estimate a period of time over which reduction to an observed 40 per cent ground cover took place.

3910   Vegetational Response

3911   *Status and Trend of Vegetation—Principles*

Distinguishing between evidence of *status* and that of *trends* in the vegetation is important both in understanding animal–vegetation relationships and in management. *Status* relates to the plants and plant remains that are currently in an area at a given time. It is an index of the prevailing relationships between the soil and vegetative cover and indicates the standing crop available to animals at the time observations were made. It is essentially a static concept, concerned with the present. It involves simply description at the occurrence level of an ecosystem, and various ways of obtaining such evidence are discussed in Section 3500.

3912 *Trends*

Trends in the vegetation indicate the dynamic responses of plants to the influences of their environment, including the man-caused influences. Some of these changes enhance ecological stability, while others are detrimental to stability. Such trends are here referred to as 'conservation trends' because emphasis is placed on maintaining a given habitat in a stable or improving state.

Trends may be downward even if the present status of the vegetation is high and there is a good plant cover. Such trends, if continuing, could foreshadow loss or serious reduction of the habitat element in the future and the trend is therefore regarded as unsatisfactory. On the other hand a recent trend may be regarded as satisfactory from the point of view of conservation, even if the present status is low, providing the trend is towards recovery of the habitat element concerned.

In referring to trend, words like 'satisfactory' or 'unsatisfactory' are perhaps not ideal terms as they may be considered by some as too emotional for use in objective surveys. Their use in this handbook is simple. If a habitat element on which an animal (or form of land-use or management) depends is stable or improving (increasing) it is considered satisfactory. If the same element is downgrading, that is becoming scarcer, it is considered as unsatisfactory.

A change in the status of vegetation depends upon the average annual rate of production and reproduction of plants in the population and whether through this process the plant species or vegetation type is increasing, maintaining itself or declining. This may be represented by changes in the number of individual plants, or by changes in the average amount of annual growth which survives the year, or by both. There are continual fluctuations in the composition of plant communities due to weather, as well as changes in the nature of the vegetation due to long-term climatic changes. The former are temporary, and the latter are usually almost imperceptible and, for most practical wild-life management purposes, can be ignored. Against this background there are often persistent, easily observable trends, which can be attributed to land-use or management practices, that cause an imbalance between the plants and their environment. It is these changes that particularly concern the wildlife manager.

Common examples of such trends are the effects of incorrect burning and

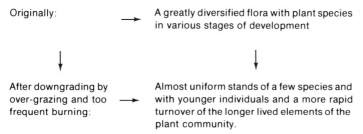

Figure 45   Response of the floral spectrum to over-use (after Riney, T., 1970)

of over-populations of browsing and grazing animals. Seperately, or together, these may lead to an overall reduction in plant production, or they may alter the nature of the vegetation by favouring certain species or classes of vegetation at the expense of others. A simple example involves burning in certain types of grassland. A fire, even at the time of the year when it does least harm, may lead to a reduction in grass growth the following season; but by the second or third season the grass has often regained its maximum vigour. If such areas are burnt every 3 or 4 years, the average plant cover remains high. Although there is a downward fluctuation after each fire, undesirable symptoms, such as accelerated erosion of surface soil, do not necessarily occur. However, if burning is too frequent and the grassland does not have time to recover fully after successive burns, then trends are set in motion towards eventual denudation of vegetation and ground cover and serious erosion of soil. The process may be intensified if, in addition to too frequent burning, the area is also subject to heavy grazing pressure, or the grazing pressure itself may initiate a downward trend. The species composition of plant communities may also change if preferred food plants are heavily used by animals. The weakened plants may no longer be able to withstand competition from less palatable forms, which gradually replace them. Figure 45 illustrates in very general terms how a previously undisturbed watershed in which wild ungulates are living can be expected to change following over-grazing and too frequent burning, and Figure 46 suggests how animal habits are thus modified.

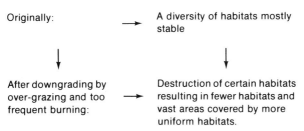

Figure 46   Response of wild ungulate habitats to over-use (after Riney, T., 1970)

3913   *Ecological Succession and Regression*

As vegetation develops it passes through several successional stages. Succession is the universal process by which formations of plants develop. This begins with the first species that colonize bare areas (pioneer species) and culminates with the final plant communities a given environment can produce, the climax vegetation. Ecological succession is thus the normal, natural mechanism by which improving vegetative trends take place.

This continuous progressive process of species overlapping each other in time is important for wildlife managers to understand. Optimum habitat requirements for some animal species may be met only in the lower successional phases; others may require higher phases or a climax vegetation and still other animals may require a mixture of successional stages. As plant communities change, populations of large mammals become automatically and consequently adjusted (e.g. as suggested in Figure 46).

Succession proceeds at slower rates under difficult climatic conditions, as in arctic or desert regions, and proceeds at its most rapid rate under the influence of the warmth and humidity of the tropics.

Both succession and regression in plant communities may be initiated or

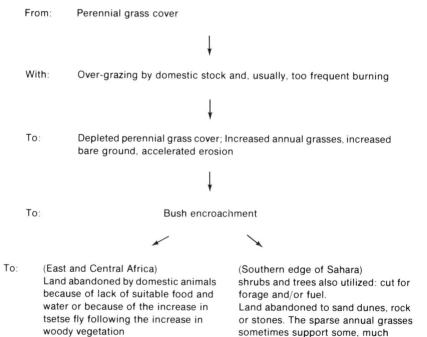

Figure 47   An example of a downgrading sequence of events starting with a woodland savannah with a good perennial grass cover (after Riney, T., 1970)

greatly modified by the type of management, or animal use, to which an area is subjected. Furthermore, either increasing or downgrading vegetation may be halted to form what is commonly called a sub-climax, and these sub-climaxes may become stable *under a given consistent form or pattern of use*. Such sub-climaxes are commonly referred to as grazing climaxes, fire climaxes, or by such other form of major use or practice as may be relevant.

Although an improving vegetative trend will normally proceed smoothly from the colonization of bare ground by plant species through a succession of successively more complex vegetative communities, downgrading is often sudden and complete as, for example, when a forest is clear-felled for timber or by fire or flooding. However, downgrading can also occur more slowly and through successive stages as under different kinds and intensities of over-grazing. The sequential stages of habitat degradation will vary in different environments. For example, such a sequence is indicated for an initial wood-land savannah with good perennial grass cover and a stable savannah watershed in Figures 47 and 48 respectively. A consequent response by ungu-lates is suggested in Figure 49.

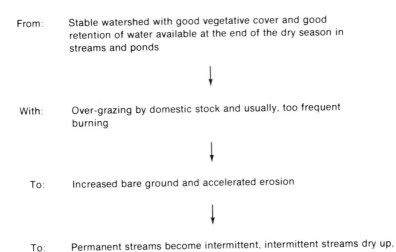

Figure 48   An example of a downgrading sequence of events starting with a stable savannah watershed (Africa) (after Riney, T., 1970)

204

A diversified fauna with many species.

↓

Decline of certain species, especially those which require perennial grass and which must drink frequently at the end of the dry season.

↓

Marked increase of those species utilizing shrubs and trees and which are not so dependent on readily available water.

↓

Large herds of a few species; in some areas large migrations become a feature of the country.

Figure 49   An example of animal response to successional downgrading in open woodland savannah (after Riney, T., 1970)

3914   *Observing Vegetative Response*

3914A   *Indicator Plants*

The most useful indicator plants in any given situation can usually be determined by the inspection of specific areas with a known history of land-use. The investigator should acquire as much knowledge as possible from old people and old records, including old photographs, of changes that have taken place in the general nature of the vegetation under a given form of land-use. Careful inspection of the vegetation then often reveals particular plants or plant forms which have benefited by this treatment. These plants may then be used as indicators in recognizing and interpreting changes which have taken place over large areas for which detailed records are lacking.

The concept of key species or utilization indicator species has been described in Sections 3731–3732. Such species can at times also be used to describe response, as indicated below. Likewise the presence or absence of age classes can be indicative of past over-use, as when no young trees or shrubs are present because browsing animals are eating all seedlings. In general comparatively few plant species are present in degraded areas compared to the same areas covered by higher successional stages. A striking example is the presence of carpets of annual wild flowers of a single colour blanketing plains or hillsides. Such displays may attract tourists but the very presence of large blocks of single species or very few species usually indicates a low successional stage.

Extreme environments provide an apparent but not a real exception to this generalization. In high mountains or in arctic regions one reaches a stage where the plant species are few and where the colonizing pioneer species and the climax species are one and the same. Some arctic environments will permit only lichens to grow on the cold rocks.

3914B *Key Areas*

Many wildlife workers are faced with the need to ensure that satisfactory vegetation–soil relations are maintained over large areas. In such cases, it is important to avoid unnecessarily detailed analysis and to develop techniques which are reliable and can be applied in the shortest time possible. It is also useful to learn to select key areas in which to concentrate measurements of the trends. These are areas which are of such importance to a species of animals that their well-being depends on the maintenance of satisfactory conditions on such places. Factors which may determine whether an area is a key area or not, include its intrinsic properties, such as geology, local topography and drainage, or the nature and depths of the soils; or they may depend on the treatment to which it has been subjected, or its use by animals and its proximity to favoured areas, such as salt licks, waterholes (including artificial waterholes) and ecotones (the transitional zone between two or more vegetation types.).

Since certain parts of an area are usually more sensitive to change than others, the interpretation of the results of trend measurements on key sites needs care. If certain classes of vegetation are disappearing from only the most sensitive areas, this is indicative of a fairly early stage of range deterioration. If, on the other hand, vegetation classes are disappearing from the least sensitive sites, this constitutes good evidence of general and more advanced range deterioration. Used in this way careful observations of key areas facilitate a more rapid definition of basic problems.

3914C *Age Classes of Plants*

It is usually sufficient for wildlife management purposes to classify plants by age into seedlings, young plants, mature plants, decadent plants and dead plants. With shrubs and trees an additional class is often needed, i.e. crown-sprouts. Foresters usually classify trees as seedlings, saplings, poles, mature trees, over-mature trees, decadent (dying) trees, and dead trees. The last two categories may consist of plants of all ages.

Each age class may have a special significance associated with its stage of development. For example, a stand of spruce trees in their first few years may be of importance to a deer population mainly in supplying a proportion of their winter food. Later, the same tree species may provide both food and shelter and a few years later provide shelter alone. The following describes briefly each of these classes of shrubs and trees as used by wildlife managers.

Seedlings

Seedlings are defined as very young plants which have survived the early die-off period. Each favourable year, thousands of seeds germinate and produce small seedlings, the majority of which succumb before the year is out. It is the fraction of these which survive and become established that are of interest, the 1-, 2- and 3-year-olds. Such very young plants can be distinguished by their relative size, simple branching and succulent bark. Note that all trees do not seed each year. It is thus necessary to know something of the pattern of seeding for a particular species before interpreting evidence based on presence or absence of seedlings.

Young plants

These are larger, and have more complex branching and more fibrous bark, than do those in the seedling class. But they do not show the size and other signs of maturity, such as the rounding-off of crown and heavy fibrous stems. With trees, the pole class falls into this category.

Crown-sprouts

Crown-sprouts develop from the root-crown of some browse species after burning or crushing. They are so classed until they begin to show the characteristics of mature plants.

Mature plants

These will be distinguished by their complex branching, rounded growth form, larger size, and heavier and sometimes gnarled stems. The crown should be made up of more than three-quarters living wood.

Decadent plants

Decadent plants are defined as those shrubs or trees which are dying from age or other influence. Their crowns show one-quarter or more dead wood.

Dead plants

These include only those dead shrubs and trees which are still rooted in place.

The relative abundance of various age classes of a plant may be an expression of its location, competition, or fire, as well as of varying degrees of use by browsing animals. But the absence of one or more age classes should always be carefully considered. The effects of soil erosion, climatic fluctuation, plant density, or wild-fire may favour some age class and some plant species at the expense of others. Stands of vegetation at full density may lack a representation of younger plants because of severe competition from established vegetation. Species of plants that reseed satisfactorily only after burning represent another group from which young plants may be lacking from a cause other than browsing pressure. Where areas are over-stocked with browsing animals, the younger age group of preferred forage plants are particularly susceptible to losses resulting from over-browsing.

Thus, the analysis of age class data is done in order to determine if enough seedlings and young preferred forage plants are present to take the place of dying and dead plants. If the older preferred plants are dying without replacement, the apparent trend in condition of the shrubs or trees must be rated as *downward* or degrading. If the survey findings show that there are far more young plants than those decadent or dead, the apparent trend (of the plant species) may be rated as *upward* or improving or as stable. Subsequent surveys will show the actual (rather than the apparent) trend.

3914D *Browse Lines and Hedge Forms*

Since it may take some time for an animal population to create a certain age structure in trees or shrubs it is useful to consider another more sensitive response in the form of browse lines and hedge forms. The act of browsing can change the shape of woody plants as easily as can hedging shears. Many browsers, feeding in a characteristic way, thus modify the vegetation into recognizable forms.

Wildlife managers find these characteristic changes in shape of shrub or tree useful, as the change in shape normally takes place before age classes are eliminated. It is thus more sensitive as a guide to trends in animal–plant relations. The frequency and type of hedge forms constitute an important kind of observation in determining areas of key use. They may provide useful information concerning competition for food between coexisting species of animals, or on the impact of one species on the cover or food requirements of another. They may be useful in recognizing past changes in population pressure of browsing animals and may facilitate and extend interpretation of size and age class distribution of plants. For example, old browse lines or hedge forms recently unused may indicate a higher populaton of browsing animals in the past than at present; or the absence of young plants that are shorter than an animal's characteristic browse line height may indicate that the animal is preventing the establishment of young plants. In another broader sense, changes in the forms of plants can permit recognition of the early stages in a downgrading process which can lead to loss of habitat. Because of the significance of this process, particularly in developing regions, this aspect has been emphasized below.

Many of the characteristic effects of particular browsing animals on plants will be known to the experienced field man. Others can be learnt by direct observation of the way animals feed and their influence on the growth forms of plants. This is usually easiest in areas with a reasonably high animal population density. Attention should be paid to such factors as whether the animal takes leaves alone, or leaves and a section of stem, the thickness of stems that are bitten off, whether trees are pushed over or broken off and how this is achieved, the maximum height to which use is limited, whether use is uniform or patchy with animal populations of different densities.

Form classes of woody plants

To make such observations more objective, clear criteria must be established for the various classes and plants sampled then recorded under the classes of their description. One may prefer to assign form classes on the basis of combining both the degree of hedging and the present availability of forage. An example of this combined approach may be described in this way.

When shrubs or trees are not browsed, or only lightly browsed, they tend to assume the natural forms, or shapes, which are normal for each species. As intensity of browsing increases, the departure from these normal shapes becomes more striking. Continued heavy browsing, year after year, results in tightly hedged or highlined, and partly dead browse plants, which stand out as evidence of poor condition and declining forage yield. And small trees and shrubs can of course be killed by over-browsing.

Degree of hedging may be broken down into arbitrary classes, for example, little or no hedging, moderately hedged, and heavily hedged. Hedging is evidence of response to past use and should not be confused with present use.

If only three classes are used the heavily hedged class is reserved for shrubs so closely cropped they are being damaged, losing vigour and are moving into decadence.

Browse plants may also be classed for availability in a similarly simple way as: all available, largely available, mostly unavailable, and unavailable. This classification will cover most conditions found in the field.

When form and availability are combined into a composite rating the possibilities are as follows:

Form/availability class
    1: All available, little or no hedging
    2: All available, moderately hedged
    3: All available, heavily hedged
    4: Largely available, little or no hedging
    5: Largely available, moderately hedged
    6: Largely available, heavily hedged
    7: Mostly unavailable
    8: Unavailable.

Availability may result from height, location or density of plants. The browsing height limit will have to be set according to the easy-reach height of the animal studied, even though some of the interior growth may not be available for browsing. Where shrubs or trees occur in very dense stands, leaving only the growth around the margins of the stand available for browsing, such stands or patches should be classified as largely available or mostly unavailable depending on the circumstances.

Browse plants that are growing out of reach, but which still offer considerable forage below the browsing height level, are classed as largely available. Where browsing, shading or other factors have killed most of the available growth below the browsing height level, or along the margins of impenetrable brush stands, the browse should be classed as mostly unavailable, or, if no available forage is present, as unavailable.

In classifying browse plants for degree of hedging, remember that those plants that exhibit a more or less natural shape are classed as lightly hedged, while those that have been hedged so severely that they exhibit poor vigour and short growth are classed as heavily hedged. All others will fall into the moderately hedged class.

It is recognized that no age or form class description will fit all species of shrubs and trees, and that personal bias has room to operate in borderline cases. It is always necessary to develop *written* standards as a guideline to field observations in order to overcome these weaknesses.

Application in the field

Measurement of trends

Trends in vegetation classes can be discovered by the periodic repetition of any of the measurements used to describe the status of the vegetation, provided these are made at similar times of the year. Even casual previous descriptions of the vegetation, including historical accounts and old photographs (especially aerial photographs) can provide valuable evidence of changes in the vegetation. However, it is often unnecessary to initiate such periodic observations, or to await their results in order to assess important trends. Assessments made at the most critical time of the year for plant survival, such as at the end of the dry season or at the end of winter, can often yield adequate answers.

Assessments of trends in vegetation classes may be made on a purely subjective basis, but it is usually preferable to introduce a measure of objectivity by standardizing procedures. This allows comparisons to be made between areas, or in different years on the same area, or between the results of different observers. This is the method described below.

A careful examination of vegetation in the immediate vicinity of points at regular intervals along a transect line can be used to assign an apparent trend to vegetation, according to the criteria outlined below. At least ten separate assessments of the trend in each vegetation class are made. If clear conclusions as to the direction of the trend are still not possible, then the sample should be extended to twenty, thirty or more assessments, until the observer is satisfied that the various classes are either stable, increasing or decreasing. It is preferable to increase the samples by multiples of ten, as this makes their comparison with other samples, or their statistical evaluation, easier.

The information should be recorded on field forms. The form can also be adopted for recording other types of relevant data along the transects as shown in Figure 141. It is also important that individual transects should sample, as far as possible, uniform habitat conditions, similar in slope, vegetation type, distance from water and the like.

Trees

Small trees may be treated as shrubs as long as the objective is to hold the stand in suppressed form for browse production rather than to allow normal growth. Some trees can maintain a vigorous supply of forage by multiple branching when cropping is as heavy as 50 per cent of the annual growth; the same trees will not make normal-height growth if the terminal leaders are cropped and overall browsing exceeds 20–30 per cent. Also once a tree achieves a height at which a significant part of its annual growth is above the reach of browsing animals, heavy browsing no longer will suppress it but will merely hasten normal die-back and self-pruning of lower branches.

Moderate browsing of young trees, including the terminal leaders, may not necessarily result in a lowered production of wood. Often such trees expand their root systems during the period of suppression above ground and if browsing is reduced for a session or two, height growth is phenomenal. Once such trees are released they may catch up in both height and in general form with unbrowsed trees of the same age. Also, there are usually more young trees growing in wooded areas than are needed for full stocking at maturity. In such instances, a portion of the young trees will be stunted or lost in any case whether or not browsing occurs. On the other hand, if trees are girdled (as by porcupines), or debarked (as by elephant, deer or bears), or broken or pushed over (as by elephants), they must be considered lost to timber production.

Terminal leaders have been mentioned above. A terminal leader is the terminal part, or tip, of the main stem or stems of a tree that makes the principal height growth. Figure 50 illustrates several sorts of terminal leaders. If the terminal leaders are cropped, the height growth is suppressed. Records may be made of the number of terminal leaders present on each tree and of the number which have been browsed.

If there is a clear need to make actual browse utilization measurements other than terminal cropping and general broad classifications as described above, methods described in Section 3721 may be used.

It must be remembered that for sustained production of browse it is essential that enough trees (at least two to four per acre) achieve release to grow to normal height and maturity. Otherwise the stand may be lost in time for lack of seed sufficient for adequate regeneration. In contrast, if the objective is to

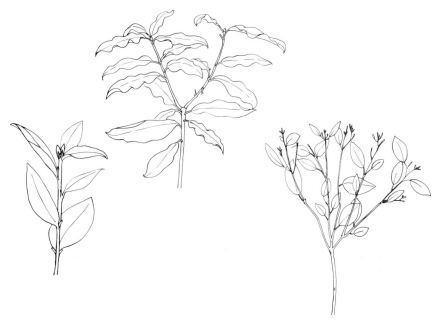

Figure 50   Three different types of terminal leaders

maintain woodland or forest habitat, it is essential that enough young stock escape prolonged suppression to fill in spaces left by dead trees. The appropriate spacing desired depends on the objectives of management.

It is normally impossible to look at and evaluate all parts of a large area; a system of sampling will be necessary. Of the kinds of sample plots described above, the circle-plot transect is the most useful for sampling stands of trees. Whether managing for browse or timber, it will be necessary to establish a series of sample plots on the key areas, i.e. the parts of the range where browsing of trees is most concentrated (either seasonally or year-long). We must do this in order to maintain a record of the effect of browsing on the stand of trees. We will be concerned principally with the trees that provide preferred browse or provide shelter to the animals with which we are concerned. Otherwise, because of heavy suppression of the trees sought by the browsing animals, and the comparative light to moderate use of all others, there will be a gradual shift in composition of the stand of trees toward the non-preferred, less desirable species. Hence it is essential that we make a record of the ages of the trees that are rooted within the sample plots. From these records we can determine whether the stand is maintaining itself or whether the species composition is changing. Whether the wildlife manager's or the forester's classification of trees is used is a matter of choice (see preceding section).

212

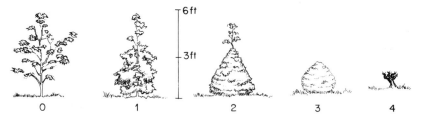

Figure 51   Criteria for classifying New Zealand beech into five hedge form classes: (0) no observed effect of hedging by animals; (1) lightly hedged, shape slightly altered; (2) hedged but breaking away, or capable of breaking away, from browse pressure; (3) growth held entirely in check by animals; (4) plant killed by deer (no other browsing mammals in this area) (from Riney, T. *et al.*, 1959)

A practical example

From the above it should be clear that there are various simple ways of classifying shrubs and trees into form classes resulting from browsing and that availability classes can be combined with form classes if required. As mentioned above standards will have to be defined in each study and animal concerned. The following example is given to illustrate how form classes were used in learning the effects of red deer on the forest and shrub vegetation in a New Zealand study area (Riney, T. *et al.*, 1959).

The trees and shrubs available to deer were classified into five groups as explained in Figure 51.

One result of the classification, shown in Figure 52, indicates that 96 per

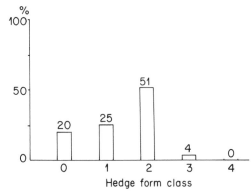

Figure 52   Percentage of shrubs and trees assigned to each hedge form class. Ninety-six per cent of the plants examined were in no danger of being eliminated by deer, although 80 per cent had been modified in shape by deer. Removal of leaves and twigs by deer equalled or exceeded annual growth for only 4 per cent of the plants examined (Class 3) (Riney, T. *et al.*, 1959)

cent of the plants examined were in no danger of being eliminated by deer. Other results not shown compare the frequency of various hedge form classes for the three key species of plant. Still another kind of information was obtained by taking a sample of plants in hedge classes (1) and (2); measuring the height of hedging (browsing height) and the total height of the tree or shrub. When the annual growth rings were counted at the browsing height this indicated an age, at that height, of between 7 and 14 years. The ages of the plants themselves, judging from counts of growth rings taken at the bases of the same plants, ranged from 17 to over 100 years (see Figure 53). This was interpreted as suggesting a drop in the deer population 7 to 14 years previous to the study.

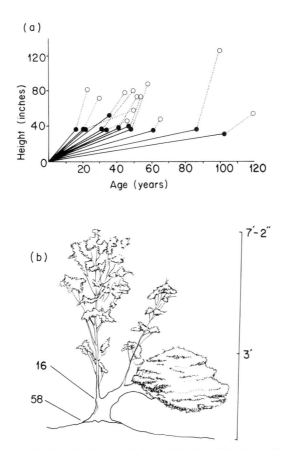

Figure 53   (a) Total height (white circles) and height of hedging (black circles) of beech saplings, both plotted against age. The age of the shoots taken at the level of hedging ranged between 7 and 16 years, suggesting a drop in the deer population in this period (see text). (b) Location of counts of annual growth rings on one of the plants (modified from Riney, T. *et al.*, 1959)

Shrubs

While many species of browse plants produce vigorously under moderate hedging, vigour declines and production falls off if heavy hedging persists. Where browsing animal populations are at a level at which there is no more than light-to-moderate cropping of preferred forage plants, the individual animals will maintain health, vigour and productivity. There can be no doubt that pound for pound, the leaves and twigs selected by browsing animals under light-to-moderate stocking, if only on the basis of lower percentage fibre, are more nutritious than are the coarser twigs and stems consumed when shrubs are browsed heavily.

It has been demonstrated that at least deer and some game birds have the ability to choose plant parts higher in crude protein than can investigators making clippings for chemical analysis. Freedom of choice is, however, restricted greatly where key areas are over-browsed; hence, heavy cropping of browse may be evaluated from two perspectives: the effect on and corresponding response of the browse plants, and the effect on and corresponding response of the browsing animals. With some species of browse plants, enough leafage is protected by their height, density, or by their cage-like growth forms to enable the plants to survive extended periods of heavy cropping. In such cases, either because of actual shortage of preferred forage or because of the lower value of the plant parts eaten, the browsing animals may demonstrate an adverse reaction before there is actual decline in the browse stand. So, either from a plant or an animal standpoint, a high percentage of heavily hedged preferred-browse plants on a key wildlife area spells trouble. Hence the investigator on shrub ranges should be interested not only in determining the age composition of the stand of preferred-browse plants, but the form classes as well, and this may be done in essentially the same ways as described in the preceding section for trees.

On even a moderately used habitat the examiner must expect to find some heavily used shrubs of preferred species. It must be determined how much one is willing to sacrifice. On some deer ranges in America, the standard has been set at 15 per cent. Thus if more than 15 per cent of the shrubs on a key area are heavily hedged, stocking by browsing animals is rated as too high and a recommendation for either improvement of habitat or reduction of the number of animals is made. As mentioned elsewhere many individual shrubs and young trees can sustain a loss of over 50 per cent of their annual twig growth and still continue to grow and to reproduce. This is only an example of the form which setting a standard may take. Standards and tolerance limits must be specifically developed for the vegetation and animals concerned.

Different animal species create different shapes of shrub and tree

Each browsing species modifies the shape of trees and shrubs in a characteristic way. Naturally even within a single plant species not all plants browsed will be the same shape for they will not only be of different ages but subject to different intensities of browsing. A range of shapes common in beech in the South Island of New Zealand, when browsed by red deer, is shown in Figure 51. Where sheep alone are involved a similar set of shapes may develop but the apex of the hedged cone shown in (2) of Figure 51 will be appropriately lower. If cattle are involved, instead of a triangular profile or cone, a hemispherical hedge form may result. Goats hedge a proportion of the same plants into columns with comparatively straight sides. One of the hedge forms produced by giraffe is similar to cattle but of course much higher. Hares and rabbits under-cut shrubs to form a characteristic profile and so on.

In Africa, horns are used by several antelope, e.g. kudu, eland and buffalo, to reach higher branches and break them, thus making more leaves and twigs available. The breaks will be of a greatly or slightly different type for each species. Elephant not only push over trees but create a pattern of broken branches which creates hedge forms of various kinds, depending on the plant species. Former heavy elephant use has been recognized on trees over 100 years old in some African study areas which now hold but few elephants. The percentage of trees showing a past major damage to the stem that could only have been done by elephant, when compared with similar damage on a sample of several younger age classes, may thus reflect changes in elephant populations.

Caution must be used in interpreting these forms and shapes until you are certain of which animal is involved. Certain shapes can be created by more than one species. For example, mopane (*Colophospermum mopane*) about 5–10 ft high can be modified by both eland and elephant to a similar hedge shape, and using this evidence alone it was difficult to distinguish hedged forms created by these two species on this particular height class of mopane.

A distinction is commonly made between maximum height of hedging and browse lines. The hedging height is the height of the apexes of hedged plants such as shown in Figure 51(2), and is formed by animals feeding at normal head height and below. Browse lines are made by animals reaching upwards to browse which, when they become intensive enough, form a low ceiling to the canopy of trees, below which there is little food available. Many species, but not all, will stand on their hind legs to browse overhead leaves and twigs, thus raising the height of the browse line above the height of the apex of the hedged plant.

To associate particular hedge forms with particular animal species, observe undisturbed feeding in areas of shrubs and trees, especially in key feeding areas, for it is here and in adjacent areas that a range of hedged forms can usually be found. It is an obvious advantage to find an area in which only the species concerned occurs. A visit to research stations where animals are kept under controlled and known conditions can be useful in the early attempts to describe a spectrum of characteristic hedged forms However, once a range of hedge forms is recognized for several animal species, even when several animal species have been feeding in the same area of shrubs or trees, their past presence can be identified by recognition of characteristic hedge forms alone. Fortunately hedge forms created by cattle, sheep or goats are easily recognized and readily distinguished from most species of wild herbivore.

3914E *Trends in Grasses*

The following characteristics are usually associated with the progressive deterioration of healthy grassland, such as may occur from too frequently burning, burning at the wrong season, over-grazing, heavy trampling, following drought, or from a combination of such factors.

Deterioration usually begins with a decline in the number of plant species. Sensitive forms or palatable forms that are over-used will no longer be able to maintain themselves. This may not lead to a reduction in ground cover, if less sensitive or less preferred species colonize the space previously occupied by declining species, but it may result in reducing the amount of food available to certain grazing animals. Sometimes the loss may be dramatic, as when a stand of closely spaced, vigorous perennial grass plants, with a high proportion of plant litter between the tufts, gives way to a sparse grass cover with few young tufts and much less litter. Animal signs, such as trails, will be numerous if deterioration results from animal use. As the process proceeds the proportion of bare ground increases, and the soil may become compacted or subject to wind or water erosion. The annual grasses tend to increase and forbs or woody plants may become more numerous.

An increase in grass is accompanied by the opposite trends. Bare ground and the proportion of annual grass and forbs generally diminish, as perennial grass colonizes more and more of the available space. This may be slow, especially if the nature of the soil has been changed or the density of competing woody plants is high (see also Section 3913).

When assessing trends in grass populations, it is advisable to pay particular attention to the edges of patches of bare ground, to see if these are being colonized by young, perennial grass plants and if any of these have survived the season. The presence of such invaders, and a mixed population of healthy tufts of all ages, indicates an upward trend. On the other hand, if the bare patches are not being invaded and have abrupt edges composed of old plants and there are signs of soil erosion, and if there are few young plants and the older ones are generally unhealthy, it is indicative that grasses are declining. Unhealthy tufts usually have exposed roots, dead or dying centres (in some species this applies even in young plants) and can often be dislodged easily by a gentle kick with the toe of a boot.

If observations taken at intervals within circular plots along a transect indicate that decadent and dead plants outnumber healthy tufts, that there are few young plants in evidence and that the amount of bare ground is increasing at the time of year most critical for plant survival, the perennial grass stand should be judged as down grading at that particular station. If this is true of the majority of the stations along the transect, the perennial grasses in the area sampled should be assessed as declining.

Often relative trends can be determined between various species of grass occupying the same area, provided the differences are reasonably well defined.

Individual assessments made in stable grassland, in which a proportion of the grass plants are approaching senescence, may have certain similarities to assessments made in deteriorating grassland, but in stable grassland the status is generally much higher and there is usually much more litter. In addition, bare patches are infrequent and limited in extent and most young plants are healthy. With a little experience there should be no difficulty in distinguishing between stable perennial grassland and range in which the grass cover is deteriorating.

If examination of grassland reveals that no clear trends are evident, the area should be recorded as stable, except where there is inadequate grass upon which to base a decision. In such a case this fact should be noted as characteristic of the transect, and it may then be necessary to seek out relict patches of perennial grass, in an area nearby, in order to recognize recent trends.

Line point transects are recommended for measuring trends in grassland. The toe point transect (see Section 8341) is particularly useful where vegetative communities can be easily recognized and characterized as more or less distinct habitat types and where the density and composition of the vegetation is apparently evenly distributed within each type.

While actual trends in range condition are best judged by comparison of findings of original with subsequent surveys, a judgment in regard to apparent recent trend can be made at the time of initial survey by observing appropriate evidence within small circular plots[1] for each of which an assessment of trend is made. If such plots are spaced at regular intervals; e.g. every other step along a transect line designed to describe present condition or status, this will facilitate later cross-checks and comparisons with other kinds of observation made along the same transect. Within each circular plot observations are made characterizing the site and may include, for example, observations on: severity of cropping, degree of trampling, vigour of preferred species, changes in species composition, presence or absence of established seedlings in bare spots, recent covering of grasses by soil or gravel, or dead or dying plants. Evidence of soil loss through wind or water may be observed within the same plots. For example, the lower lines of lichens on rocks, or grasses growing on top of small pedestals of soil, may indicate the depth of soil recently removed, or the raised bases of tussock with depressions in between may reflect soil loss during the lifespan of the tussocks observed, caused by wind or water (see Figure 54). Small erosion rills may be observed forming even within circular plots of 1 square metre or less. Observations such as these can, in combination, form the basis for judging whether perennial grass is increasing or

---

[1]For example, cut a measuring stick as a guide: 3 square metres = a circle with a radius of between 97 and 98 cm. 1 mil-acre (1/1000 acre) = a circle with a radius of 44.7 inches.

218

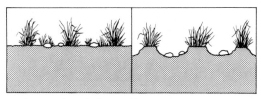

Figure 54   The way plant pedestals form. Plants may appear to rest on pedestals when
unprotected soil around them erodes (after Beeson, R. W. *et al*., 1940)

decreasing. Since the downgrading of vegetation is but one early phase of a
downgrading habitat, consult also Section 3923.

Best sampling period

The act of deciding on an appropriate sampling period can be an important
step in realizing management or research objectives. For example, if one's
concern is over the maintenance of perennial grass as an essential habitat
element, sampling should be near the end of the worst season of the year for
grasses. In semi-arid areas this will be the end of the dry season before the
rains commence. In temperate regions it will be the end of winter before the
spring growth masks the ground. At such times it is easiest to recognize relations between soil and upgrading vegetation. Then it is quicker to establish
consistent results between different observers or when making subsequent
assessments in the same area. The same conclusions may be reached even
when the rainy season is well advanced or in spring, but only with much more
time, and this often requires at least twice as many samples (see Figure 19).

The particular checklist of observations you may use in assessing trends in
perennial grasslands will vary with the environment. However, it is not difficult to set criteria for assessing trends in a particular study area, for areas
where recent trends are known. Such areas provide opportunity for recognizing and defining relevant criteria. For example, if a farm or ranch formerly
holding sheep or cattle was abandoned 5 years ago one would expect recovery
of the vegetation used by sheep or cattle to be taking place. If a borehole has
recently been opened for the first time on a cattle ranch, the area in the near
vicinity of the new watering point can be expected to be downgrading.
Another way to become familiar with a range of relevant observations to
make in assessing trend is to visit research stations where numbers of cattle,
burning and other elements of the history are precisely known. If possible run
a transect inside the enclosures. A word of caution may be helpful. Care must
be taken in using such areas to avoid confusing desirable pasture research
standards or desirable trend from the point of view of the research officer in
charge of a particular experimental area with an assessment of trend for your
own purposes. While the objectives may in fact be the same, this should
never be assumed.

In principle, recent trends in grasslands should be no more difficult to recognize than recent trends in forest or shrub lands, if considered in terms as general as perennial or annual grasses. There is, however, an important present difficulty; that of estimating the age of individual perennial grass tussocks. It is to be hoped that this will be less difficult in future, but, at present, criteria either for assigning ages or for placing grasses into age classes have, for most species, yet to be devised. There is an obvious need to develop local criteria for ageing at least the key perennial grass species. However, even without precise ageing criteria, trends in recent years can usually be recognized and demonstrated.

The series of diagrams and pictures shown in Figures 55 to 91 may assist in recognizing improving or down-grading trends in perennial grassland.

Forbs

Most woodland, shrubland, and grassland areas support a normal complement of scattered forbs: broad-leaved herbs. Many of these forbs (especially the legumes) are nutritious and form part of the diet of grazing animals. Some are higher in protein in the cured stage than are grasses and are much sought after during dry seasons. Furthermore, many kinds of forbs are prolific seeders and possess deep tap-roots which make them better adapted for survival on deteriorating sites than are grasses. Some species of forbs are aggressive invaders of grasslands of low density such as result from heavy grazing. Hence a high density of forbs and/or annual grasses is often indicative of an undeveloped, disturbed, or deteriorating soil with fertility, tilth, or moisture too low to support a full stand of perennial grasses.

Like other kinds of plants, the species of forbs most preferred by grazing animals will be replaced by those of lower preference under excessive grazing, especially when plants are cropped so severely they cannot store enough food. For maintenance or improvement of a stand of forbs, grazing ordinarily should not be so close as to leave a slicked-off tightly cropped appearance, but rather an uneven, patchy type of cropping which allows some plants to mature and cast seed.

The methods of measuring condition and trend, and utilization, described above for grasses may be used also for management of areas with stands of forbs.

3914F *Bush Encroachment: an interplay between trends in woody plants and grassland*

An increase in the frequency of woody plants often accompanies a decline in perennial grasses and vice-versa. Bush (shrub or tree) encroachment may be a gradual process or it may take place rapidly. It may involve one, a few or a whole range of species. Gradual encroachment can result in a wide dispersion of age classes and may be difficult to distinguish from a normal population without historical evidence. Where the process has been rapid a high proportion of the plants of the same or different species may be of a similar age. This often results in comparatively uniform stands of woody plants.

'Bush encroachment' may refer either to the colonization of grassland, or to a thickening of woody vegetation in already wooded areas. In the first case patches of open grassland are generally fringed by an area in which there are progressively younger woody plants towards the centre. However, some trees and shrubs can disperse more or less evenly throughout a clearing. Where the transition is short, it may be difficult, without historical evidence, to distinguish encroachment from a normal woodland/grassland edge. Indicator plants and plant forms may help to reach a conclusion in such a case.

Encroachment resulting from certain factors may be characterized by typical species or growth forms. Over-grazing by domestic stock may lead to the spread of certain species. Frequent burning may result in typical growth forms, such as ground-level coppicing, which may be apparent in more than one woody species.

The most reliable basis for recognizing bush encroachment under most conditions depends upon an estimate of the relative ages and frequencies of woody plants. Age can often be determined from the sizes of the plants, either in terms of height or girth. It is useful to examine the edges of areas where woody plants are established, in order to determine if these are spreading or receding. This should be done at regular intervals along a transect.

A note of caution is necessary. Bush encroaching into grassland areas may be stunted, or the growth form modified in some way, so that there is not a smooth regression in over-all size into the area. Agents, such as fire, can modify the form of the plants so as to produce a distinct gap in smaller size classes suggesting a shrinking population. The relative density of the plants may then provide the clue as to whether the population is spreading or not. When in doubt, one can usually determine the age of the plants from growth rings.

3920   Ground Cover and Erosion

3921   *Principles*

When one basic objective of management is to continue with a form of use on a stable long-term basis, one prerequisite is clear: that form of use or management cannot afford to destroy the resource base on which its existence depends. The minimum conservation requirement is thus to maintain the resource base required to achieve management objectives on a long-term basis. Bare ground and erosion are of importance to the wildlife manager in so far as they may be associated with downgrading habitats, and these two kinds of evidence may conveniently be considered together. In terms of conservation, a downgrading trend involves an increasing proportion of bare ground and, if the process continues, accelerated erosion.

When comparing two areas, if the status of ground cover in one area is demonstrably less than in another, from a purely conservation point of view the area with the lowest ground cover would be regarded as in the poorest condition. If the ground cover can be demonstrated as in the process of decreasing over the past few years, that area could be regarded as downgrading.

Stability, from the conservation point of view, normally requires both vegetation and litter, the two combined providing the ground cover. Ground cover protects the soil from raindrop impact and splash. Its presence allows the water to soak slowly into the soil and protects the soil surface from rapid soil removal by washing and retards run-off. It reduces water loss from the soil by evaporation. A minimum amount of litter is required before many species of plants can establish themselves. Thus, in assessing recent trends, observational evidence of recent changes in plant cover and litter, compared with evidence of recent changes in the proportion of exposed bare ground, is used to indicate improving or downgrading trends. This approach is simpler, quicker and requires less training than assessment of recent trends based on observations of vegetation alone.

If stable areas are to become unstable and downgrade, the process is often as follows. As litter disappears from the surface and as vegetation becomes scarcer, bare patches of soil are exposed to the direct impact of raindrops. Through the effect of direct impact and splash when raindrops hit the soil and through increasingly unimpeded run-off of water, the amount of surface soil removed considerably increases. As this process continues more soil is displaced and the formation of rills, erosion pavements or gullies takes place. Eventually the large-scale conservation problems of intensive gullying, flooding and silting become apparent.

Ground cover provides an easily measured index to allow recognition of downward trends at an early stage in this downgrading process, for even when vegetation is at an early stage of downgrading, ground cover may also be decreasing measurably.

3922 *Evidence for Assessing Condition of a Catchment*

Evidence for assessing the condition of a catchment may be obtained from several sources and expressed in terms of characteristics of soil, vegetation, animal, or water run-off. Furthermore, trends in soil and vegetative condition may be apparent before correlated trends are reflected in animal condition or water run-off or accelerated erosion. Therefore observation of evidence of the present status of soil and vegetation should be an early prerequisite to the arresting of deteriorating habitats.

The following list is by no means complete but is submitted to exemplify the type of evidence useful for determining present status and trend and for providing clues to past history of the vegetation–soil complex in a specific environment (modified from Riney, T. and Dunbar, G. A., 1956).

3922A *Soil*

Bare patches due to rill, gully, sheet and other types of erosion.
Soil profiles (normal, truncated, buried, inverted).
On rocks, former soil lines, lichen lines, moss lines.
Soil pedestals and exposed plant roots.
Erosion pavement.
Fresh gravel or soil covering, or partly overlying vegetation.
High-density drainage patterns.

3922B *Vegetation*

Composition of vegetation (diversity of species, density of desirable or un-
 desirable species).
Ground cover provided by living vegetation
Vigour of vegetation (e.g. abundance of seed heads on grasses).
Amount of litter.
Indicator plants.
Regeneration.
Presence of different age classes.
Dead plants.
Vegetation colonizing exposed soil or gravels.
Stability of stream flora and fauna.

3922C *Fires*

The following list exemplifies types of evidence useful when the major causes
 or contributory causes of present status and trend has been fire.
Charred woody material; ash.
Charred bases of tussock.
Ash and charred wood in soil profiles.
Increase in species resistant to fire.
Relict species present, indicating previous forest or grass cover.

3922D *Animal Use*

Occurrence and density of sign of various species of animal:
Sheep camps (beds, rubs) bedding areas, deer or buffalo wallows.
Rabbit warrens, burrow systems.
The presence and numbers of characteristic spoor of various animal species
 and the equally characteristic trail patterns, particularly in feeding areas.
Browse lines, hedging effects on browsed plant species.
Heavy utilization of preferred plant species; utilization of less preferred
 species; over-grazing.
Light use of preferred plant species; virtual non-utilization of less preferred
 species; under-grazing.

3922E *Other Forms of Human Exploitation*

Logging (cut poles, barking, stumps).
Mining activities (sluicing, dredging).
Access roads.

3923 *Illustrating Evidence of Recent Trend*

Two general environments have been selected to illustrate the assessment of
trends: semi-arid savannah and woodland savannah areas in Africa and temperate mountainous areas in New Zealand. In both of these apparently different environments the kinds of evidence most useful are: evidence of recent
trends in the relationship between bare ground, ground cover and vegetative
cover and evidence of accelerated erosion. Such evidence seems best
described in conjunction with drawings and photographs.

Each area with its combination of soil, slope, and other factors requires a
certain density of plants and litter to hold the soil. Figure 55 shows the
relation of density of plant cover to erosion. The different layers of soil are
not of equal value. The darker-coloured uppermost humus layer will produce
not only the greatest volume of plants but also the more desirable ones. As
succeeding soil layers are exposed the volume of plants is reduced and less
desirable species occupy the area. Figure 56 shows this in a digrammatic way.
But when soil has recently been removed there is usually evidence to show
that this is so. Figure 57 shows the loss of soil and the formation of soil
remnants, and Figure 58 shows exposed roots of small shrubs due to soil loss.

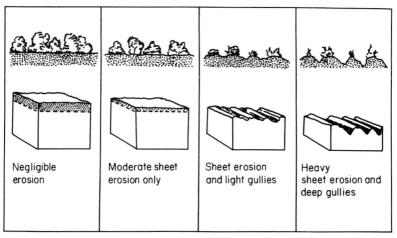

Figure 55   The relation of density of plant cover to erosion. Each area with its
combination of soil, slope and other factors requires a certain density of plants and
litter to hold the soil (after Beeson, R. W. *et al.*, 1940)

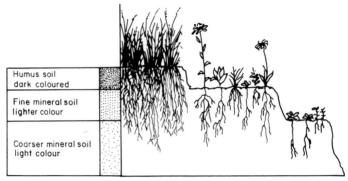

Figure 56   Relative fertility of soil layers. The fertile topsoil will produce not only the greatest volume of plants but also the more desirable ones. As succeeding soil layers are exposed the volume of plants is reduced and less desirable species occupy the area (Beeson, R. W. *et al.*, 1940)

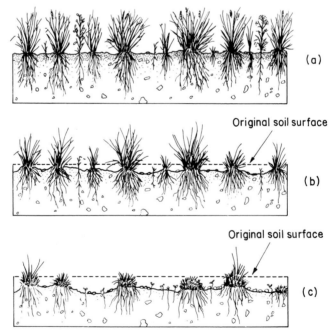

Figure 57   Formation of soil remnants. (a) Thinned stand of plants and a scarcity of litter permit erosion by sheet wash and raindrop splash; (b) more soil has been carried away, plants are left on pedestals and small stones accumulate at the surface; (c) plants are weakened and dying not only because of heavy grazing or fire, or both, but because of drying from root exposure and undercutting by erosion. Accumulation of stones on surface shows development of an erosion pavement (after Ellison, L. *et al.*, 1951)

Figure 58　Exposed roots of small shrubs due to soil loss

Figure 59　Exposed tree roots generally indicate that erosion has taken place (modified from Beeson, R. W. *et al*., 1940)

226

Soil loss is also indicated by exposed tree roots (Figure 59) and by so-called 'lichen lines' on rocks (Figure 60) which show a previous soil level. In many areas an erosion pavement may be the most easily observed evidence of past soil loss (Figure 61). Figures 62 (a) and (b) show a more normal and an extremely advanced type of erosion pavement respectively. Figures comparable to results of astronomical calculations can be easily obtained by preparing a shallow tray 1 ft square, with sides 1 in. deep, filling it with dry soil several times and weighting each time to obtain an average weight of 1 sq. ft of soil 1 in. deep, estimating or measuring the minimum depth of soil loss, then calculating the weight of soil lost per acre, square mile or square kilometre.

3923A *Bases of Perennial Grasses as Indicators of Trend*

Tussock-type grasses, when healthy, demonstrate firm growth throughout their basal area. The series of illustrations in Figures 63–66 were photographed after a light burn which facilitated photography. Figure 63 shows a

Figure 60 Old soil lines, or 'lichen lines' on rocks (A) provide evidence of the extent to which soil has been removed, and (B) shows tussock partially covered by fresh gravel

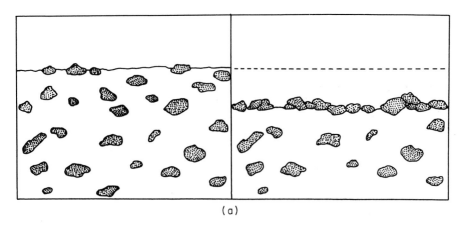

(a)

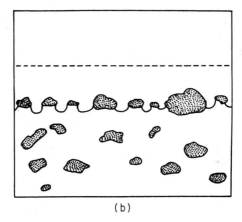

(b)

Figure 61   Erosion pavement. (a) Raindrop splash and flow of water over the soil surface removes fine soil particles and leaves first small then larger stones scattered on the ground (Beeson, R. W. *et al.*, 1940). (b) Under certain circumstances stones may be perched on and partially protecting pedestals of soil (Ellison, L. *et al.*, 1951)

healthy plant. When the tussock dies through too-frequent burning it tends to die in the central part first, as shown in Figure 64. In Figure 65 some remnant shoots can be seen arranged in circles while in other tussocks parts of the circles have died out and only arcs of small tussocks remain to indicate the former presence of a tussock with a much larger base. Where trampling is heavy dead shoots with roots can often be seen lying on the surface of the ground and hoof action at the bases of the remnant tussocks is easily recognized (Figure 66). When in doubt pry open the bases to see which parts are still living.

228

Figure 62   (a) A common type of erosion pavement and (b) an extremely advanced type of erosion pavement where most of the small stones have been washed away and only the larger boulders remain

Figure 63 A recently burned but healthy tussock, showing living shoots coming from all parts of the base

3923B *Assessment Using Burned and Unburned Areas*

Assessment of recent trend is no more difficult in burned areas; it simply takes longer. The two pictures shown in Figure 67 were taken in adjacent areas inside a recently established West African national park. Although there is still a large amount of bare ground, within the past 5 years new perennials had become established and were spreading. Thus the recent trend was judged as improving in both burned and unburned areas.

3923C *Bush Encroachment*

Encroachment of shrubs and trees into a grassland community is a common type of succession following either too frequent burning or over-grazing. Since bush encroachment is a common environmental response and evidence of a trend in terms of changing habitat it is useful for wildlife managers to recognize it when they see it, to appreciate its significance and if necessary to demonstrate the time taken for the bush encroachment to have reached its present stage. Figures 68 to 70 exemplify this process in a Zimbabwean national park. First the composition of the grasses changed following over-

230

Figure 64 (a) A tussock similar in size to that shown in Figure 63 but with the central portion dead and with most of the base missing. (b) A few remnant shoots at the edges of an otherwise dead base

Figure 65   Some remnant shoots can be seen arranged in a circular pattern while elsewhere parts of the circles have died out and only arcs remain to indicate the former presence of a tussock with a much larger base

Figure 66   Where trampling is heavy dead shoots with roots attached can often be seen lying on the surface of the ground (below A)

Figure 67 Transects are more quickly run on burned (a) than on unburned (b) areas, but assessments of trend can be determined equally well in both. The above pictures were both taken in the same habitat and inside a recently established West African national park where some reduction in the frequency of fires had occurred over the previous 5 years. The trend was improving

Figure 68   The composition of perennial grasses changes with over-grazing or too-frequent burning or both. Patches of bare ground, small at first, increase in size and provide increasing opportunity for young shrubs or trees to become established

grazing and too frequent burning (Figure 68). Then woody vegetation became established in the bare spaces and grew. Eventually large areas became largely occupied by shrubs, trees, annual grasses and forbs (Figure 69(a). In this national park, in some areas, burning was less frequent, which allowed a larger accumulation of burnable material and thus hotter fires. Consequently in such areas bush encroachment was less advanced, but the trend of encroaching bush still continued (Figure 69(b)).

In other areas the dangers of losing the perennial grasses as a habitat type were appreciated shortly after bush encroachment started, and with the removal of domestic animals (cattle) and the fairly effective enforcement of a no-burning policy the trend was reversed and within 5 years such areas were characterized by increasing perennial grass cover as shown in Figure 70.

Two main methods are used for dating a stand of bush encroachment: historical records, usually by consulting local inhabitants (see Section 2200) and by counting growth rings on the oldest woody plants. The latter is always dependable and can provide invaluable cross-check questions against information gathered in conversation with locals. If the oldest plants are a fairly consistent age, as they often are, this information is useful in making local

234

Figure 69   (a) An advanced stage of bush encroachment in a Zimbabwean national park. (b) Although less frequent, hotter fires had retarded the rate of bush encroachment in some areas

Figure 70   An area where, although bush encroachment had started on a large scale, in the previous 5 years grazing by domestic animals had been eliminated and the frequency of fires reduced with a consequent increase in the colonization of bare ground by perennial grasses. Trend, improving

enquiries about the nature of the changes that took place locally, the number of years ago as indicated by the ring counts.

A word of caution. In areas where burning has occurred for many years ring counts may have to be taken slightly below ground level as former shoots may have been burned out. Thus counts of growth rings, both at the top of the root and at the base of the stem above ground can often throw interesting light on the history of an area.

3923D   *Steppe and Savannah Woodland*

A downgrading trend in recent years was indicated in the area shown in Figure 71, taken in the Tchadian Sahara and showing an initial reduction in larger perennials and an increase in the size of the bare patches of soil. Figure 66 shows more detail of a zone of contact between the more palatable (larger) and the less palatable (small) perennial grasses. Regarding the larger perennial, all evidence is of use and none of regeneration. At the edges of tufts some grasses had been killed by trampling; others were pulled out by the roots with grazing. Along the edge between grass and bare ground this down-

236

Figure 71   General view of a downgrading area in the Central Sahara (Tchad). Although there is still abundant grass, the larger more palatable species had been reduced due to grazing by domestic animals and patches of bare ground have been increasing in size in recent years. The patchy pattern of grazing was done by nomad-owned cattle, camels, horses and sheep during a seasonal migration. Traditionally nomads have not burned this area and no signs of burning were seen either at tussock bases or on the infrequently encountered shrubs and small trees

grading area (Figure 72) showed no sign of young grasses colonizing the bare ground. The surface of the bare ground is about $1\frac{1}{2}$–2 in. below the bases of the small tussocks and tiny miniature gullies of this depth make inroads into the grassland. Action of hoofs breaks back the edges and already a small erosion pavement has formed, probably by wind erosion removing the smaller particles. These observations repeated many times along transects in this area indicated a downgrading trend in recent years.

By contrast transects were run in another area (Figure 73) similar to that shown in Figures 72 and 66 but 50 miles distant and with a different grazing history. Although nomads have moved their animals through the second area in past years, they indicated that this particular migration route had not been used for the previous 4 years. While the size and distribution of patches of bare ground were similar to that shown in Figure 72 their edges differed. Although observations in both areas were taken at the end of a dry season in a year of lower than average rainfall, observations in the first area (Figure 72) indicated a downgrading trend, while the recent trend in the less recently grazed area was assessed as improving (Figure 73). Clearly the small tussocks in the latter area were gradually spreading into the bare patches. There were no signs of small rills developing into the grassland at the edges of the bare patches, as was observed in the first area, and these observations were consis-

Figure 72   Detail of a contact zone between grassland and bare ground. Note the lowering of the soil level even in the small inlets into the lower part of the grassland (upper centre), the presence of a small erosion pavement and absence of any signs of recent colonization of the exposed area. Several small tussocks along the grass edge had been destroyed by a combination of under-cutting and hoof action. Transect assessments falling on spots like these were assessed as downgrading

tent throughout the transect. Finally, although not clear by comparing the figures, a slightly greater and more uniform accumulation of litter characterized the area improving in condition. Obviously such observations are a mixture between those concerned with present status and with recent trend. The significant observation relating to trend, in this example, is the presence or absence of grass colonizing bare patches of ground.

South of the areas mentioned above, in an open grassland savannah zone of higher rainfall and different vegetation, the kind of evidence useful in assessing trends is essentially the same. Figure 74 shows an area downgrading in condition. Here the sandy soil has eroded away from the bases of large tussocks leaving them somewhat raised. Colonization of bared areas is lacking and the bases of the individual tussocks have been receding in recent years. By contrast, Figure 75 illustrates an improving condition in the same rainfall zone. In this second area, although its present status is poor the trend is clearly improving as smaller-diameter tussocks have in recent years been colonizing and surviving in the bare patches of ground. Both Figures 74 and 75 were taken in areas recently burned to better illustrate these trends photographically.

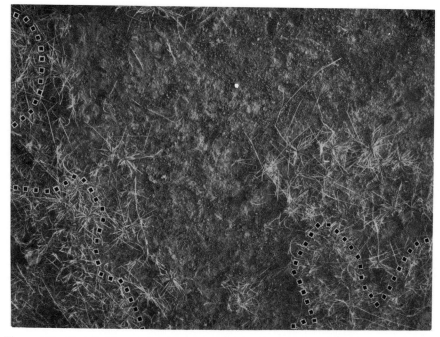

Figure 73 Detail of an edge between grassland and bare ground on a transect assessment plot judged as improving in condition. Clearly the small tussocks were spreading into the bare patches and surviving several dry seasons. There were no signs of small rills developing in the grassland at the edges of the bare patches, as shown in Figure 72. A previous edge between grass and bare ground is indicated by a dash-line

Figure 76 shows a sandy area similar to that shown in Figure 74 but in a more advanced stage of vegetative deterioration. Only one large tussock is present in the upper part of the photograph and the bare areas have been partially colonized by a small melon (a lower successional stage) which has some effect in stabilizing the sand. However, in this region, if reduction of vegetation continues, one can expect the formation and growth of sand dunes as a final result, as had already occurred in nearby areas.

The usefulness of making observations of trend at the edges of bare patches is apparent. However, in areas where patches of bare ground are particularly large, care must be taken in interpreting the trends. Even large bare areas have edges and to obtain a better perspective of the significance of recent trends it is always useful to observe the contact zones between bare ground and vegetation.

Figure 77 shows an area adjacent to an abandoned village in northern Senegal. Here at the end of the dry season there is no sign of re-colonization of grassland vegetation. The nearest grassland edge is nearly 2 km distant and by looking at this edge, Figure 78 (a) and (b), evidence of dying tussocks and recent erosion confirms the original impression of a downgrading condition.

Figure 74   In this area, assessed as downgrading, the soil has eroded away from the bases of large tussocks leaving them somewhat raised. Bases of the tussocks themselves were mostly dead or dying and no evidence of young grasses colonizing bare ground was seen

Transects run in the vicinity of all three photographs indicated a low present status and downgrading trend. If, on the other hand, the edges had been stable and vegetation moving into the bare areas, transects run along the edges would then have demonstrated an improving trend while the centre of the bared area might well have remained bare.

Even in areas clearly downgrading it is common to assess a few of the plots sampled as improving in condition. Whenever this happens and the cause is apparent it should be recorded in such a way that, should the observations proved important and relevant to your study objectives, they can be expressed in areal terms or as a proportion of percentage of plots examined. As an example, Figure 79 shows a Zimbabwean area in which several hundred transects were assessed as downgrading. Across the centre of the picture lies a dead tree pushed over by elephant several years previously. Perennial grasses regenerate and grow under the mechanical protection of the branches and several years later these mini-centres of regeneration become available to animals as the protective dead wood generally disappears. Along thirteen transects, 3–5 per cent of plots in this area were classed as upgrading, due to this by-product of elephant feeding behaviour. Ungulates mainly dependent

240

Figure 75 (a) In this area, assessed as upgrading or improving, patches of bare ground were being colonized by young tussocks which had survived previous dry seasons and were growing in diameter. The bases of the larger tussocks were firm and healthy. (b) An unburned portion of the same area several hundred metres away. In spite of heavy grazing in this particular season the percentage grass cover had been increasing in recent years

Figure 76   A downgrading sandy area similar to Figure 74 but in a more advanced
stage of deterioration

Figure 77   Area adjacent to an abandoned village in northern Senegal. No sign of any
kind of perennial grass colonizing this area was found. Trend in the vicinity of the
village, downgrading

242

Figure 78 Photographs taken at two typical places along the edge of the bare area
shown in Figure 77 (2 km away) confirm the impression of a recent downgrading trend
with dead and dying tussocks (a) and (b) and recent active erosion (b)

Figure 79 A mini-centre of colonizing and regenerating perennial grass tussocks, created by the protective branches of a small tree broken by elephant while feeding

on the availability of perennial grass at the end of the most critical dry season were few in number. But for the elephant-created focal points for regeneration there would have been even fewer.

The specific criteria used for judging improving trends in grassland vegetation will differ as the details of succession differ. For example, in a Zambian area (Figure 80) bare ground was first colonized by a ground-hugging Bermuda type (Couch) grass (*Cynodon* sp.) then by larger perennials. In a 35 in. rainfall area in New Zealand (Figure 81) earliest colonizers were small perennial forbs, followed by various species of tussocked perennial grasses.

3923E *Recognition of Trends in Gullies*

Although the presence of accelerated erosion gullies in itself may be the result of a degraded condition in the past, at any time after the gully is formed it is often useful and convenient to assess the recent trend in essentially the same way as described above where patches of bare ground were involved. If the gully is large enough, and if the study objective warrants, transects may be located within the gully. Otherwise sample assessments may be made along the upper edges and sides of the gully in the same way as at the edges of bare patches of soil.

Figure 80   In this Zambian area bare ground is first colonized by a low ground-covering Couch grass (*Cynodon* sp.), then by larger perennials

Figure 81   In this New Zealand area earliest colonizers were small perennial forbs, followed by various species of tussocked perennial grasses

Figure 82   Miniature 'badlands' in a national park in Kenya still show a downgrading trend. Note the exposed tree roots, lack of colonizing perennial vegetation at or near the edge and within the gully system. Grasses shown are annual grasses

Figure 82 shows a complex of accelerated erosion gullies in a national park in Kenya, already forming miniature 'badlands' and still downgrading. Note the lack of recently colonizing vegetation in the left foreground and the exposed roots of a small tree which, unless the trend is halted, will eventually fall into the gully as have others in the distance. The grass shown in the picture is annual grass and no signs of the edges stabilizing were observed. Figure 83 exemplifies an earlier stage of gully formation through over-use of a national park tourist road. Note the straight sides to the gully and no sign of stabilization by colonizing vegetation. The head of the gully (insert) was still actively cutting back at a rate estimated by park officials as 75 m in the previous wet season. Taken in a Dahomey national park, Figure 84 shows the edge of a depleted area similar to that shown above (Figure 82). Although unnaturally exposed tree roots are clearly in evidence, the present trend is clearly improving. In recent years grasses have increased in bare areas adjacent to the gully and a good cover of stabilizing perennials has been developing within the gully itself, even on the slopes below the break point.

3923F   *Mountainous Areas*

Figures 85–91 were taken in the New Zealand watershed shown in Figure 85 and in similar watersheds. In such terrain erosion can be spectacular and

246

Figure 83   An earlier stage of downgrading by gully formation, along a tourist road in a national park in Tchad. Note the straight sides and lack of evidence of stabilization within the gully. The head of the gully (insert) had cut back 75 m in the previous season

erosion gullies that form on the upper parts of slopes may very quickly transfer incredible tonnages of soil and gravel down to the streams at the base of the slope. Fans of transported gravel situated at the bases of the gullies, which are mainly without vegetation, may indicate recently active erosion. But this is not necessarily so for, as mentioned above, one must always be conscious of the distinction between status or present condition (what occurs on the land at present) and trend, or the direction in which the vegetative cover has been moving in the past few years.

Looking closer at the heads of several gullies with nothing but loose rock remaining (Figure 86), it is easy to see how the same principles mentioned for bare patches of ground and gullies described above for comparatively gentle savannah areas may also apply in mountainous areas. Transects run in an area of the sparse vegetation remaining adjacent to one gully (lower right) showed the downgrading trend to be continuing. Historical records indicated that denudation had taken place over the previous 80 years, following the destruction of the forest by fire and the introduction of sheep.

Consider the process by which this transformation took place. When vegetation is removed the thin topsoil quickly follows and there remain patches of

Figure 84 The edge of an erosion gully improving in condition in a recently formed national park in Dahomey. Note that the colonizing grasses are approaching the edge of the gully and note the good stand of perennial grasses present and developing within the gully itself, even on the slopes below the break point

Figure 85 The lower part of a watershed in the South Island of New Zealand. Photographs in Figures 86–91 were taken in this and in similar watersheds

248

Figure 86   Heads of several erosion gullies with nothing but gravel remaining. Transects run in an area of sparse vegetation remaining near an edge (lower right) showed a recent downgrading trend. Historical records indicated that most of this denudation had taken place during the previous 80 years

exposed rock and gravel (Figure 87). Since on slopes of this kind stabilization by colonizing vegetation proceeds normally first from the base of the bared slope upward, the base of the bare area is worth checking first. The photographic insert shown was taken at the base of the central bared area in Figure 87 and indicates a recent movement of gravel over the grass tussocks. The assessment is thus clearly downgrading. When man-made structures such as buildings, bridges, roads or fences (Figure 88) are located at or near the base of active fans, knowledge of the rate of forward movement can often be obtained by questioning local residents.

In Figure 87 the focal points where erosion first started were the sheep beds or 'camps'. With the conservation of the hillslope in mind, sheep beds are thus appropriate places on which to check recent vegetative trends. Figure 89 shows a previously used sheep bed, now almost completely colonized and covered with several kinds of perennial plants. Recent trend on such sites was assessed as improving.

Hillslopes as shown in Figure 90 must, from a conservation viewpoint, be regarded as satisfactory. In this instance observations made on the gentler hillslopes confirm the general impression. A closer look at the steeper slopes still shows an excellent ground covering and in addition that tussocks are

Figure 87   Bared area (right centre of photo) where topsoil has washed entirely away and which is still actively downgrading. The insert, taken at the base of the bare area, indicates recent gravel movement over the grass tussocks. The bare patch was thus assessed as expanding, or downgrading

growing through and replacing an earlier successional stage, the lighter-coloured vegetation. Figure 91 shows a hillside in a heavily used sheep run occupied by both sheep and rabbits and where tussock grasses were almost entirely removed by over-grazing. There even the early successional vegetation was assessed as downgrading.

3924   *Demonstrating Accelerated Erosion*

The value of such demonstrations should be obvious. Although not all accelerated erosion need be spectacular, in its extreme form it can be a dramatic visual demonstration of the process by which productive ecosystems may become extinct. In the terms of this handbook, accelerated erosion is the ultimate environmental response to mismanagement.

3924A   *Principles*

First there are several concepts basic to developing simple and rapid methods of demonstrating accelerated erosion as one aspect of environmental response.

Figure 88   Fence being destroyed by movement of gravel at the base of an actively eroding gully. The insert shows surface detail. When buildings, bridges, roads or fences are affected, accurate dating of such movements can usually readily be established with the help of local residents

Figure 89   A recent trend of vegetation increasing over and almost covering a previously bare sheep bed

Figure 90   An excellent vegetative cover on a well-managed New Zealand sheep run. In terms of soil conservation this must be regarded as entirely satisfactory. The gentler slopes shown above are entirely covered with tussock grasses. The steeper slopes are covered with perennial vegetation but since tussocks were observed growing on and replacing vegetation of an earlier successional stage (insert) the assessments of trend were 'improving'. For comparison a downgrading trend in a nearby valley is shown in Figure 91

Areas whose ground surface is covered with grass or other vegetation and litter can be expected to have only little erosion by surface run-off since the vegetation and litter reduce the impact of falling raindrops and impedes run-off. The infiltration capacity of the grass-covered soil is high and the mat of vegetation is in itself physically resistant to erosion.

Erosion increases with increases in slope.

As erosion increases patterns of rilling become more intricate.

It is the last two concepts that have proved most useful in demonstrating accelerated or abnormal erosion without actually having a figure for the normal rate of erosion and without expressing the loss in terms of change in depth of soil over known periods.

3924B   A Method for Mountainous Areas

First consider the implications of erosion increasing with slope. As an example, consider a section through a mountain. The higher elevations in this

252

Figure 91   General view of a sheep run with the perennial tussocks being eliminated
through over-grazing by sheep and rabbits, and with even the lower successional stages
downgrading, as shown in the insert. Figure 90 shows a similar area but with recent
improvement in the vegetative cover

example have steeper slopes and the slopes become progressively gentler as
one moves down through the foothills towards base level. A profile showing
the line of a typical stream bed is represented in Figure 92. Even with all
elevations covered with undisturbed natural vegetation one would expect
natural erosion to occur at a higher rate in the higher steeper elevations and,
because of the higher rate of erosion, one would also expect to find more
intricate stream patterns. Thus, if one were to divide the stream into arbitrary
units of length, count the number of tributary alluvial fans entering both sides
of the main stream and, for each stream, express the number of fans on a
per-mile or per-kilometre basis, for example, one could predict a set of results
as shown in Figure 93. In fact when fans were actually counted for the upper
12 miles of the main watercourse in a mountainous valley comparatively
undisturbed by man in the South Island of New Zealand (Cobb Valley) the
resulting diagram was not, surprisingly, as shown in Figure 94. All this naturally assumes that there are no major changes in soil, rock, frequency of fault
zones, etc. in the watershed and in this case there were none.

If, in another valley, this normal pattern was found to be radically altered
by tributary fans becoming more frequent in the lower elevations, the pattern

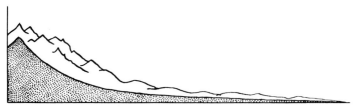

Figure 92   A longitudinal profile of a typical mountain stream bed (upper edge of shaded portion) (modified from Lowdermilk, W. C., 1934)

of occurrence of tributaries would be abnormal and accelerated erosion would be indicated if no other geological or geomorphological factors could be found to account for the abnormality. Accordingly similar counts were made in the upper 10 miles of another South Island watershed, the Avoca, which differed from the Cobb Valley in that its rainfall was somewhat less but mainly because of its past history of forest burning and over-grazing by sheep. When these Avoca sections were adjusted to a per mile basis for comparison they produced the pattern shown in Figure 95(a) and when only actively eroding fans were counted, as shown in Figure 95(b) the abnormal pattern was even more striking. This was interpreted as a demonstration of abnormal erosion or accelerated erosion having occurred sometimes in the past in this drainage. In this example there was little difficulty in being clear as to the main causes to the problem. Where forest and alpine vegetation was comparatively undisturbed in the upper parts of the drainage, even with good populations of red deer and chamois, the pattern remained normal (see sections A to C, Figure 95).

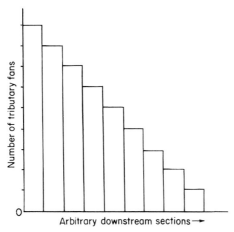

Figure 93   Theoretically the numbers of fans per mile should decrease as one proceeds downstream from the headwaters

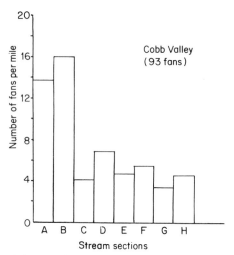

Figure 94   Counts of detritus fans, converted to number of fans per mile, along the comparatively undisturbed Cobb River in the South Island of New Zealand

In the lower two sections of Figure 95(a) and (b) (sections G and H) burning of the forest had been followed by over 50 years of free-run sheep grazing.[1] The erosion problem had become severe and was becoming worse (see also Figure 86).

It took one man 2 and 3 days respectively to obtain the information shown in Figures 94 and 95. To my knowledge this technique has not been used elsewhere than in steep mountainous terrain. The angle of hillslope was about 45°. The average difference in elevation between the stream bed and the top of the slopes was about 3000 ft (900 m). Figure 85 is typical of this area.

Uses

When compared drainages show striking enough differences this method, perhaps elaborated and refined, may prove useful. At present it seems most useful in contributing to a syndrome composed of various types of inter-related evidence and in diagnosing a land-use problem. The method is well worth considering where the severity of the land-use or management problems in question have not been sufficiently appreciated by administrators within the departments or organizations, or individuals responsible for the land-usage concerned.

3924C   Drainage Network Analysis

More useful and more widely applicable is the second technique which involves recognizing changes that occur in drainage networks when erosion is

[1]The sheep were simply turned loose on the range.

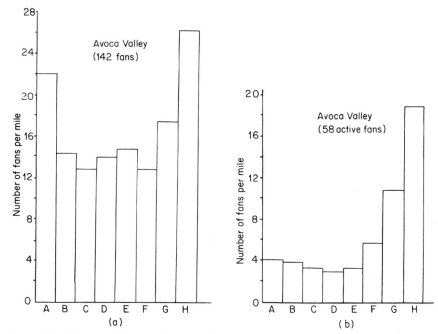

Figure 95 Counts of (a) total alluvial fans, converted to number of fans per mile, along the upper 10 miles of the Avoca River. This area has a past history of forest burning and consistent grazing by free-run (unshepherded) sheep; and (b) those fans assessed as unstable or actively eroding. Counts from the last two sections (G and H) would seem to indicate an abnormal rate of erosion in this area. The major difference between this area and that shown in Figure 94 was a difference in the intensity of land-use

accelerated. Characteristic drainage patterns thus created may indicate that accelerated erosion has taken place in an area. Since care must be used in interpreting this evidence a brief review of relevant principles is presented first, to enable you to modify the approach to your particular study areas and objectives.

Principles

Drainages are composed of a network of stream channels which assume characteristic patterns. These basic or normal patterns are referred to as dendritic, rectangular or trellised, as shown in Figure 96. The commonest drainage pattern is dendritic, which, for this reason, is used in the illustrations.

In describing drainage patterns in a watershed, geomorphologists have found it convenient to rank the streams in order. Thus the small tributaries at the head of a stream are called first-order streams. Two first-order streams join to form a second-order stream segment; two second-order segments

256

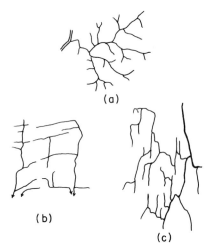

Figure 96   Types of drainage patterns. (a) A dendritic pattern developed on homogeneous, flat-lying strata. (b) Rectangular drainage pattern developed on jointed rock. (c) Trellised pattern developed on folded rock strata. Drainage patterns are drawn to different scales (modified from Morisawa, M., 1968)

join forming a third-order, and so on. It takes at least two streams of any given order to form a stream of the next highest order, as shown in Figure 97.

As mentioned above, as erosion increases patterns of rilling become more intricate. As rills grow they not only become larger but coalesce with other rills. Although many rills are lost in this way the net effect of increased erosion is an increase in the intricacy of the network by considerably increasing the actual number of first-order streams in the area of concern. Remember the drainage

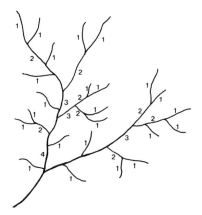

Figure 97   System of stream ordering. Numbers and pattern indicate order of respective segments. Because the largest order shown is four (see text for explanation) this is called a fourth-order watershed (modified from *Streams: their dynamics and morphology*. Copyright © 1968 McGraw-Hill Book Company. Used with the permission of McGraw-Hill Book Company)

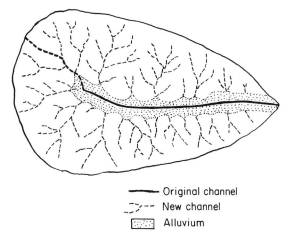

——— Original channel
⌐⌐-- New channel
[::::::] Alluvium

Figure 98   Drainage basin transformation from a low density with but a single first-order channel to a high drainage density with seventy-one first-order channels in the same area. Severe gullying and badland development on slopes was accompanied by filling in the main valley (modified from Strahler, A. N., 1956)

courses do not have to have water in them all year round. Intermittent watercourses count just as much as water-filled streams, for they form an integral part of the drainage pattern. Figure 98 illustrates the transformation of one drainage basin from a low density with but a single first-order channel to a high drainage density with 71 first-order channels in the same area.

Although there may be difficulties in interpretation, to be mentioned later, patterns of accelerated erosion are usually easily recognized and they seem best illustrated photographically for there are several variations.

Recognition of patterns

In practice the simplest way of recognizing accelerated erosion is by noticing abnormally intricately dissected parts of an otherwise normal drainage network. Such patterns may easily be recognized by inspecting air photographs, observing from an aircraft or simply by inspecting the countryside as one moves from place to place. Figures 99–102 were taken in central Kenya, while flying at elevations varying between 1000 and 3000 ft above ground (about 300–900 m).

In photographs (a) and (b), Figure 99, the pattern of accelerated erosion is clear, although since the erosion occurred perennial grasses and a few shrubs and trees have colonized the slopes and bottom of the newly formed drainage network (a). In (b) the bed of the aggraded watercourse has not yet completely healed. Figure 100 shows another variation on the pattern of accelerated erosion which became stabilized while still at the stage shown in Figure 83. In Figure 101, in an area of somewhat steeper terrain, a somewhat differ-

258

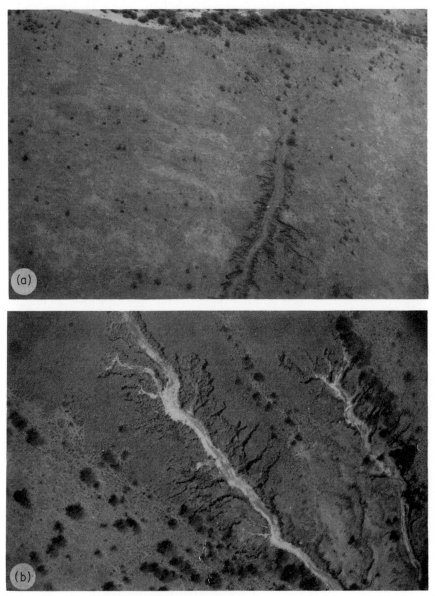

Figure 99   Two accelerated drainage networks in savannah; (a) completely healed
with vegetation, and (b) not quite recovered

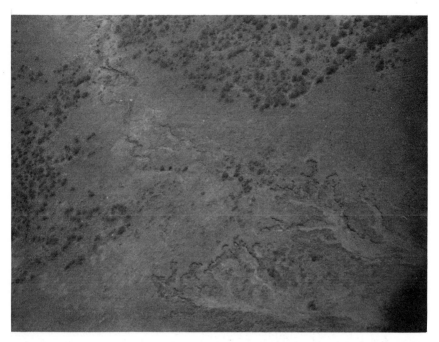

Figure 100   Another abnormal erosion pattern. The rounded edges of the eroding gullies suggest that there must have been a sudden relaxing of the stimulus to erosion at a time when gullies were actively cutting back and in a stage similar to that shown in Figure 83. At the time the photograph was taken the area was fully recovered. Fairly even-aged stands of shrubs and small trees (bush encroachment) nearby are of the same age as the oldest shrubs within the accelerated gully system

ent pattern of accelerated erosion can be seen in the lower half of the photograph. Even when covered or nearly covered with shrubs and trees patterns of accelerated erosion may still be recognized, as shown in Figure 102. In each case the drainage network has become far more intricate, that is with a much higher proportion of first-order streams or rivulets per unit area, than before erosion accelerated.

Within all the patterns shown trends may be assessed by inspecting the ground surface inside the boundaries of the new watercourses. It is surprising how frequently an approximate date can be fixed through counting growth rings in the oldest woody plants, then cross-checking the historical records with old residents, or both. This is particularly relevant when accelerated erosion has been recent or is still in its early stages. On the other hand abnormally intricate drainage networks can persist for hundreds of years and long past the lifespans of trees. They are still associated with many Bronze Age camps and Roman hill forts in Scotland. Stone Age sites in Zambia have been first located by recognition of intricate drainage patterns in an otherwise normal hillslope.

Figure 101   Another variation of an accelerated erosion pattern, this time in an area of somewhat steeper terrain. Note the comparatively normal drainage pattern in the background

It is of course the recent and current trends in this connection that most concern the wildlife manager, for accelerated erosion almost always involve basic habitat changes and in semi-arid lands this involves not only changes in vegetation but associated changes in the amount of water available throughout the dry seasons.

Use of accelerated erosion patterns in interpreting cause and effect

Demonstrating accelerated patterns of erosion from the air can be useful as evidence of past trends but should at present not be used as evidence of very recent trends (e.g. within the previous 5 years) without first making careful cross-checks between aerial and ground observations. However, as part of a syndrome of signs and symptoms all consistent with one conclusion, or as a means of developing hypotheses involving past changes in habitats, such evidence may be useful and sometimes invaluable. As with preceding parts of this chapter this can perhaps best be illustrated with photographs.

Figures 99–101 were taken in Kenyan semi-arid grazing areas. Similar erosion patterns were observed not only throughout this region of several thousand square miles but in comparable areas in neighbouring countries.

Figure 102 Even when covered, or nearly covered (camouflaged), with shrubs and trees patterns of accelerated erosion may still be recognized. Although this erosion pattern is still active at some parts of its edges the recent trend has been improving as the gullies have been largely colonized with vegetation. Growth rings of oldest trees place the age of this pattern at about 80 years

The presence of abnormal[1] drainage patterns suggests some kind of over-intensive use due either directly or indirectly to man's activities. The only kind of use known previously in this area was by graziers and, as it is part of traditional grazing practice here, one can assume in addition, some pattern of burning. In general these erosion patterns, developing at some unknown time in the past, had either completely or nearly healed at the time the photographs were taken. This much could be deduced from observations in flight and a general knowledge of the published history of the area.

Ground checks included counts of growth rings on the oldest trees growing inside the healed drainages. Although samples were taken in widely separated parts of the large region the ring counts, surprisingly, were consistent at 85–90. This was interpreted as placing a minimum age on the erosion gullies.

Local enquiries were then made as to what major event took place throughout the region 85–90 years before the photographs were taken and which may have accounted for less intensive use of some kind. Anticipated was an answer indicating a major change in type of land-use due to shifting

[1]Abnormal, in the context of this discussion, refers to comparatively small areas with vastly more intricate stream patterns and patterns of gullying than exist in nearby areas with the same climate, soil and slope.

Figure 103   A cause–effect relationship between a 5-year-old dam and the development of an accelerated erosion pattern in Kenya. (A) marks the upper limit of an old
pattern of accelerated erosion which had healed after the removal of cattle 80–90
years previously (see text for exaplanation). (B) indicates the upper limit of recent
accelerated erosion within the previous 5 years, since the construction of the dam
made over-grazing possible once again

populations, to some political changes or tribal conflict, or from depleted land
having been abandoned. The growth ring data in fact coincided perfectly with
a widespread epidemic of rinderpest which devastated cattle populations
throughout eastern and southern Africa.

In the same region other more recent causes were even more obvious.
Figure 103 is interesting in that it contains an old and stabilized accelerated
erosion pattern (see A), the gully heads being similar to those shown in Figure
100. In addition, 5 years before the photograph was taken an assistance
agency, designed and organized to help third-world countries, provided permanent water in the dry season by constructing a small dam. This resulted in
over-grazing to the extent that a new pattern of accelerated erosion (B) was
actively superimposing itself upon the older healed pattern (A). The dam had
obviously tremendously accelerated the destruction of vegetation through
over-grazing and reactivated accelerated erosion.

In hilly country Masai manyattas[1] are commonly accompanied by acceler-

---

[1]Dwelling places enclosed by a thorn fence within which their livestock shelter at night.

Figure 104   An old accelerated erosion gully to the right of the Masai manyatta has healed largely with woody vegetation. The lighter areas at its edges and in the foreground indicate bare ground caused by over-intensive use by livestock belonging to the present occupants, thus setting the stage for the development of additional gullies in future

ated erosion gullies; some old, some recently forming and others with bared areas downgrading and susceptible to further erosion. In Figure 104 an old erosion gully has been stabilized largely with woody vegetation while on the foreground slope transects indicated a downgrading trend which, if continued, would inevitably develop into another erosion gully. Figure 105 (a) and (b) shows two more variations of patterns of accelerated erosion: (a) in the vicinity of grazier's dwellings and (b) in a Zambian area occupied by a family engaged in subsistence agriculture and hunting and with no domestic animals. The active erosion in (b) is due entirely to over-frequent burning, the local reason for burning being to reduce numbers of snakes and to improve visibility for hunting. Similar patterns may be associated with any form of land-use which becomes intensive enough to maintain downgrading conservation trends. In addition to over-grazing, particularly susceptible are areas along recently opened roads, near new villages, recently cleared agricultural lands or clear-cut forests.

3924D   *Special Patterns Associated with Particular Animal Species*

This is a phenomenon well worth exploring as more detailed standards and criteria are needed than the few that now exist. Here the subject is simply

Figure 105   Two more variations of patterns of accelerated erosion: (a) in the vicinity of grazier's dwellings in northern Kenya, and (b) in a Zambian area occupied by a family engaged in subsistence agriculture and hunting and with no domestic animals. Here over-frequent burning was the principal cause of the depleted ground cover

introduced as a likely avenue for future development by stating the principle on which the evidence depends, then giving a few illustrations.

Principles

In principle, every domestic and wild animal has a range of ecological niches that is unique to that breed or species. In wild animals this involves a combination of differences in habitats preferred, combined with differences in movement and social behaviour, anatomy and so on. Domestic animals differ between each other in the same ways as do wild animals, and in addition their movements and numbers may be strictly controlled by man. In short, each wild and domestic animal will occupy a drainage in a somewhat different way in terms of, for example, trail systems, feeding areas and bedding areas. It follows, then, that if any species of large herbivore becomes so numerous as to be the principal cause of accelerated erosion, the focal (or starting) points of erosion will develop in a way characteristic of the species concerned and attributable to it.

Sheep in mountainous areas

In mountainous areas the principal focal point of erosion is the sheep bed or sheep rub. These may occur on hill slopes irregularly distributed in areas of somewhat better exposure or shelter from the wind or in lines along the upper edge of the first escarpment above a river bed. A close-up of a sheep bed in a fairly stable condition is shown in Figure 89. Bared areas which initiated with one or more sheep beds are shown in Figure 87 (a) and (b). As one observes from ever-greater distances, large areas of the hillside may be completely bare, but the focal points of erosion may still be recognized as small arcs (the sheep bed) below which eroded soil and gravel has shifted down-slope in long thin spills (see Figures 106, 85 and 86). Viewed from even greater distances it may be difficult to see the individual sheep beds but the pattern of thin strings of erosion may still be diagnostic. These sheep-induced erosion patterns have been recognized from high-flying commercial jet aircraft in temperate mountainous areas over New Zealand, Hawaii, Scotland, Western United States and Africa.

When sheep beds or rubs occur in a line—as for example, just below the top of either a main or side ridge—habitual use ensures active erosion and produces the typical sheep-induced erosion patterns shown in Figure 106. The pattern shown can sometimes be due to sheep trailing along the upper edge of the escarpment, but more commonly to bedding just below the upper lip of the slope. A variation of this pattern occurs on hillsides where very large bare areas have already been exposed or erosion gullies are well advanced. Beds are commonly found at the upper edges of such areas and accelerate the upward spread of the bared areas.

Figure 106   Aerial view of an over-grazed sheep run in the South Island of New Zealand. A characteristic pattern of focal points of erosion can be seen above the large bare area in the lower centre of the photo

Other domestic animals

Similarly the past presence of other domestic animals can be recognized by the signs they have left behind and which last much longer than hoofprints, droppings or even characteristically hedged plants. In distinguishing between sheep and cattle, for example, if accelerated erosion is occurring due to excessive numbers of either, this can be recognized without seeing an animal and from a height of about 10,000 ft above ground, or even less. Two kinds of evidence will be relevant: trail systems and erosion patterns.

When ungulates feed on a hill slope they tend to follow a system of main and subsidiary trails, the latter forming an intricate network. In cattle, even the feeding trails approach the horizontal more than do corresponding sheep trails. In time small trail-wide terraces may be formed which may last for many years. Figure 107, taken on one of the outlying Hawaiian islands, shows a pattern of cattle trails still identifiable under a solid covering of fern about 6 ft high and 20 years after the cattle had been shot and buried in this part of the island as a result of the virulent disease anthrax.

In addition to differences in trail patterns the focal points of erosion will

Figure 107   A favourable angle of the sun makes it possible to identify old cattle trails
20 years after cattle had been removed from this Hawaiian island

differ. As mentioned above the focal point of sheep-induced erosion in hilly and mountainous areas is commonly the sheep bed. On the other hand cattle-induced focal points of erosion usually start on one of the more heavily used trails. Erosion progresses by the following stages. From a distance the main trails first become more pronounced. Erosion starts immediately above the trail and thin bands of erosion develop into long thin arcs. If erosion continues material from the eroded strips flows over the terraced trails and eventually one can have a pattern of large exposed patches much as shown for sheep in the preceding photographs. But at the edges of the large patches the earlier stages of developing erosion will be recognized as characteristically caused by cattle.

In comparatively flat areas as in semi-arid savannah lands, focal points of erosion may still be along cattle trails leading to and from villages or water (Figures 108) and in areas providing shade from the midday sun. In parts of southern Africa the latter can be recognized from the air first as arcs of bare soil at the edge of the tree canopy furthest from the sun. In time the bare patches spread into larger and larger areas.

As mentioned above care must be taken before assigning a human cause to an accelerated erosion pattern for such patterns may develop, for example, along active fault lines or glacial spillways. Furthermore, ephemeral streams in desert areas are often aranged in similar patterns. Since detailed exposition and illustration of exceptions is outside the scope of this section it is suggested that, before adapting or modifying such techniques for one's own purposes, close collaboration should be established with a local geomorphologist to

Figure 108   Continual trailing of cattle along the same route creates a strip of bared
soil vulnerable to accelerated erosion

assist in cross-checking such features as mentioned above, or others that may
complicate interpretation.

Wild ungulates

Wild ungulates have not been used as examples, simply because in my experience examples of focal points of erosion and patterns of accelerated erosion
attributable to one species of ungulate are difficult to find. Although smaller
mammals such as rabbits and ground squirrels may develop characteristic
erosion focal points with their warrens and burrow systems, the high density
of these small mammals is often the result of previous or of continued over-
use by domestic animals. It is suggested that the main reason wild ungulates
are less apt to develop the extreme environmental response of accelerated
erosion is because under free and uncontrolled conditions natural environmental controls become effective in reducing numbers before such an
extreme response becomes apparent. It seems well worth warning that, if wild
ungulates are to be domesticated and brought under the same kind of management regimes as cattle or sheep, one can confidently expect new kinds of
erosion patterns to develop as characteristic of the newly domesticated
species, whether they be red deer or musk-ox; Cape buffalo or eland.

Figure 109   The Luangwa River, Zambia, exemplifies an aggraded river with annual deposition of silt and frequent changes of course, which helps maintain a mixture of habitat types along the river

3924E   *Aggradation*

In its extreme form the deposition of silt in a river and the deposition of silt overflowing the banks in flood periods can make periodic, marked changes in habitats available to animals (Figure 109). Although an increased severity of flooding, silt deposition and changing river channels can make life difficult or impossible for humans living alongside such rivers, it is often beneficial to wildlife, for a mixture of different successional stages is in effect an interspersion of various habitat types. The creation of a dam upstream which eliminates flood periods also eventually reduces the variety of species living near the river.

Accelerating aggradation is invariably associated with some form of accelerated erosion upstream and it is there one must look for the cause, whether natural or man-induced.

Chapter 4000

Collection and treatment of materials

4100   INTRODUCTION

Three kinds of materials are briefly described: the museum specimen, plant specimens and parasites and disease organisms. For countries well supplied with active museums and veterinary organizations it is well worth becoming acquainted with the services they offer and with their special requirements for collecting, recording and preserving specimens. This chapter is included especially for those for whom such services are still absent or developing. It is intended to present the point of view of a museum worker and to give enough information to save specimens which may be important enough to place in the hands of a specialist.

Quite apart from the desirability of preserving material accruing from research projects as material evidence, there is an ethical responsibility on workers in the wildlife field to make the best use of any material which they may handle. Wildlife workers have many opportunities to make useful contributions to science by preserving material resulting from research projects or management practices. This material can be invaluable in directions other than those for which it was originally collected. It should, wherever possible, be preserved for future reference.

4200   THE ROLE OF THE MUSEUM

Natural history museums have a vital role to play in the field of wildlife investigations. They can do this not only by their own contributions to the studies in question but also as a cooperating agency in providing specialist services to wildlife organizations.

The contributions which they can make can be summarized under the following headings:

Technical services

Consultative, educational or direct services in the preparation and processing of natural history material are supplied. Professional and technical staffs of natural history museums are qualified to give advice on the collection, handling and preparation of animal and plant material in the field or on the various processes later employed to render it suitable for study purposes. Museums in several Asian, African and Latin American countries loan technicians to wildlife projects to assist in training wildlife technicians in the methods of preparing specimens.

Storage and curatorial service

Museums provide proper accommodation for research collections and the specialist knowledge for maintaining and keeping in good order such materials, and their safe packing and transport.

Material prepared for research purposes requires to be properly housed to avoid damage by insects, light and other agencies. It is the function of museums to provide the specialized type of storage required for this purpose.

Collections

Research collections of both plants and animals are maintained in such a manner as to make the materials, and the data related to them, readily available to research workers.

Once provided with proper storage, natural history collections require continuous maintenance to ensure their maximum life. Museum research collections are a source of named comparative material. Museums either employ specialists in identifying, classifying and naming animals (taxonomists) or have access to them to assist in the proper identification of material accruing from field investigations.

Research

A range of professional services of their scientific staff is offered, especially in the field of taxonomy and ecology.

Libraries

Scientific libraries, including photostat or microfiche services, are commonly provided.

Education

Educational services for schoolchildren and adults are normally provided based on display or research collections and carried out either by specialist staffs or preferably by trained teachers seconded to the museums by educational authorities. Museums should take an active part in conservation education, especially at the school level. There are many obvious advantages to wildlife, national park and museum staff working together in close harmony.

4300   THE MUSEUM SPECIMEN (LARGE MAMMALS)

4310   Collecting

The methods and equipment required for the collection of material from large mammals vary with the size and nature of the material to be collected.

Large mammals may be divided into two groups: (a) the large and dangerous species such as elephant, buffalo and lion, and (b) the non-dangerous species such as the larger and smaller antelopes. For inexperienced personnel the larger and more dangerous species should only be collected with the assistance of trained and experienced operators. Where such experienced assistance is not available, heavy-calibre magazine rifles of not less than .375 bore or double-barrelled rifles of not less than .450 should be used. Individual collection should be done only after consultation with experienced personnel and the achievement of proficiency in use of heavy-calibre rifle on a proving range. As a general policy inexperienced personnel would be advised not to collect dangerous species themselves.

For the collection of the larger antelopes a heavy rifle of not less than .375 calibre, using soft-nosed bullets, may be used. Lighter calibres are inclined to increase losses by wounding.

For the smaller antelopes a rifle of approximately 30.06 calibre, again using soft-nosed bullets, should be used for collecting during daylight hours. If collecting is done at night, where the species can be shot at closer ranges with the aid of dazzling lights, a 12 bore repeating or double-barrelled, well-choked shotgun using not less than $2\frac{1}{2}$ in. cartridge $1\frac{1}{8}$ in. loaded with AAA

274

or SSG pellets may be used. Where the species are shy and difficult to approach at night, a light rifle of a maximum bore of 30.06 with a telescopic sight, used in conjunction with the light, is very effective. When rifles are used, the heart shot is usually the best. Experienced operators may use the neck shot, which causes less damage. If collection is done with shotguns, the head should be the target.

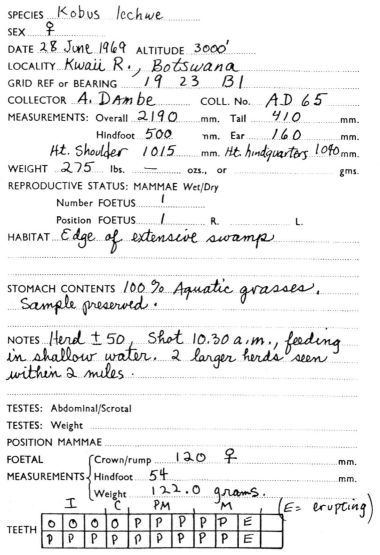

MAMMALS

SPECIES *Kobus lechwe*

SEX ♀

DATE *28 June 1969* ALTITUDE *3000'*

LOCALITY *Kwaii R., Botswana*

GRID REF or BEARING *19 23 B1*

COLLECTOR *A. Dambe* COLL. No. *AD 65*

MEASUREMENTS: Overall *2190* mm. Tail *410* mm.

Hindfoot *500* mm. Ear *160* mm.

Ht. Shoulder 1015 mm. *Ht. hindquarters 1040* mm.

WEIGHT *275* lbs. — ozs., or gms.

REPRODUCTIVE STATUS: MAMMAE Wet/Dry

Number FOETUS *1*

Position FOETUS *1* R. L.

HABITAT *Edge of extensive swamp*

STOMACH CONTENTS *100 % Aquatic grasses. Sample preserved.*

NOTES *Herd ± 50, Shot 10.30 a.m., feeding in shallow water. 2 larger herds seen within 2 miles.*

TESTES: Abdominal/Scrotal

TESTES: Weight

POSITION MAMMAE

FOETAL ⎧ Crown/rump *120* ♀ mm.

MEASUREMENTS ⎨ Hindfoot *54* mm.

⎩ Weight *122.0 grams.*

(E = erupting)

		I		C	PM			M	
TEETH	0	0	0	0	P	P	P	P	P	E
	P	P	P	P	P	P	P	P	P	E

Figure 110   One type of printed form used by national museums, Zimbabwe (from Smithers, R. H. N., 1973)

4320 Recording

The proper recording of accurate and adequate data is by far the most important aspect of the collecting of zoological material.

Owing to the nature of the mammal material, certain data have to be obtained from the specimen prior to skinning. For ease of future reference these and other data should be recorded in a standard form. It is recommended that printed forms are used for this purpose (for example, see Figure 110). These forms should be printed on a good-quality durable paper and the data entered on them in black waterproof ink. If this is not available a medium-hard pencil may be used. Do not use other types of writing media as these are prone to fading or obliteration by dampness and other causes. If printed forms are not available, data should be entered in a good-quality notebook in a standard itemized sequence.

The first step in handling a specimen is to give it a collector's number. This is appropriately prefixed by the collector's initials, the numbers running serially, as: Collector, Albert Dambe: AD1, AD2, etc. This collector's number, AD1, is entered on the form and is applied to all material collected from a specimen, such as skin, skull, stomach contents, reproductive tracts, as later outlined under Section 4340, Skinning and preparation. The data needed may be classified into two categories:

(a) *essential data*, which should be obtained from all material handled;
(b) *desirable data*, which should if possible be obtained at the time of collection.

The above data are in addition to any specialized data which may have to be obtained in connection with the specific work in hand.

The information may be summarized as follows:

ESSENTIAL	DESIRABLE
Date	
Locality	
Sex	Stomach contents
Measurements	Measurements
Overall length	Height at shoulder
Tail	Height at hindquarters
Hindfoot	Girth
Ear	Weight
Habitat in which specimen was collected	Altitude. May be ascertained later and added
State of udder in females	Weight of testes
Wet or dry	
Number of embryos, designating their position in right or left horns of uterus	Embryo measurements
	Time and method of collecting baits, traps, etc.
	Notes

276

Attention is drawn to the following aspects of the data to be recorded.

Date. This is the date of collection of the specimen. In the case of known-age specimens kept in captivity, both the dates of birth and of death should be recorded. Owing to the diverse ways of presenting a date it should be expressed as '2 August 1969' or '2.VIII.1969'. In the case of trade skins or specimens purchased where the date of death is not known, the date of purchase should be recorded under NOTES, and not on the date line.

Locality. Give this as accurately as possible in relation to some known nearby map locality, e.g. '5 miles NE Gaberones, on Francistown Road', or in the form of a bearing to the nearest minute '28° 14′ S, 15° 12′ E' or by a quarter-degree grid location as explained below. Vague and inaccurate localities are a source of confusion.

The quarter-degree grid system is becoming more frequently used as more detailed maps become available, and is worth special mention. Using this system mammal localities are conveniently given by the use of a 'quarter-degree latitude–longitude grid'. This is a subdivision of a one-degree square of latitude–longitude into sixteen equal parts (see Figure 111), each of the sixteen parts representing an area of approximately 15 miles by 15 miles.

In the example shown in Figure 111 a locality occurring within the shaded section of the figure can then be described as 20 18 Da, making it readily traceable by workers outside the area of recording. For further details of this system, see de Meillon *et al*. (1961).

Plotting maps, already gridded, are available for many parts of the world

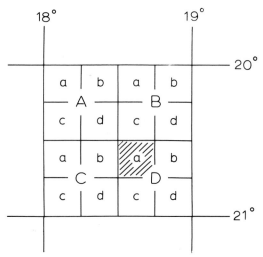

Figure 111 The latitude–longitude quarter-degree grid system of recording specimens. The shaded portion is referred to as 20 18 Da

and their availability may be checked with a local museum, university or government mapping office.

Sex. If there are difficulties in externally sexing the specimen, this may be checked after skinning by dissection of the body.

Measurements. See Section 4330.

Weight. Wherever possible a weight should be recorded as soon after collection as possible, in terms of pounds and ounces or kilograms.

Habitat. A note on the habitat from which the specimen has been collected should be made in general terms, as grassland, Mopane woodland, *Brachystegia* woodland; plains, watercourse, broken hilly country, rocky outcrops; dry, well-watered; and the soil or rock type. If habitat requirements are being studied a precise description of the location of the collection should be made.

State of udder in females. The udders and teats should be pressed to see if the specimen is in milk, and the condition reported on the form as 'wet' or 'dry'.

Number of embryos in females. Examination for the presence of embryos in females is carried out on the body of the specimen after skinning. The total number and position of embryos should be recorded in the appropriate section of the field form, indicating in which horn of the uterus these are implanted, e.g. 2R 0L (Figure 112). The crown–rump length and length of

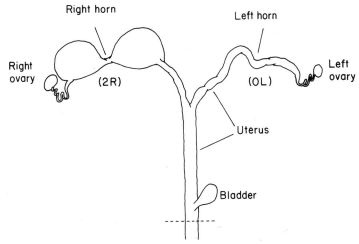

Figure 112   Method of recording embryos in uterus. In the above example there are two embryos in the right horn of the uterus; none in the left (recorded as 2R 0L). Note that if the entire uterus is removed for preservation the bladder should also be included, to ensure re-identification of right from left horns of the uterus

278

the hindfoot (Section 4330) and weight of embryo should also be recorded on the form. Where there are no embryos, the whole of the reproductive tract may be preserved in about three times its volume of 5 per cent formalin in a plastic bag for later microscopic examination. It is important that the bladder remain attached to the tract as this allows the orientation of the uterus to be determined even after removal from the body, since the bladder lies under the uterus. Thus, the position of the embryo in the left or right uterine horns can be determined. Always place the tract directly from the body cavity into the plastic bag. If it gets gritty, wash it in the body fluids but never in water. Label it with the collector's number so that it may be correlated with the specimen. The label should be written on a good-quality paper in black waterproof ink or with pencil.

Stomach contents. About a double-handful of the stomach contents, in the case of the larger species, is taken from the front end of the stomach and preserved in 5 per cent formalin in a plastic bag. This is labelled in the same manner as the reproductive tract. Correspondingly less stomach material may be preserved in the case of smaller animals.

4330   Measurements

All mammal measurements should be presented in millimetres. The measurements should be taken from fresh specimens. If measurements are taken from specimens in any other condition this should be stated.

(a)   Overall length

With the specimen lying on its side, stretched out in a natural position, the measurement is taken between the tip of the nose and the end of the vertebrae at the end of the tail (Figure 113(a)).

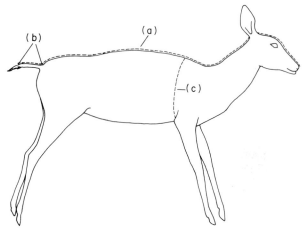

Figure 113   Method of measuring total length (a), tail (b) and girth (c) (modified from Funaioli, H., 1972)

(b)   Tail

The tail is usually measured at an angle to the line of the body, from the base of the tail to the end of the vertebrae in the tail (Figure 113(b)).

(c)   Hindfoot

In ungulates the hindfoot measurement extends from the tip of the hoof to the midpoint in the curve shown in Figure 114(a).

(d)   Ear

Measure from the notch of the ear to its tip (Figure 114(b)).

(e)   Head

Measure from the tip of the nose, along the contour, to the furthest point at the back of the head, as shown in Figure 114(c).

(f)   Height at shoulder, height at hindquarters

With the specimen lying on its side, measure from the base of the hoof or with the foot in the walking position, to the top of the shoulders or hind-quarters, to give a measurement of the specimen when standing in a natural position (Figure 115(a) and (b)).

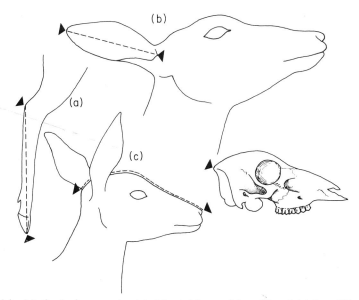

Figure 114   Method of measuring hind foot (a), ear (b) and head (c) (modified from Funaioli, H., 1972)

280

(g) Girth

If a girth measurement is required it is taken behind the front legs as shown in Figure 113(c).

(h) Embryo measurements

As embryos normally lie curled, a crown–rump measurement is taken. The hindfoot is measured as indicated in Figure 114(a).

(i) Measurement for museum exhibit

A specimen to be mounted as a museum exhibit requires extra measurements, as shown in Figure 115.

4340 Skinning and Preparation (for museum specimen or as a trophy)

The first step in the handling of a specimen is to give it a number (see Section 4320) and fill this in on a standard form or notebook.

Two labels are then prepared; one for the skin, the other for the skull and numbered with the specimen number. Waterproof linen or an embossed lettering machine make good labels. After the necessary external data have been taken from it, the specimen is then processed as follows.

Lay the specimen to be skinned on its back. If a hoofed animal, cut from the back of the hoof down the back of the leg to the knee (front leg) or hock (hind leg), then turn the cut to the inside of the leg and continue to the centre of the body. Proceed in the same manner for each leg making sure the cuts

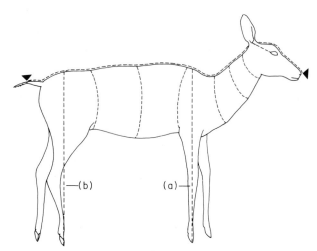

Figure 115 A specimen to be mounted as a museum exhibit requires extra measurements, as shown (modified from Funaioli, H., 1972)

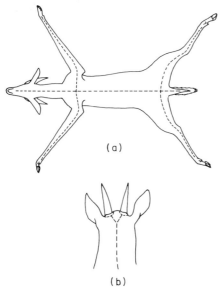

(a)

(b)

Figure 116   Dashed lines show location of cuts for skinning (a) and detail for skinning animals with horns (b) (modified from Smithers, R., 1973 and Funaioli, H., 1972)

meet at the centre of the body (Figure 116(a)). Slit the underside of the tail and carry this cut forward along the midline of the belly and chest to near the chin, making sure to cut the skin only, and not through to the intestines.

If the animal is not hoofed, then cut from the middle of the centre pad, along the inside of the leg to the mid-belly, the longitudinal cut being as in the case of the hoofed animal.

Now work the skin off the hindlegs and tail as far as possible with use of the hands and fingers. Use the knife only where the skin muscle attachments are too tough to be released by pressing. Sever the leg bones at the joint just above the hoofs and by inserting the point of the skinning knife between the bone and the skin and pushing downwards, open up so that the salt can later penetrate as deeply as possible into the base of the hoof. Continue to remove the skin from the sides and back right forward to the base of the ears. Where the knife has to be used to assist in this process, endeavour to leave the very minimum of muscle or fat attached to the skin.

Now cut through the cartilage of the ears close to the head and continue to skin over the head, cutting deeply round the eyes so as not to detach the eyelids from the skin. Continue forward to the tip of the nose, severing the skin free at the tip of the cartilage of the nostrils. If the specimen has horns, a further cut is required from a few inches behind the base of the horns down the back of the neck. (If the ungulate has a mane, cut down one side of this, the length depending on the length and spread of the horns.) Two cuts are then made from the forward end of this cut to the base of each horn and continued round the base of each horn (Figure 116(b)). The skin is then freed from the bases of the horns with the aid of a sharp screwdriver or the point of the knife. This allows the skin to be removed over the horns.

Attach a second numbered label to the skull by tying it through the lower jaw or the arch of bone below the eye. Apply a temporary label to the carcass, corresponding in number to the skin and skull, as this may later be examined. Now lay the skin flat, flesh side uppermost, and as far as possible remove any gross excess of fat or muscle. Cover the whole with a plentiful amount of salt, making sure that the salt is worked onto the whole surface of the skin, more especially the feet, tail and its edges which tend to curl over. Roll up from the tail end leaving the head free for work on the ears.

With the skin of the head turned inside out, the cartilage of the ears is now freed from the skin of the ear, by catching hold of the visible base of the cartilage and, with the aid of the thumbnail or a blunt tool gradually freeing it and at the same time pulling it through until it is free to the tip of the ear.

If the lips are thick they may be split with the knife without cutting the skin so as to allow penetration of the salt.

Now salt the whole of the head skin liberally, attach the label through the nostrils and fold into the centre of the rolled skin. Leave the skin with the salt rolled up in the shade in a cool place (never in the sun) for at least 24 hours. If a large thick skin, it is as well at the end of this period to open up the skin, shake off the water and excess salt, and re-salt and leave rolled up for a further 24 hours. One salting is sufficient for small or thinner skins.

At the end of the salting period open the skin, shake off the excess salt, and hang the skin up to allow it to partially dry. Before it is completely dry, sprinkle the hair side with 5 per cent DDT powder or naphthalene to avoid insect damage and fold it into a neat flat bundle, hair side inside, with the legs and tail inside, the head outside, and allow to dry thoroughly. A properly folded specimen of a large antelope should be about 24 in. square; smaller antelope, such as a duiker, 12 in. square. Improperly folded skins are very difficult to pack and transport. In this form mammal skins are ready for tanning.

4350   Skeletal Material

The skull of a mammal specimen should be recovered whenever possible. Even if the skin for any reason is so badly damaged that its recovery is impossible, a great deal of valuable information can still be gleaned from examination of the skull alone.

After skinning, the skull (with numbered label attached) should be carefully removed from the body and the eyes and as much of the flesh as possible cut or scraped from the bone. The brains should be removed by scraping through the aperture at the back of the skull where it is attached to the vertebrae with the aid of a thick piece of wire, the end of which is broadened out by hammering the point, and the tip bent at a right angle to form a scraper. Alternatively the brains may be broken up in the skull with a stick and as much as possible shaken out. In both cases the addition of water into the skull cavity with shaking will greatly facilitate the complete removal of the brains.

If the skull is to be handed over to a museum for final cleaning it is then allowed to dry out in a safe place. Alternatively, if a suitable vessel is available the skull may be further cleaned in the following manner.

The skull should be simmered (not boiled) in water or in a solution of 1 lb sodium perborate per 3 gal of water until the remaining flesh can be completely removed. The length of cooking depends on the size of the skull and the amount of flesh remaining on it. Avoid over-cooking or teeth may fall out and in young animals some of the skull bones may separate.

If the skull is horned, the horns should be raised out of the water during the cooking process as otherwise they will split. If the cooking vessel is large enough to contain the horns then a sack should be used as a lid so that the horns are held in the steam. In this case horns may often be removed after steaming by carefully screwing from side to side.

If the horns are removed, the bone core may be sawn off, leaving just sufficient to hold the horns in position later. The cleaned skull should then be put in a safe place to bleach in the sun. If it shows signs of fat exudation it may be recooked and the process continued until no further signs of fat show.

Owing to the conformity of certain horns (e.g. buffalo, springbok, etc.) they can never be removed from the skull. In such cases clean the skull in the above manner and bleach in the sun for a considerable period when the gristle between the bone and the skull will liquefy and eventually dry out. Poison the aperture between the horn and the skull from time to time with formalin or a solution of approximately a teaspoonful of corrosive sublimate (very poisonous) in a quart bottle of methylated spirits. This solution if recovered may be used for several such treatments. As other alternatives, sheep or cattle dip containing arsenic, or one of the water-miscible DDT powders, may be used.

Skeleton

A collection of skeletons of a representative adult male and female and young is used by some wildlife departments in training field workers to identify skeletal material that may be found in the course of a field investigation.

After removal of the skin the skull should be removed from the body and treated in the above manner. As much flesh as possible is then removed from the other bones by cutting and scraping without separating the bones, although the leg bones may be removed and treated separately. Fine salt rubbed in at this stage will prevent putrefaction. If the final cleaning is to be done in a museum, then the well-scraped skeleton is hung up in a safe place to dry.

In the case of very large skeletons, for convenience of transportation, it may be necessary to partially disarticulate the skeleton. If so it is convenient to separate the leg bones and the vertebrae just at the back of the rib case, packing the loose pieces within the rib case and, finally, securely tying all together. Each separate item should be numbered with the collector's number.

If it is desirable to clean further the skeleton the whole should be treated as outlined under the processes used for cleaning the skull. Several cookings are normally necessary in the case of large skeletons, and a lengthy period of sun-drying to get rid of the fat within the bones. Holes should be drilled in the ends of the larger bones and the marrow fat syringed or blown out. During this process the individual bones of the skeleton may be completely separated and retained together in a container which is numbered with the collector's number.

Care must be taken here not to lose the smaller bones. In cleaning smaller skeletons it is desirable to carry out the disarticulation over an appropriate sieve so that the smaller bones are not lost.

4360 Packing

A great deal of damage can be caused by improper packing of material. The following points are worthy of note.

Flat skins properly folded are comparatively easy to deal with. They should be packed tightly in solid wooden boxes and liberally dusted with 10 per cent DDT powder (which will kill any insects present) or with naphthalene flakes (which will not kill insect pests but will discourage their presence). Para-dichlorobenzene will also kill, but quickly evaporates and is only really effective if boxes are reasonably airtight. It is a good solution as a quick killer but once the pests are dead, the skins should be protected from further infestation with naphthalene flakes.

Never pack both field-cleaned skulls and skins in the same container, as they almost invariably attract insect pests which, while they may continue the cleaning process on the skull, constitute a danger to the skins.

Large skulls should be protected by layers of sacking, grass or any soft material bound tightly round them, the lower jaws in separate packing attached to them. Grass packing is not permitted for export to Europe or the United States of America, in which case hessian (burlap) or woodwool (excelsior) should be used. If lower jaws are left on the skulls, movement may cause damage to the teeth unless a pad of grass or sacking is tied between the teeth to protect them.

4400 PLANT SPECIMENS

Game departments, field stations, game management areas and national parks that include large mammals will have, or wish to have, a correctly identified reference collection of at least the main plants utilized by these mammals. Even when good handbooks for plant identification exist small collections are invaluable to newly arriving biologists and wildlife managers and for training purposes. In building up such material normally plants must be collected and submitted to a botanist for specific identification.

4410 Collecting

One good specimen is preferable to a large collection of scrappy poor material.

The minimum amount collected of each species should be sufficient to fill a herbarium sheet, which measures 16½ in. (40.6 cm) by 10½ in. (26.7 cm). Whenever possible the whole plant, including the underground parts, should be collected, and a number of plants to show the range in size is much better than a single plant. In many cases, where the plant is longer than the sheet, the upper part of the plant can be bent down and the whole thus accommodated on a single sheet. Sometimes, as in the case of larger herbs, it is not possible to collect the whole plant, but it has to be cut up to occupy two or more sheets. On the other hand, with very small plants a number must be taken to fill even one sheet. For woody plants or very large plants only representative material of essential parts should be taken; for example, a branch with some leaves, flowers and fruits. Sufficient material should be collected to give a fair representation of the species, at least enough to fill one sheet.

For reference collections material should always be collected in flower or fruit, otherwise it is of little value. The only exceptions are bamboos, for these are only occasionally found in flower or fruit. The ideal is to obtain both flowering and fruiting specimens of the same species. It is particularly valuable if both flowers and fruits can be obtained from the same plant where possible, as in trees and shrubs. In some cases it is useful to include a small piece of stem, bark or parts of the roots, bulbs, tubers, or other underground parts. Fleshy parts can be cut into slices to facilitate drying. Mosses are the easiest to collect for they do not have to be put into a press, one simply puts them in a stout envelope and leaves them alone, apart from keeping the package dry.

There are generally two steps involved in producing a plant specimen: collecting the plants in the field and placing them in a portable field press or portfolio and, later, transferring them to a press where the final drying takes place.

4420 Collecting and Transporting in the Field

The field press is simple and easily constructed. It consists of two pieces of hardboard cut to the size of the press and as light as possible while still being fairly rigid, with two webbing straps round them and passing through slits at either side. Alternatively, rings cut from the inner tubes of tyres will serve to hold the boards together. Between these two sheets of hardboard are some (say fifty) folded sheets of thin paper of the same size as the boards; pieces of newspaper will do. One folded sheet is used for the specimen or specimens destined for one herbarium sheet. The specimens are spread out and put in the folded sheet and this is then put in the field press and the straps drawn tight. In this way quite a large number of specimens can be collected in a day. It will be obvious that plants already partially pressed in a field press can be quickly and easily transferred to the final press. Delicate specimens need not be removed from their field folder, the whole folded sheet being put in the press without disturbing the specimen at all.

4430   How to Press Plants

The press consists simply of folded sheets of stout absorbent paper (blotting paper for example) in which the plants are placed and held between two wooden lattice frames by means of webbing straps. Two or three sheets should be placed against the wooden frames of the press to prevent specimens being damaged by the wood. The object is to flatten and dry the plants at the same time, and this is done by keeping the straps tight and changing the paper as it becomes damp. The original paper should be spread out in the air and allowed to dry off. Paper changing can be greatly reduced if the press is kept in a warm dry place (hot sunshine is excellent, otherwise near a fire but it must not be too near or the plants will be cooked). The gentle heat to which the press is exposed helps to evaporate the moisture as the plants lose it, and the paper does not then become too wet. In this way the plants gradually lose their moisture, which is absorbed by the paper, and die, becoming flattened by the pressure exerted by the tightly drawn straps. Specimens should be spread out when put in the press, so that the leaves are flat and do not overlap more than is necessary. Likewise flowers should be spread out as far as possible. Very thick stems can be split lengthwise, so too can thick roots, tubers, etc. Bulbs normally dry well without splitting. To overcome the unevenness which exists when there is a thick part at one end of a specimen, a pad of paper can be put on top of the rest of the material to even the press.

When the specimens are dry and rigid they are removed from the press and put into folders of thinner paper (newspaper type) and eventually packed into bundles for sending for identification.

4440   The Field Record

A most important part of the technique of plant collecting is the making of notes. General information required may conveniently take the form of a field form to ensure that nothing of importance is omitted. The notes for each specimen will include the collector's name, date, specimen number, local name if known, correct scientific name if known, locality, altitude, a description of the plant and notes on the plant habitat.

The collection number should start with No. 1 and only one series of numbers should be used by each collector, the sequence being maintained no matter where or when the collections are made. The number associated with the notes is entered on the sheets containing the specimen.

Local names should not be attempted unless the collector really knows the language and is sure that the plant actually has a local name.

Under *locality* give the name of the place, if it has one, and also the distance and bearing from some place large enough to be shown on maps of the scale 1 : 1,000,000. An entry like '25 miles S of Mto-wa-Mbu' is much better than 'Maji Moto', which is the actual name of a place too small to be shown on most maps. It is better still to give both names. Some herbariums use also the quarter-degree grid system of locating collection sites, as described in Section 4320 above.

Under *habitat* list the conditions under which the plants are growing—for example, marsh, grassland, alpine meadows, thickets, deciduous woodland, crevices of rock, limestone, clay etc.; on the general aspect of the terrain (steep slopes, flat bottom of valley, etc.); whether wet or dry; whether the plant is in shade, or full sun, and so on. The idea is to present a picture of the conditions under which the plant is growing, information which is valuable and greatly enhances the value of the specimen.

In *describing* the plant start with a brief description of the plant and its variation in size unless this is evident from the specimens: e.g., 'erect herbs 6–18 inches high', 'prostrate herbs forming patches 1 foot across', 'densely branched, rounded bush 4–6 feet high and as much through', 'erect openly branched shrub 12–20 feet high', 'tree 30–60 feet high with wide-spreading branches', and so on. Then a note about colours of leaves, flowers and fruits, and particulars about the structure of the plants which might be of interest, bearing in mind that much which is obvious in life is going to be difficult or impossible to recall when the plant is flattened. If sketches can be made to supplement the notes, they will be most useful. Another point on which information is useful is the degree of variation in the species in the locality of collection. It is most useful to have a selection of plants to show this, but if only one specimen or very few can be taken, then notes should be made on how the species varies in size, colour of flowers, for example. Notes should also be made on the relative abundance of the plants, whether they grow gregariously in large numbers or singly and whether the species is common, occasional or rare. For collections in national parks and wildlife management areas the extent to which the species is used by animals is important—for example, 'heavily hedged on south facing slopes in winter'.

4500   PARASITES AND DISEASE ORGANISMS

4510   Introduction

No wildlife field worker is expected to know all the parasites and diseases of wild animals; but the field worker should be able to collect from living or freshly dead animals the relevant material that is required for laboratory study and diagnosis.

There are many kinds of parasites. Some are on the animal's body (ecto-parasites) and some inside the body, as those in the blood or in the alimentary canal (endoparasites). They vary in size from microscopic blood parasites to very long tapeworms in the gut. Although parasites can be found almost anywhere inside or outside the body, each *species of parasite* will usually prefer a *certain type of host* (the host is the animal supporting the parasite), and *a particular place* in or on that host.

The structures of parasites differ greatly and vary from the small hard body of the flea to the long soft watery body of the tapeworm. Methods for collection and preservation must therefore vary accordingly.

4520   Collecting, Labelling and Dispatch of Parasites

The following are general points on the collection, labelling and dispatch of all species of parasites.

(a)   Collection

Collections from different animals (even of the same species) or from different localities, *should never be mixed* in the same container.

(b)   Labelling

Each tube of specimens should be *carefully labelled*; otherwise, they are useless. The label should give the following information:

1. Date of collection.
2. Locality: accurate map references or nearest town or village.
3. Species of host (scientific name if possible).
4. Sex of host.
5. Part of the body of the host from which collection was made.
6. If not found on a host, the place where it was found should be noted.
7. Name of collector.

Data recorded on the label placed inside each tube should be written with a lead pencil (or waterproof India ink) preferably on both sides. *Do not use a ballpoint pen.* Suitable plain or plastic containers may be supplied by laboratories or through a game department head office.

(c)   Preserving

Table 8 gives a general indication of the collection and preservation for most kinds of parasites and diseased tissues. It is intended only as a rough guide

Table 8 Collecting and preserving parasites and diseased tissues

Species	Host	Occurrence	Collection	Preservative
EXTERNAL PARASITES				
Ticks	Vertebrates Large mammals	Grass stems, nest or lair of the host, body of host. Eyes, ears, neck, belly, body sides, genitalia, tail, hooves, etc.	Remove ticks gently and avoid damage to mouth parts. Remove entire tick. Females can be bloated with blood, resembling fat peas. Male ticks never become blown up in this way. Alcohol may encourage ticks to release their hold. (Special instructions needed for sending live ticks for laboratory work. Contact local lab.)	70 per cent alcohol or gin (undiluted or obtain more sophisticated preservatives from nearest veterinary or game research laboratory.
Fleas	Mammals	*Eggs*: in the bedding or nesting place of the host. *Larvae*: feed on faeces, dried blood, etc. *Adult*: anywhere, but particularly on abdomen, may be found on captured or killed animals. The flea leaves the host shortly after its death.	One way, when local conditions permit, is to hang the animal over a water-container and the fleas will drop off and can be collected from the water. Also try to find a few hairs with the egg-cases attached to them and include these.	As above.

Table 8 Continued

Species	Host	Occurrence	Collection	Preservative
Lice	Specific mammals and birds	The entire life cycle is spent on the host.	When lice are collected try to find a few hairs with the egg-cases attached and include these, as they facilitate indentification.	As above.
Sore-producing flies	Vertebrates	The larvae (maggots) live as parasites in or under the skin or in inner organs. Heavily infested ungulates may stand with heads bent downwards, sneeze and be so inattentive that an observer may approach relatively close. Wild animals which look sick are usually heavily infested with all kinds of parasites including fly maggots.		As above.
	Antelope.	In or under the skin, nasal cavities, pharynx.	To collect mature maggots from the nasal cavities cover mouth and nasal openings with a bag for several hours or overnight. The maggots will crawl out and drop into the bag. Maggots para-	As above.

sitizing the gut leave it through the anus and may be found in freshly deposited dung or in the hindpart of the rectum.

→ Rhinoceros	Skin along sides and all parts of the gut.		As above ←
→ Zebra	All parts of the gut.		As above ←
→ Elephant	Skin, pharynx, stomach, hindgut.		As above ←
→ Hippopotamus	Nasal cavities. For above, larval development is completed in the host, but the pupation occurs in the soil.		As above ←

INTERNAL PARASITES

Roundworms Flukes Tapeworms Thorny-headed worms	Any large mammal or bird	Intestine, stomach, kidneys, liver, body cavities, blood-stream, trachea, lungs, large intestine. (Some worms may be quite large but some are barely visible to the naked eye.)	Cut the intestine into suitable lengths. Put the pieces in a dish. Carefully cut the intestine open. When gently washed out in a tray the larger worms may then be seen floating in the water.	70 per cent alcohol or gin (undiluted). In all cases specimens must be completely immersed in the preservative.
Various virus or bacterial diseases	Any animal	Especially lungs, internal organs, intestinal tract, glands.	Collect sample from affected part	70 per cent alcohol or 10 per cent formalin.

and for use in the absence of more detailed instructions. There is no substitute for establishing close working relations with local veterinary officers who will advise on collection and preparation of specific kinds of material for the particular diseases of local concern.

(d) Dispatch

Packages sent by post should be *properly preserved, packed and addressed.* Bottled specimens, because of possible leakage, are best packed in wooden boxes well filled with sawdust.

4530   Blood Smears for Microscopic Examination

Certain parasites, such as those causing sleeping sickness and African horse fever, can be detected in the blood or lymph by means of a smear of the fluid on a glass microscope slide (available from laboratories). Blood smears may be made from the living animal or from an animal which has recently died.

Precautions

When a dead or sick animal has been handled, do not eat, smoke or scratch yourself before washing at least your hands thoroughly, preferably in disinfectant. This is not easy without conscious thought or effort. A protective apron or a change of clothes is always desirable.

Use clean slides

The glass slides must be *absolutely grease-free and clean.* New slides appear quite clean, but there is always some grease which must be removed with spirit. Immediately before use, polish the slide with a clean handkerchief or a clean piece of soft paper.

Fresh blood is best

A blood smear is, however, worth sending even if the animal has been dead for several hours. In that case, use blood from the ear as this is the last to decompose.

Technique for thin blood smears

1. *To obtain blood*, clean the skin over an ear vein and prick it so that a bead of blood wells up. Only a very small drop is required, because too much blood makes microscopic examination impossible. *One cannot make the smear too thin.*
2. Touch one slide on the bead of blood, so that the bead is transferred to the centre of the slide, about 1 cm from one end.

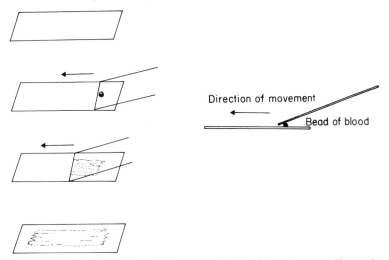

Figure 117 Method of making a thin smear of blood, lymph, pus, diseased organs etc.: (1) take two clean slides, (2) place small drop of blood on one slide, with second slide just touching it as shown, (3) move second slide forward, (4) leaving a thin smear on first slide

3. Place this slide on a firm flat surface. Hold a second slide at an angle of about 20° and move this second slide onto the bead of blood so that it spreads out along the line of junction of the two slides. Push the top slide along the lower, keeping the angle the same, and dragging the blood along the lower slide (see Figure 117).
4. *Dry* the smear rapidly by waving in the air.
5. After drying, the smear is *fixed* by pouring a few drops of methyl alcohol onto it and leaving the alcohol for 1 or 2 minutes before again drying.

Technique for thick dry blood smears

In certain cases (e.g. sleeping sickness) it may be desirable to make a *thick* blood smear. As with thin smears, it is particularly important that the slide should be completely free from grease.

Spread three or four drops evenly over an area of about 2 cm diameter in the centre of a slide using the corner of another slide or a clean pin. Leave it to dry in a place free from dust and insects and *without fixing*.

Labelling

Each slide should be labelled directly with a grease pencil or with a special adhesive label. If this is impossible, the information should be written on a piece of paper in which the slide is then wrapped.

Dispatch

Slides can be posted if securely packed in cigarette boxes. If only a small number are sent, they travel well if tied together in pairs, back to back, wrapped in a piece of paper and placed in an ordinary envelope. However, in most countries the latter practice is risky.

4540   Field Report

As *an aid to diagnosis* by wildlife disease specialists, the 'on-the-spot' examination of a sick or dead animal, made by the field worker, is a very important step towards determining the cause of the illness. The observer should always attempt to describe fully the condition *as he actually sees it*. It is more helpful to the specialist to be told 'the mucous membranes were pale' than to be told 'the animal was anaemic', unless anaemia has been established by the proper tests.

Include reference in your description to the general appearance, expression and condition of the animals, the state of the skin, mucous membranes, eyes and other external symptoms such as prominent lymph glands, swellings, etc. Also give figures for body temperature. Note any deviation from normality of bodily functions with respect to appetite, thirst, respiration, faeces and defecation, pulse and heartbeat, urine and urination, abnormalities of milk or other discharges.

Here are some conditions recorded by Tanzanian animal health officers when inspecting sick animals or examining dead animals:

Live animals
(a) *Activity*	What is the animal doing?
(b) *Posture*	Moving about naturally, lying down, standing still or practically immobile.
(c) *Feeding*	Feeding naturally, not feeding or feeding abnormally (e.g. eating unusual things).
(d) *Drinking*	Normally, very much, or not at all.
(e) *Respiration*	Normal, slow (few long breaths) or fast (short and fast breaths). Try to record the number of breaths in 60 seconds.
(f) *Temperature*	If it can be taken by inserting the thermometer in the rectum the information will be valuable.
(g) *Urination*	Normal, a little urine or a lot of urine passed. Colour of urine if possible.
(h) *Defecation*	Normal, hard, soft or liquid. Any signs of blood present. Smell normal, or bad and unpleasant.

Dead or live animals	Describe any abnormality.
(a) *Skin*	Any observable damage.

(b) *Swellings on body*	Size and location of any abnormal swellings on body; soft or hard; lymph glands or not.
(c) *Mucous membranes*	Colour, normal pink, very light pink or white, reddish, reddish blue or yellow; moist, dry or wet.
(d) *Faeces*	Defecation normal, back legs soiled with liquid faeces; colour of faeces or any abnormal contents (e.g. worms).

Finally, labelled specimens, preserved and accompanied with a field report, should be taken into your nearest cooperating veterinary laboratory as soon as possible.

Chapter 5000

Determining habitat requirements

5100 INTRODUCTION

Animal–habitat relationships and interactions provide the central core to the ecology of a species. An understanding of habitat requirements is thus basic to any form of management; it is essential whether one's interest centres on the protection of rare species or the production of optimum numbers for commercial utilization. It has been assumed that the practical effectiveness of management may improve, following increased understanding of the significance of various habitat elements. The present chapter suggests several ways of moving forward with this kind of understanding.

5200   HABITAT REQUIREMENTS DEFINED

Habitat requirements may be defined in two main ways. In terms of the individual animal this term refers to the minimum requirements for the existence of that individual. In terms of populations a qualitative dimension is added and it is convenient to consider habitat requirements as the optimum requirements for production of highest stable populations or the highest sustained yield. These are biologically somewhat different concepts, important for managers of large mammal populations to understand.

The fact that a species is invariably associated with a certain habitat or habitat elements automatically means that the minimum requirements for existence are present. But this does not imply that habitat is best suited to sustaining high numbers. The minimum requirement for keeping an individual alive involves a perspective summarized by Figure 118; while the optimum requirements for sustaining high densities will involve the interaction of various habitat elements with a population, as indicated in Figure 119. The presence of the essentials determines the survival of an individual. The size and distribution of the essential habitat elements superimposed against a background of environmental changes determines the richness or poverty of an environment in terms of the numbers of animals it can now safely support.

There is no one way of studying habitat requirements. The approach used

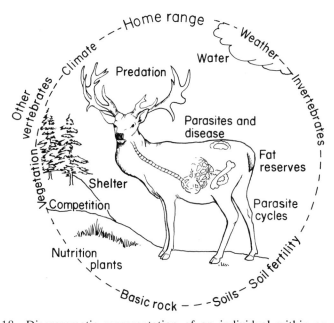

Figure 118   Diagrammatic representation of an individual within an ecosystem, bounded by a rather fixed home range and emphasizing its relations with various components of an ecosystem (modified from Riney, T., 1956b)

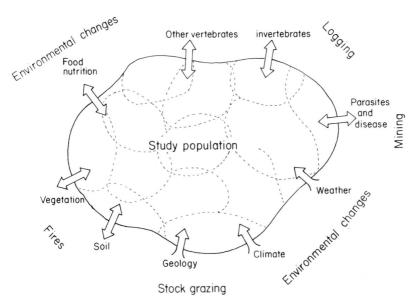

Figure 119   Extending the perspective of Figure 118, a population tends to maintain a dynamic balance within the sum of its individual home ranges and can be understood only when placed against a background of recent historical events which may have modified the ecosystem (modified from Riney, T., 1956b)

here is to discuss a range of possible approaches and to emphasize that an increasing understanding of habitat requirements can (and I believe should) become an integral part of any wildlife research or management organization's activities. The subject represents to me the biggest gap in present-day wildlife research on large mammals and, for this reason alone, seems worth the emphasis of a separate chapter.

Before discussing habitat requirements it is convenient to briefly consider several ways of relating animals to their environment. This will be done by defining and discussing the terms 'distribution', 'habitat', 'habitat elements', 'home range' and 'ecological niche'.

5201   *Distribution*

The most general of all ways of relating a species to its environment is to consider its global distribution, e.g. 'circumpolar, in northern tundra regions above 60° N latitude'; or 'Equatorial rainforests in Asia and southern islands extending eastwards to Borneo'; or 'occasionally found on the eastern slope of the European Alps above 2500 metres'.

Normally this is not useful in terms of identifying habitat requirements; those parts of the habitat most important for the existence of the individual and the health of the population. However, as a guideline for selecting study areas it may prove useful. For example, comparisons may be made of habitat elements and ecological niches used at the extreme edge of the distribution range of a species, with similar observations made in more favourable areas nearer the centre of its range. The differences these comparisons reveal often produce questions or theories, which can then be tested more precisely by using some of the methods suggested below.

5202 Habitat

Habitat is usually described in a general way in terms of a vegetation type (see also Section 3500). In the course of a day's movements an individual may use, or pass through several habitat types, for example, grassland, croplands and regenerating forest land.

5203 Home Range

As mentioned earlier, *home range* is the basic areal unit of the ecosystem and this area is normally more or less fixed during an individual's adult life. It is this fixed character of the home range that not only makes it the basic areal unit but provides a basis for critical studies of specific habitat elements occurring within the home ranges (see also Section 3200).

5204 Ecological Niche

Two Australians (Andrewartha, H. G. and Birch, L. C., 1954), in referring to 'niche' note that it is convenient to have some term to describe the status of an animal in its community, to indicate what it is doing and not merely what it looks like. The term was first proposed by an American (Grinnell, J., 1917) as the specific place in an ecosystem occupied by a single species of animal, and he included adaptations of structure and habit that fit the individuals of the species to the local environment.

   In distinguishing between habitat and niche the word *habitat* is here used to mean the kind of place where an animal lives (as, for example, within one or including several vegetation types), and the term *ecological niche* means the role that the animal plays in the ecosystem. Most ecologists referring to niche have had very much in mind the use which the species concerned makes of particular physical and biotic features of the ecosystem and of the dependence of the species on these features (Dice, L. R., 1952). It is these specific uses that we are concerned with in further definining those parts of the home ranges that the spcies concerned occupies for shelter, for breeding sites and for other activities, the food that it eats and all the other features of the ecosystem that it utilizes (Dice, L. R., 1952 and Allee *et al.*, 1949).

   By defining niche in this rather loose, broad way it is only natural that some workers, recognizing that it has an infinite number of dimensions, have felt that 'we cannot completely determine the "niche" of any organism' (Krebs, C. J., 1972).

This may be true, but the concept of 'niche' is an obviously important one to have in mind in assessing the relative importance of various habitat elements and, as noted below, in defining niche one automatically identifies elements of the habitat which can by other means be then assessed in regard to their importance to a species in a particular area.

Figure 120 shows diagrammatically the distinction between habitat, habitat elements, home range and ecological niche. The general habitat may be described as forest edge. Dashed lines indicate the boundary of the home range of an individual ungulate. Within the home range several habitat elements occur: forest, patches of shrubs and young trees, grassland, special riverside vegetation and water. In this example the home range is used throughout the year. Examples of seasonal differences in use are indicated by

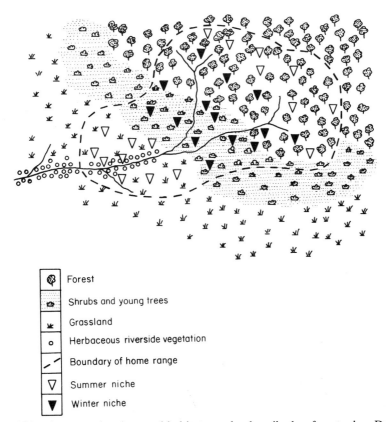

❀	Forest
🌰	Shrubs and young trees
⍦	Grassland
∘	Herbaceous riverside vegetation
⸌	Boundary of home range
▽	Summer niche
▼	Winter niche

Figure 120 An area whose general habitat may be described as forest edge. Dashed lines indicate the boundary of the home range of an individual ungulate. Within the home range several habitat elements occur: forest, patches of shrubs and young trees, grassland, herbaceous riverside vegetation and water. The particular seasonal uses of the habitat are indicated as winter and summer niches. See text for further explanation

solid and open triangles to distinguish between winter and summer niches respectively. In winter, bedding at night occurs up to 400 m inside the edge of the forest. Feeding is mainly on shrubs, particularly of two species, at the edges of the shrub zone closest to the forest and in small areas particularly sheltered from the wind. In summer, bedding occurs inside the forest but on higher elevations. Feeding is mainly in grassland and on herbaceous plants in moist areas bordering the stream. Drinking from the stream occurs at least once, sometimes twice, daily.

Two other aspects of niche are worth discussing in relation to identification and evaluation of specific habitat elements.

E. P. Odum (1966) observers that: 'the same species may function differently—that is, occupy different niches—in different habitats or geographical regions.' As an example he notes that in some region man's food niche is that of a carnivore (meat eater), while in other regions it is that of a herbivore (plant eater); in most cases man is omnivorous (mixed feeder). As part of the same function I have observed that in two different regions, densities of red deer may be stabilized at about the same levels but by entirely different combinations of environmental factors. This serves to further emphasize the need for continuous improvement in identifying and evaluating strong and weak elements in the particular habitats under management.

Finally one must recognize that species vary enormously in the breadth of their niche, in the range of different habitat elements and combinations of these in which they can exist. As E. P. Odum has said (1966): 'Nature has its specialists and its generalists.' Some large mammals feed on a very restricted number of plant species, while others may thrive on almost anything. Naturally the less adaptable a species (the narrower the niche), the more vulnerable it is to large changes in habitat due to changes in weather or to drastic modifications of habitats by man. Most environments have both specialists and generalists and if more than one species is being managed it is important to be aware of the comparative degree of specialization of the various species. As mentioned in Chapter 8000 this is an important and early question to resolve in dealing with problems involving competition between two or more species.

5300 DESCRIBING HABITAT REQUIREMENTS

For all species, defining habitat requirement always involves at least two kinds of evidence: habitats available (Occurrence–Environment, Section 3500) and animals present and associated with these habitats (Occurrence–Animal, Section 3400). For more detailed evaluation additional evidence may involve evidence of utilization or of response.

Biologists in developing countries can easily become discouraged on recognizing that so little usable information on habitat requirements is available on the animals of their country or region. It is especially difficult when they are under political pressure to advise on management, while recognizing that an improved understanding of limiting factors would considerably modify their advice. This situation seems to require an attitude or approach that would let the biologist and wildlife manager get on with the job of improving his understanding but at the same time avoid being caught in the plight of the poor unhappy centipede, . . . 'to whom the frog, in fun, said:

> Pray, which leg comes after which?
> This wrought her mind to such a pitch,
> She lay distracted in the ditch,
> Considering how to run.'

The basis of an approach I have found useful in several developing countries is to recognize and accept that: (1) there are a number of ways of improving understanding of habitat requirements—some crude, rough and over-generalized and some elaborate and sophisticated; and (2) as a wildlife organization grows in competence, its understanding of habitat requirements should keep evolving and improving through a series of different stages or levels.

A simple way to start is to use the following procedure: (1) assemble existing knowledge; (2) evaluate, in terms of how this existing level of understanding of habitat requirements can best be applied to achieve the objectives of management; and (3) proceed to improve understanding of habitat requirements, selecting or devising usable techniques with both current and long-term objectives of management firmly in mind.

5310 Assembly of Existing Knowledge

5311 *From Published Material*

Published material in semi-popular accounts or mammal books may be the only information available at the start of a study. Although unsatisfactory, this is at least a start; a beginning which future observations may extend and modify.

Such information usually takes the form of general statements about distribution as, for example the following, taken at random from several general regional works on animals: 'Central and South Western Sahara'; 'throughout the woodland savannah areas of Africa'; 'the forested zone of the Himalayas up to 9000 feet (2740 m)'; 'oak forests'; 'in riparian vegetation along steams'; 'rocky outcrops and high alpine meadows in summer and at the upper edge of the forested zone in winter'; 'lowland areas on the fringes of human settlements'.

This kind of loose description is sometimes as obvious and crude as saying that a certain species of fish is always found in fresh water. Nevertheless, even this crude approach is important for management, as the fish would die if the stream dried up. Many species of large mammals are equally dependent on, for example, some mixture of habitat elements associated with forests.

Descriptions of habitat found in more specialized mammal or game books are, on occasion, somewhat more precise but are often almost as generalized. A few such definitions taken at random from several books of mammals for a region are:

Giraffe: 'Light woodlands. In Barotse also on plains' (Ansell, W. F. H., 1960).

Banteng. (Wild Cattle) 'It occurs in herds of some size, and may be found high on mountains' (Harrison, J., 1964).

Impala. 'Tree and Bush savanna, in Rhodesia closely associated with Mopane' [a species of tree] (Smithers, R. H. N., 1966).

Reedbuck. 'Open grassland or extensive vleis [damp grassy water-courses], flood plains or montane grassland with sufficient tall grass or reed beds to provide cover, in association with water' (Smithers, R. H. N., 1966).

Black rhinoceros. 'Dry bush country and particularly thorn scrub. Also mountains up to 11,500 ft. (Mt. Kenya) in the cloud-soaked moorlands' (Dorst, J. and Dandelot, P., 1970).

White rhinoceros. 'Grassland and open savannas with scrub' (Dorst, J. and Dandelot, P., 1970).

Roe deer. 'The roe is a woodland animal, and is mainly nocturnal in its habits, generally lying up in thick cover during most of the day' (Matthews, L. H., 1952).

White-tailed deer. 'White-tailed deer show a preference for timbered areas. However, the pattern and distribution of the timber in a given area influence their presence and abundance to a great extent since they utilize the borders or edges more than dense uniform stands' (Schwartz, C. W. and Schwartz, E. R., 1959).

Red deer. 'The species is really a woodland animal, and in the forests of central and eastern Europe it reaches its largest size and greatest development of antler' (Matthews, L. H., 1952).

'Mainly woodland; chiefly deciduous, but also coniferous woods. Originally in very open woodlands, dotted about in flat country. The deer in Scotland are completely adapted to living outside the woodlands, and are found on the higher moorlands' (van den Brink, F. H., 1967).

Sambar. 'The forests of India and Ceylon, especially those in hilly districts, form the habitat of the typical sambar' (Lydekker, R., 1907).

'Forested hill-sides, preferably near cultivation, are the favourite haunt of the Sambar. Their food consists of grass, leaves and various kinds of wild fruit.'

'Sambar in New Zealand are in areas, usually swampy, which hold good stands of toi-toi, flax, or both, and are never far from clearings containing coarse grasses. The species uses native and introduced forest for shelter, but apparently does not penetrate far through exclusively forest areas' (Riney, T., 1955b).

Many books on mammals include no notes of any kind on habitat and are concerned only with identification and general distribution.

Habitat descriptions found in such texts may provide a useful starting point but it is necessary to insert a strong word of caution in using this material. Two commonly observed factors may operate to make such habitat descriptions highly inaccurate in your particular study area. First there is a strong tendency for modern books to quote older books. Sometimes the older descriptions have been made many years earlier and based only on the small amount of knowledge then available. Another kind of difficulty arises from modern books or papers which describe an existing association between animal and habitat from one area; however, such associations may be quite different in different areas. Often an implication is that the habitat is a preferred one because that is where a species is most frequently observed. This can be misleading if, over the previous century, the preferred habitat has dwindled and the species exists now only on marginal habitat. Or hunting pressure may have increased to the extent that animals are now only seen in areas of dense cover, for example, when previously they preferred more open areas.

Existing understanding of habitat requirements should never be taken entirely for granted. Some form of cross-checks of the suitability of various habitats available to the species in question is always desirable. One of the first and easiest cross-checks is to compare published assessments of habitat from more than one source, as exemplified above for red deer and sambar. Such comparisons may be useful in raising pertinent questions regarding the comparative value of specific aspects of the habitat. After considering the above quotations on red deer one might question more precisely the role of woodland: precisely what aspects or combination of elements in woodland are important and if woodlands are essential in Europe, how is it possible for them to live without woodlands in Scotland? In considering the above-mentioned sambar examples as a basis for planning future investigation one should question the importance of the role of both forests and the significance of coarse grasses.

5312 *Technical Articles*

Many technical articles are written dealing with the ecology of individual species, or with one aspect of an animal's biology, such as food habits. When appropriate articles are available these may provide some basis for the start of management. When starting, a good way to get acquainted with such articles is to consult *Wildlife Review*[1] or contact local universities or museums for appropriate journals and other publications.

However, in my experience, it is rare that the literature is completely satisfactory with regard to the significance of particular habitat elements in the study areas and for the species of one's concern. When it comes to making

---

[1] *Wildlife Review* is an indexing and abstracting publication of wildlife literature issued to appropriate organizations by the Fish and Wildlife Service, USDA, Colorado State University, Fort Collins, Colorado 80523, U.S.A. The coverage is international.

practical decisions based on an understanding of relevant requirements in our own areas, we are apt to be rather more ignorant than informed. This has been a very real hindrance to the process of improving wildlife management in many developing countries.

Three factors seem to combine to maintain this state of comparative ignorance.

The first, or 'trendy', factor seems to combine history and custom: it is, at present, simply not a popular subject for study. In North America in the 1920s and 1930s, for example, it was much more fashionable for field biologists to be concerned with describing habitat and in identifying specific ecologic niches than it is for the wild-life biologists of today. Now, while habitat requirements are recognized as important, little effort is spent in defining or in refining our understanding of them.

The second factor, like many aspects of life in the twentieth century, involves over-specialization. When habitat requirements are studied, emphasis is too often on one element in isolation as, for example, a study of nutritional requirements or of shelter requirements. This over-specialized approach can but delay progress in understanding interactions between animals and various habitat elements. A proper understanding of the role of any one element will require some understanding of the other elements as well, for they are normally inter-related in so far as they affect populations. Commonly, as one element changes so may the requirements for another change in direct or indirect response. For example, deer living in areas offering excellent available shelter during a cold winter season will require less food than deer whose home ranges are in areas poorly supplied with shelter.

A third factor might best be posed as a problem with two faces. Where a species is adapted to survive over a wide range of environmental conditions there is often the question of deciding to what extent understanding of animal–habitat relations in one area may be applied to another area, region or country. Failure to profit from existing knowledge can obviously slow down progress in understanding and in terms of the identification of relevant habitat elements. This problem commonly resolves into one of two forms.

For example, studies in one area may not be immediately recognized as relevant to another area. A certain habitat element may be demonstrably limiting populations in one area. In another nearby area this same element may be abundant and its importance masked by a deficiency involving an entirely different requirement. Problems of this kind are common, especially in the early phases of studying a species.

Another form of the problem arises when local biologists charged with studying a species dismiss all work outside their country or state as being irrelevant simply because the work was done elsewhere. This problem is difficult, not because of animals and their environments but because of people and their attitudes. In my experience the difficulty commonly rises under circumstances where biologists have not learned enough of local animal–habitat relations to enable them to recognize the relevance of work elsewhere. Although few, there are some species which are well enough understood to enable large-scale improvement in the level of management to proceed based on existing knowledge and experience, and it would be obviously advantageous for developing countries to take advantage of this knowledge.

Where the suspicion exists that 'outside' studies of habitat requirements *may* have relevant and important bearing on the management of the same species in another area, two obvious approaches are worth considering. Experts with much experience in other areas can be brought in for short seminars, training courses or consultations with local biologists, and visits made to the field to determine the extent that essential habitat elements differ or are comparable. Alternatively, the studies may be made by local biologists, after visiting other areas and eventually returning and testing locally various elements recognized as important elsewhere. In the latter case it is still useful to be in close touch with workers with experience elsewhere to ensure the information gathered is comparable.

But whatever the reasons for little recent emphasis, there is every reason for a developing country to press for priorities in research and management relevant to their own country and for the populations found therein. Almost always this will involve priority attention to improved understanding of habitat requirements. Sometimes the words are arranged differently and concern is over 'which are the most suitable habitats for a given species', or 'where is the best place to locate a reserve to protect a rare species', or 'can we really tolerate wildlife in production forests', but all these questions and many more lead inevitably to the need for a better understanding of habitat requirements relevant to the issue of the moment.

Several obvious links between understanding of various animal problems and a knowledge of these requirements are elaborated in Chapter 8000.

5320 Evidence of Historical Changes in Habitats and Associated Animal Populations

5321 *Historical Documents*

Published or unpublished early diaries or journals of early settlers, explorers or hunters, or regional histories and other historical documents, can be of value in reconstructing an understanding of former animal habitats sparsely or abundantly occupied. Pay particular attention to descriptions of the countryside and to any mention of animals seen or hunted for meat. Visiting the same localities today can be most revealing.

5322 *Questioning Old Men*

This technique has already been discussed in Section 2200, but is mentioned here as a possible first step in learning more of habitat requirements than is available in publications, or where publications are lacking. Question old residents familiar with the study area, especially those who have spent much time in wildlife areas, as hunters, foresters, honey-gatherers, etc. This method may provide an especially useful starting point in developing countries whose environments have undergone major changes since the turn of the century.

5400   FIELD METHODS

5410   Recording Detailed Observations of Activity, Time and Place

Some of the earliest good work in identifying specific habitat elements was done by taxonomists, while studying ways in which the inherent anatomy and physiology of animals was related to the function of a particular species or sub-species in the environment. It is understandable that the concept 'ecological niche' naturally developed from considering thousands of such detailed observations as were made as early as 1917 (Grinnell, J., 1917).

The method consisted basically of recording precise details of the occurrence of every animal collected. These involved an exact geographical location and note of what the animal was doing at the time of collection, as well as a description of the precise part of the ecosystem being used. For example, certain species of birds were to be found feeding on insects only on the inner side of the outer canopy of certain oak trees; others foraged only on trunks of trees and on branches greater than 4 in. in diameter; others only on the ground under a shrubby canopy. Where these conditions did not exist, neither did the species. Nesting requirements were equally specific. For small mammals the exact location of the trap was described and, by comparing the sites of successful sets with nearby unsuccessful trap locations in a different part of the habitat, gradually it was possible to understand the extent to which species occupied a specific or more generalized niche and to describe this niche rather accurately. For species of large mammals which can be positively identified in the field this method is still useful in increasing understanding of animal–habitat relations, particularly if undisturbed activities can be recorded. When an animal or group is disturbed by an observer it is often possible to record two observations. For example, a deer may be observed feeding on an identifiable species of grass, in perennial grassland 100 m from a forest edge, on a certain date, at a certain time of day. On being disturbed it dashes quickly out of sight into the forest. In this situation, both the use of the grassland as a feeding area, and the use of the forest as shelter or screening cover, would be recorded. Key feeding areas, specific bedding sites, connecting trails, breeding grounds, watering areas and many other habitat elements can be precisely described providing they can be unmistakably related to a positively identified animal species. As these observations accumulate it soon becomes clear that in a given study area a species will be either invariably, or commonly, or occasionally, or never associated with this or that habitat element or combination of elements. Comparing different areas will of course increase understanding by identifying habitat elements in common between them. For a given species all areas will have certain elements in common and these can be regarded as the key habitat elements; the basic requirements for the species' existence. It is an easy way of identifying elements of the available habitat which may then be further evaluated in other ways, some of which are described below.

Figure 121 shows a simple form that has proved useful in accumulating this kind of observation. Separate pages should be used for different species. Note under 'habitat' precisely in which part of the habitat or habitat element the animal was observed. Under 'activity', record exactly what the animal was doing when first seen, preferably in an undisturbed condition. If animals are observed for long periods of time subsequent changes in habitat and activity can be recorded as they occur or at frequent periodic intervals.

One old, but still useful, way of presenting such material is to list in vertical columns the main habitat elements available and then enter the accumulated information on a number of species in horizontal bands, as illustrated in Figure 122. It will be noted that although these categories are of a general nature they are sufficiently precise to recognize that most of the species can be associated with a particular combination of even roughly described habitat elements. From the standpoint of management to maintain the existing number of different kinds of animal, this information can be useful.

With a management aim of making conditions better for a particular species, or of sorting out more precisely the extent of apparent overlaps in habitat, such rough assessments are often inadequate and more detailed observations of a different kind may be required. An example involving the warthog in southern Africa will illustrate this. It also illustrates how the significance of more than one important habitat element may become apparent by observing differences in patterns of animal use in different areas.

For a 4-year period a 10-mile road transect was run in Wankie National Park, Zimbabwe, during the dry season. There all animals seen were recorded and accurately located on a map, to the nearest tenth of a mile (80 m). On a 1-mile section of this drive warthog were never seen feeding, although occasionally they were seen crossing to known feeding areas. A detailed vegetation map was prepared but was not sufficient to explain the feeding distribution for warthog, although it proved adequate for several other species. Finally an excellent correlation was obtained between the location of feeding warthogs and several species of perennial grasses; those associated with seasonally flooded areas. In the wet season, standing pools 3 ft or more in diameter were mapped, and an almost perfect correlation was obtained between those temporary 'pool sites' and several hundred observations of warthog feeding in the dry season. The above-mentioned area where warthog were not observed feeding was an area of deep sand into which rainwater sank quickly. Consequently the above-mentioned grasses were absent.

It was concluded that in that area, in the dry season, one essential habitat

Combined Classification, Habitat and Activity Form

		Young			Adult					Species	
Date	Time and location	C	Yr.	2 Yr	M	F	U	Total	Habitat		Activity

Observer _____

Species _____

Figure 121 A simple but useful form for accumulating observations of classification and animal activity in relation to particular parts of a study habitat

Figure 122 Major habitats preferred by large mammals in Loskop Dam Nature Reserve, Transvaal. (X = habitat preferred O = occasionally occurring; –X– = into both habitats indicated but principally along edges (from Riney, T. and Kettlitz, W. L., 1964)

Species	Hill slopes Lower	Hill slopes Middle	Hill slopes Upper	High country, tops of ridges	Low flats	Grassland	Open woodland	Dense woodland	Near (or in) water	Rocky outcrops
Ostrich	O				X	X	X		X	
Chaema baboon	X	X	X		X	X	X	X	X	X
Black-backed jackal	X	X	X	X	X	X	X	X	X	
Brown hyaena	X	X	X	X	X	X	X	X	X	X
Leopard	X	X	X	X	X	X	X	X	X	X
Chapman's zebra	X	X	X	O	X	X	X	O		
Hippopotamus									X	
Grey duiker	X	O	O		X		O	X		
Steenbuck					X	O	–X–			
Oribi				X		X				
Klipspringer				X						X
Mountain reedbuck	X	X	X	X		X	X			
Reedbuck					X	X			X	
Waterbuck			O		X		–X–		X	
Impala	O	O	X		X	X	–X–			
Sable antelope	X	X	X	O	O	O	X	O		
Blesbok	X				X	X				
Blue wildebeest	X				X	X	X–			
Bushbuck								X	X	
Nyala		O	–X					X		
Kudu	X	X	X	O	X	O	X	X		

element for warthog was perennial grass associated with seasonally inundated areas, even though these areas might be very small and hold standing water for only about 1 week in the wet season.

In another dry season, in Gorongoza National Park, Mozambique, warthog were seen feeding on similar grasses at the edge of a large seasonally inundated plain which extended away from a shrubby edge and occupied several square miles. In other words the habitat element limiting in the first study area was, at the same time of the year, virtually unlimited in the Mozambique park. But only a small band of this grassland was used by warthog for they were never more than about 100 yd (91 m) from the shrubby edge of the open grassy plain. Whenever disturbed, either by other animals or by man, they ran back to the protection of the cover, disappearing from sight in the shrubs and woodland.

Even these crude observations suggested the importance of cover as another important element in warthog habitat. Several years later, again in Mozambique, I was to see an extremely high concentration of warthog, this time in the Zambezi delta region. Although no counts were made, over a distance of several miles one was never out of sight of warthog; as one group flushed, another was seen ahead. This too was a border between shrubs and trees and a very large seasonally flooded grassland, but with an important difference from the area mentioned above. Instead of the edge between shrubland and grassland forming a fairly smooth border, and extending in a long line, as in the first Mozambique example, the border at the second area consisted of numerous fingers of grassland 30–50 yd wide, extending into the shrubland for a distance of a mile or more. This interspersion of food and cover thus set the stage for exceptionally high populations of warthog. Conversations with local residents suggested that the high numbers seen were part of a stable population, as this area was said to have supported these high populations for many years.

5420 Introductions

Studies of introduced animals can provide valuable insight into the process of habitat selection and avoidance. The opportunities are usually easily exploited. They are essentially the same for animals reintroduced after a period of local extinction, whether the introduction is man-made or is part of a natural dispersal into newly occupied habitat from adjacent occupied areas. In each case the challenge is to observe habitat selection in a growing and expanding population.

In some cases it is clearly easier to learn the essential requirements of the species by study of an introduction into a 'strange' environment than by study of the species in its natural environment. This is apparently the case of the sambar (*Cervus unicolor*) in New Zealand (Riney, T., 1957b). In India the sambar is commonly described as a creature of the forest, occurring in forested foothills and up to elevations as high as 9000 ft in the Himalayas. In New

Zealand, in one of the main areas of introduction, rather than penetrate the forest-clad mountains, sambar remain on the sandy, gently undulating, coastal plain. The search for a common element between these two environments (so different that one might feel that this was an excellent example of quick and great adaptation to a new environment) revealed that in both cases the deer were closely associated with broad-leafed grasses and sedges. In India these grew along the streams that ran through the forest; in New Zealand in the swampy areas between the sand dunes. They were not present in the New Zealand mountain forest areas adjacent to the coastal dunes.

Lupins and small scattered (introduced) conifer plantations suffice as cover in New Zealand, while in India nearly the entire environment was under cover. Cross-checking with other sambar introductions in New Zealand confirmed that the broad-leafed grasses and sedges associated with swampy areas were always an important element of the sambar habitat.

In identifying habitat elements of importance to introduced large animals the stage of the initial eruptive oscillation should be kept in mind. In the early stages of the oscillation it is useful to determine which particular habitat elements are favoured by the colonizing animals. Comparisons with original habitats can be revealing, as mentioned above for sambar. As animals increase in number differential use of various parts of the available habitat will be readily observed and in this way key seasonal areas of use may be identified. In the long term, once the initial adjustment has taken place, the introduced population will be adjusted with its habitat in the same way as native species.

5430 Analysis and Comparison of Home Ranges

As discussed in Section 3200, adult animals' movements are not over an unlimited area but are normally restricted to a home range which has more or less fixed boundaries. With this in mind studies of home ranges of individuals can contribute to an understanding of habitat elements and their importance, for individuals respond by living in areas of different sizes and shapes depending on the occurrence and interspersion of different habitat elements.

In the case of red deer, where various major elements of the habitat (e.g. edible shrubs, grass, water, shelter from the elements, places comparatively free of insects, etc.) are closely interspersed, home ranges are small, often under a mile in greatest diameter. On the other hand, where these elements are widely spaced, home ranges are larger, by many times.

This method involves comparing the occurrence, spacing and interspersion of habitat elements available inside individual home ranges with the size and shape of these ranges. A simple way to start is as follows:

1. Map the outlines of several individual homes ranges in the same study area, either on a yearly basis or during a critical season, e.g. the winter period or dry season.

2. Identify and record common habitat elements that occur within each individual range.
3. Within each mapped home range note the distance between habitat elements and their degree of interspersion. Identify habitat elements common to all ranges.
4. On this basis predict the comparative sizes of home ranges in other areas whose habitat elements are interspersed to a greater or lesser degree than in the first area where a few home ranges are known. Then check the sizes of the home ranges in the new areas to see how well they fit your predictions.
5. Compare this kind of information between two areas where populations are stable; one with a high and one with a low density of animals. Normally the closer the habitat elements are spaced (i.e. the greater the degree of interspersion) the smaller the home ranges and the greater potential for higher numbers.

In addition to the occurrence and distribution of habitat elements the size and shape of an animal's home range may also be modified by territorial behaviour or by intolerance behaviour from another species. However, as long as the home range is fairly fixed, and it usually is, this should not affect the use of such home ranges as a basis for better understanding habitat requirements. Regardless of the reasons for its size and shape it still contains therein all the elements necessary for the animal's existence.

When studying animals migrating between two or more seasonal ranges the seasonal range occupied during the most critical season may have priority. As noted above (Section 3200) there is good evidence to support the claim that the origin of migration lies in the species' need to develop patterns of occupancy that satisfy all requirements for existence throughout the year. Where migration is necessary it occurs; the completely separate seasonal home ranges being connected by migration routes. With this in mind it may be useful to use this approach in comparing the seasonal home ranges of migratory species with the yearly home ranges of non-migratory populations of the same species.

5440   Associating Different Habitat Elements with Differences in Amount of Use

Another method of identifying habitat elements and contributing to an understanding of animals interacting with them is to map the habitat elements available in your study area, then determine the extent to which either indices of animal density or evidence of use or both are correlated with these elements. Thus, for example, along a series of transects one may compare differences in density of faecal droppings with differences in vegetation type, slope, aspect or degree of shelter. The strongest correlations will indicate a relationship which may be critical in that study area, thus suggesting particular strong or weak links in the chain of requirements. For example, a strong positive correlation might be obtained between some index to available shelter and animal droppings and a negative correlation between faecal droppings and grazing intensity as measured on plants. This would by no means suggest that food was not important. It might simply suggest that, in this area, for the period studied, shelter of a particular kind had an overriding influence on the distribution of the feeding animals.

In addition to faecal droppings other kinds of evidence of animal use differentially associated with available habitat elements may be equally relevant, such as concentrations of heavily hedged shrubs, browse lines, and concentrations of bedding areas.

This method is good for identifying habitat elements and in defining areas of use, but it has certain limitations. One limitation is that under certain circumstances such evidence can be misleading. It is most appropriate in areas where either a single large mammal species is involved or, if several species are present, where there is litttle competition with other species. If the extent of competition is not known, then similar information should be simultaneously obtained for all species. If hunting occurs the influence of hunting on the distribution and behaviour of the species should be understood before such information can be safely interpreted. And finally, where several species are involved, in an environment of unusually uniform habitat, the home ranges of individuals or groups may be influenced by social interactions between species. This usually involves intolerance of other species closer than a certain 'threshold' distance, even when the home ranges are not defended on a territorial basis.

To understand the effect of the habitat elements on the population, some evidence of animal response must be taken, as suggested in the final method described below.

5450   Animal Response as a Measure of the Suitability of Animal Habitat

Indices of animal response are most useful in comparing the suitability of two or more areas which exhibit some clear difference in at least one habitat element. Such indices can also be used in defining and in assessing the significance of critical seasons and key areas; the weak links in the chain of essential requirements. A common study objective served by this method is to determine if differences in animal response exist between two or more clearly defined habitats, each with a different balance of habitat elements.

The underlying principle in using this approach should be clear. Where more detailed knowledge of habitat requirements is desired one must have some standard or criterion for judging if one area is in fact better than another. For this there is but one authority and this authority is neither man nor his publications (including this one) but the animal itself. Differing levels of physical condition and differing mortality rates associated with different habitats speak eloquently—on the animal's behalf—regarding the comparative suitability of different habitat elements. Either their presence, the quantity of the elements available or their pattern of interspersion may be factors, either prohibiting the occurrence of a population, or strongly contributing to determining its ecological carrying capacity.

But life can be much more complicated than this.

While the basic or ultimate cause of differences in productivity is likely to be differences in the quality of the environment, the mechanisms by which productivity is lowered are many and varied. However they always involve a chain of progressively deteriorating and linked characteristics. A thread common to all such links is the declining, or lower, physical condition of the animals involved; a consequence of environment becoming, or being, less suitable.

For example, if one accepted perennial grass at the end of the dry season as an essential habitat requirement for sable in southern Africa, a chain of responses in a deteriorating habitat may be organized in tabular form as shown in Table 9, assuming that all other requirements remain the same.

It should be clear that when one worries about productivity for some management purpose, one may be involved in a considerable chain of linked cause–effect events. The kind of evidence produced by each method usually represents only one link in the chain: for example, fewer females conceiving (lower fertility), fewer females with young (size of annual crop), or higher winter mortality (mortality).

Such 'single links' often represent much good work. The estimates may be accurate. But they are often irrelevant to the priorities of management and at worst may be misleading. For example, in poor environments it is especially noticeable that females without young at heel during the previous year are in better physical condition than females with young and are therefore more fertile. These differences are often striking, as was found for red deer in Scotland, most of which is at best poor environment for this species. There in one study area the fertility rate of 655 non-lactating females was 84.15 per cent, while 170 lactating females averaged 40.33 per cent, being less than half as fertile (Mitchell, B. et al., 1977).

It does not particularly help management to select two such links from a chain of cause–effect events, interesting though they may be zoologically. Their explanation that the low calving rate is due to lower fertility is like saying that hens that do not lay many eggs do not produce many chicks. And of course if the concern is over productivity one cannot stop with the chick stage for some areas may produce few chicks, most of which survive. Other areas may produce many chicks, most of which die. The number of eggs laid in the first place will be less if the hens are in poor physical condition. With wildlife, one is ultimately forced back to understanding more of the basic determinant of sustainable levels of productivity—the quality of the habitat within the individual home ranges, or in the area occupied by a given population.

A glance at Figure 126 may re-emphasize that within a given environment, as populations increase, the suitability of the habitat decreases thereby exerting progressively more restricting effect on the population. As populations

Table 9 Response by sable (*Hippotragus niger*) to perennial grass near end of dry season

Change in available perennial grass	Initial, basic response (first level of response)	Response by sable	
		Second level of response	Third level of response
Unlimited amount available	Healthy populations, a high proportion of animals in 'fair' condition	Highest sustainable productivity (highest fertility, highest calf crop)	Highest sustained numbers
Half the amount available	Animals in poorer physical condition	Fertility down; less young/100 females	Fewer numbers
Very little available	'Very poor' condition	Lowest productivity (lowest fertility, lowest young/100 females)	Lowest sustainable numbers
None available	No sable	No sable	Even animals dispersing into the area from nearby suitable habitats cannot successfully colonize

reduce, habitat conditions for the remaining animals improve and animal response improves accordingly. With this in mind another irrelevancy is to spend much time on relating population density to poor productivity for the two are, respectively, twice and once removed from the basic cause, as indicated in Table 9. Such relationships are often easily demonstrated but eventually the basic cause should be understood in order to develop the soundest possible long-term management procedures.

Condition ratings or indices are here emphasized as the principal tools used because of their sensitivity in reflecting changes in environmental conditions: measurable within a few weeks of a marked seasonal or other habitat changes. Either mortality in a critical season, or mortality of young as reflected in ratios such as calves per 100 females, may be used on occasion. But where the time-lag between cause and effect is more than 6 months it is difficult to interpret such data. In young : 100 female ratios, causes may have operated at any time between conception and 2–3 months after birth; and since females in poor physical condition are less fertile, a cause of the lower numbers of young may even extend back to several months before the mating season. The safe procedure is thus to rely on condition indices not only as the main evidence of response but as easily applied cross-checks against, and as a way of adding new meaning to, other kinds of evidence already obtained for other purposes.

For example, it may be suspected from general observations, or statements from hunters[1], that shelter of a certain kind is necessary to minimize loss of animals over a winter or a dry season. Other conditions being equal, animals wintering in areas with demonstrably better shelter would be expected to have lower mortality and to survive the winter period in better condition than animals whose home ranges were more deficient in shelter. The measure of differences in mortality in this example might confirm this initial suggestion.

On the other hand even if differences in mortality are clearly demonstrated they may be basically due not to differences in the winter environment but to differing summer habitats. That is to say, animals from one area may be entering the most critical winter season in poorer physical condition. Where this is suspected, it may be tested by obtaining indices to the physical condition of animals from each of the different summer habitats.

I am always surprised at how quickly this approach increases understanding and awareness of the significance of particular habitat elements. The following example has been selected from many such examples taken from African savannah areas.

In 1959, in western Zimbabwe, in mid-dry season and within 1 week, three different populations of sable (*Hippotragus niger*) were observed in three separated and apparently different habitats.

---

[1]For the benefit of European or American readers, hunters in developing countries bear little resemblance to the socially ritualistic trophy hunters of Germany, for example, or the multitudes of recreational hunters after both sport and meat that have evolved in North America. Hunters in developing countries, who supply their own villages with meat as their fathers did before them, are often the best existing sources of local information about wild animals.

The first habitat was in a forest and woodland area managed by the Forestry Commission. There strips of perennial grassland in good condition bordered a small stream. This was locally known as a 'vlei habitat'. Several transects were run which provided an index to the available perennial grass. At the edges of the grassland was a typical 'miombo' woodland. Sable were numerous and all of the fifty adult females classified were assessed as 'fair' by the method described in Section 3864.

On the following day sable were observed within a national park about 50 miles distant. Although the 'miombo' woodland, rainfall and soil were similar, the national park habitat differed from the first area in two main ways: there was less perennial grass available along the transects and water was much less plentiful and less accessible. Water was found only in widely spaced artificial ponds which were maintained by park personnel as a basis for tourist viewing. Sable were equally tame and approachable by car, but conspicuously less numerous. Adult female sable in the park area were in noticeably poorer condition, only three being classed as 'fair' while the remaining twenty were clearly in the 'poor' category.

Observing these differences in availability of both water and perennial grass, as well as the differences in physical condition, led to a quick cross-check in another national park. This second park was characterized by poor perennial grass but with unlimited water, as the vlei-type grassland bordered the large and perpetually flowing Zambesi river. Within a week of observations made in the first two areas similar transects run in the Zambesi park showed the available perennial grass to be lower than in either of the other two areas. Sable occurred at about the same density as in the first national park and were much less numerous than in the forest area. The twenty adult females classified were in 'very poor' condition, using the same method mentioned above. Viewed from the side the lateral processes of their vertebrae, ribs and two points on the pelvis were all not only visible but prominent. The ears of several sable were starting to droop, indicating they were approaching the poorest condition possible. They were visibly weak.

Table 10 is included to illustrate one way of summarizing these preliminary observations. Obviously it would be useful to classify more adult females in the last two areas and more transects should be run in the second park. But even these preliminary observations serve to emphasize one habitat element. Whatever other factors may be involved, lack of perennial grass in vlei areas at the end of the dry season (where sable in each area were mainly feeding) may be an important limiting factor for sable populations.

5500 UNDERSTANDING HABITAT REQUIREMENTS . . . A CONTINUING PROCESS

Almost every technique mentioned in this handbook can be used to contribute in some way to a better understanding of habitat requirements. From the earlier sections there are two kinds of information that are particularly valuable in this respect.

Table 10 Differences in available perennial grass and condition of sable at end of dry season in three Zimbabwean study areas[1]

Study area	Comparing habitats	Total stations	Percentage perennial grass (1)	Average height (2)	Index to available grass (1) × (2)	No. of adult sable classified	Condition class			
							Good	Fair	Poor	Very poor
1	Well watered, best grassland (forest area)	400	26	7.0	182	50	0	50	0	0
2	Poor water, less grass available (national park 1)	700	16.7	9.4	156	23	0	3	20	0
3	Best water, least grass (national park 2)	100	25	2.76	69	20	0	0	0	20

[1]The additional class 'very poor' included animals where (b), (d) and (e) (Figure 43) were each easily observed.

First there is the understanding of the extent to which the population is restricted and localized, which enables one to know to what extent one can relate the observations of habitat elements in one area to those taken in another nearby area.

Two considerations that may be involved: (1) the sizes and shapes of individual home ranges and the niches used therein; and (2) the equally useful indirect evidence of localization, as when two adjacent areas with different habitats reveal different levels of physical condition between samples of the two populations. If animals in these nearby areas were freely moving back and forth a given sex and age group would be in similar condition. The demonstration of significant differences in physical condition is thus good indirect evidence of localization.

The second kind of consideration, always re-emphasizing the importance of understanding habitat requirements, concerns the eruptive oscillation (Section 7100). Recognition of the existence of a particular phase of an eruptive oscillation depends on the ability of the field observer to recognize changes in at least some of the easily recognized main elements of the habitat.

Learning about habitat requirements is a process that will probably continue as long as we are interested in managing animals. As noted in the above section, learning can proceed in many ways.

It is difficult to imagine any management programme or management-oriented study that would not profit from increased understanding of habitat requirements, for they represent one half of the essential core of animal–environment interrelationships (ecology). They provide the basis for progressing with the environmental side of management. Populations can of course be managed with information taken from animals alone (the other half of the core), but past experience has shown this to be a dangerous numbers game. Whether we like it or not, every man-stimulated manipulation of numbers of large mammals produces a corresponding response in terms of one or more habitat elements. And when we manipulate essential parts of an animal's habitat the population must respond as certainly and inevitably as night follows day. This is the way life, including wildlife, is. Eventually, to be effective in achieving most of our management objectives, we must improve understanding of both animal and habitat in an integrated way.

To move ahead in our understanding, the following approach seems worth encouraging. First there is the need to be aware of our ignorance, of the need to know more of, and to take every opportunity to improve the precision of, our understanding of the habitat requirements of our study species. As we become increasingly aware of the significance of one or another habitat element in our particular study or management area we can consequently be more precise in our definition of key areas, in our knowledge of how to improve the conditions of living of our study species in their critical seasons. Thus gradually we may improve our standards of management for whatever purpose.

Chapter 6000

Management of special areas

6100   INTRODUCTION

Among the areas which have been completely or partially dedicated to wildlife, there are some where complete protection of animal and plant life is the rule and where populations are allowed to fluctuate unhampered and unaffected by the activities of man and his technology. Scientific research is the only activity permitted, under strict supervision, in these natural reserves and wilderness areas. We do not need to discuss these strict reserves where management activities are excluded. Rather, this chapter will discuss several types of wildlife areas which require some management measures: controlled hunting areas, national parks and game reserves, sanctuaries, controlled hunting areas and game cropping areas.

6110   The Management Plan

There is, throughout the developing countries, an urgent need for plans for management of lands where wildlife is considered a primary resource or an important form of land-use. Such lands will include parks, reserves, areas for controlled hunting or for the production of meat from large herbivores. Although the needs are different, management plans are as important for the well-established parks and reserves as for newly developing wildlife management schemes.

A management plan is a practical tool designed to further the major objectives of a given wildlife management area. As these objectives in turn relate to an overall wildlife policy, it is essential to have a stated national wildlife or national park policy and clearly understood national objectives. Appendix 4, which reviews guidelines in establishing national legislation on wildlife and national parks, has in it many of the elements of such a policy statement.

Management plans are essential in that they can: (1) concentrate and coordinate limited funds and personnel on the really major objectives (objectives that should be well known to every person charged with any phase of activity in a managed area); and (2) aid the authorities responsible for the administration of the resource in educating the public and in meeting criticism. A general plan for a wildlife management area should include:

1. A brief review of the origin and history of the area to be managed and of the policy which led to its establishment, and including a record of former land-use in the area.
2. A general description of the area and details of management techniques formerly applied.

3. A list of basic needs and objectives, in order of priority, towards the achievement of which the management is directed. (The realization of these objectives will provide a foundation on which a more specific and elaborate plan may be built at a later date.)
4. The general approaches, guidelines, and specific methods for accomplishing the major objectives of management. (In the case of a wildlife area in its very early stages, this may simply be a plan for development. In the case of established areas this may mean monitoring to detect problems or for routine management decisions and it may involve a periodic evaluation of the effectiveness of management techniques currently in use.)
5. A programme and schedule of short-term and long-term jobs to be accomplished.
6. Periodic review of the over-all working plan, to make sure that activities within the managed area are consistent with the major objectives. This review should initially be done on an annual basis.

I encountered a strange idea in one country where an organization director was not interested in management plans because he felt there was not enough knowledge to prepare a plan and, furthermore, that the preparation of a management plan could not be met by the department as it would entail a team of costly foreign experts. Although the department concerned administered some of the best-known wildlife areas in Africa, there was no national policy. Conversations with several senior wardens revealed that they were not at all clear as to the major objectives for the areas they supervised and thus it was impossible for them to evaluate the effectiveness of their daily efforts.

Management plans can be simple and they can be prepared by existing staff, with existing knowledge: a plan to do the best job possible of working toward the long-range objectives with existing knowledge and personnel. Since such plans are never final they should be modified as more knowledge is accumulated.

Lack of money and manpower is often given as a standard reason for failure to develop or improve management and it is, of course, sometimes true. However, in my experience a vastly more important cause for concern is failure to make the most appropriate use of existing staff, coupled with failure to identify critical priority issues. Lack of money and manpower may even be an advantage in assigning priorities by stimulating an awareness of the need to make the most effective use of these limited resources.

6200 MANAGEMENT OF CONTROLLED HUNTING AREAS

The basic laws enacted for national wildlife conservation and management cannot be altered except by the legislative body that enacted them. But the essence of good game management lies in flexibility of control. Such a flexibility is needed: (1) to allow for different conditions in different parts of the country; and (2) to meet the ever changing local abundance of different species of animals resulting from hunting pressures, seasonal migrations, movements in response to drought or rain, outbreaks of disease and disturbance by man, whether from tribal hunts or migrations or from rural development schemes. For this reason it is usual for wildlife conservation laws pertaining to controlled hunting areas, hunting zones (or similar titles) to authorize the chief game warden (or sometimes the minister concerned) to make separate rules or regulations applying to each of them. Such regulations elaborate locally the general hunting policies laid down in the main body of the law. For instance, it is made an offence for any person to hunt in a controlled hunting area unless in possession of a special permit to do so, or to commit a breach of any of the provisions of that permit (consult Appendix 4 for other examples).

6210 Objective

The usual objective in management of controlled hunting areas is to obtain an optimum possible sustained yield through hunting. However, that objective is difficult to achieve for several reasons:

1. As shown in Chapter 7000 there is, even for one species in a particular environment, a wide range of ecologically acceptable sustained yields.
2. Because of changes in understanding, politics, other social factors, the cost of management, or an unusually severe winter, an optimum yield for one year may considerably change several years later.

As mentioned elsewhere optimum yield is rarely if ever the same as the highest possible (maximum) sustained yield. As used in this handbook optimum yield is the yield which represents the best compromise towards meeting management objectives, while taking realistic account of existing local knowledge, or lack of it, and other dampening social and economic factors.

It is usually impossible to develop long-term plans because game departments are given responsibility for wildlife management only and have no control (and often no advance knowledge) of other land-use activities, such as agricultural schemes, cattle production projects, or disease-eradication programmes. Particularly in countries where management is in its infancy there is the problem of sustaining an unknown amount of illegal hunting which can by itself prevent effective management.

6220 Establishment of Blocks

In setting out to operate a controlled hunting area system the first thing to do is to divide the usually large regions into blocks of manageable size. The dimensions of these blocks will vary not only with the relative abundance of game, but also with the nature of the country. They may range in size from the traditional 100 square mile block of the state forests of India to those of 5000 or 6000 square miles found in the sub-desert country of northern Kenya. Generally speaking, in open savannah country carrying abundant game such as is found in many African countries it would be unusual for blocks to be less than 750–1000 square miles in size. In heavy mountain forest, where animals such as bongo or leopard form the main quarry, blocks of only 100–150 square miles have been found adequate, provided they are free from human disturbance. Where possible, the boundaries of blocks should be ridges, valleys, or tracks as these can be easily recognized on the ground. Rivers, except in very well-watered country, do not make good boundaries as they tend to draw game from neighbouring blocks to the common border, thus causing interference, and sometimes friction, between hunters.

Another, and most important, point bearing on the size and limits of a block is whether all hunting can be reserved for the permit-holding hunter, or whether local people also have hunting rights in the area at the same time. The latter circumstances are dealt with in a subsequent paragraph. For the moment, the simpler case will be assumed in which the block is either uninhabited or inhabited by non-hunting people, and no-one but permit holders has the right to hunt.

From the hunter's point of view the main advantage of the controlled hunting area system is that it affords him the exclusive use of a block even for a short period. It is for that exclusiveness that he is usually willing to pay high fees. Therefore not more than one party of hunters should ever be allowed in a block at any time. It is also normal to limit the number of actual hunters in each party.

Disturbance usually has a serious effect on game even if little shooting is done, so each block should be rested as much as possible between hunts. Conditions will vary enormously, but it is suggested as a guideline that not more than one party of hunters should ever be permitted in a block in a fortnight (irrespective of the number of days out of the maximum 14 they actually spend there). Where a fortnight of use proves to be too much, the block should be open for booking for only the first fortnight in each month.

Detailed knowledge of a block and of the home ranges and movements of the species concerned will enable knowledgeable managers of these blocks to increase numbers of hunts by developing a rotational hunting system where hunting takes place only within selected blocks during a given month, while other previously hunted areas are rested. In some cases a block may be closed to hunting for several months on end, as in countries that have a heavy rainy season, or where an over-all closed season for game is in force.

6230 Application of Yield Determinations to Hunting Areas

The next step is to determine the annual offtake of each species of game animal that each block can afford to yield (see Chapter 7000). Once the number to be harvested for each block is determined, that number can be divided among hunters in two ways: (a) by limiting the number of permits that are issued to hunters, or (b) by limiting the number of animals that each permit-holder is allowed to hunt.

To fix the number of animals that each permit-holder may take, one has merely to divide the maximum permissible offtake of each species for each block by the number of permits it is proposed to issue. It is most unlikely that every hunter will succeed in killing all the animals he is allowed. With rarer species other systems may be appropriate; for example, only one permit may be issued to each party, leaving it to the members to decide amongst themselves who the lucky hunter is going to be (and who must take out the necessary game licence). With even rarer animals, the chief game warden may hold a secret ballot in advance, resulting in the permission to hunt an individual of that species being offered to, for example, only the third, sixth or eighth party to book that specific block.

6240   Regulation of Hunting

It cannot be over-emphasized that the conditions of a permit relating to any block should govern the number and species of game animals that the holder may take in that block irrespective of whatever other game licences he may hold for additional areas, numbers of animals or species. If the wildlife organization considers that an otherwise common animal requires special protection in any particular block, it may remove that animal from the list of species that may be taken there. The holder of a game licence for that species will have to go to a different block where its hunting is permitted, if he wishes to fill his licence.

Having decided the number of hunting parties to be permitted in a block during the course of the year, and the number of each species of animal that may be taken, the next step is to decide if there is need for any special restrictions. Such restrictions may take the form of:

1. Holding one part of the block as an absolute sanctuary.
2. Protecting females of some species not generally protected throughout the country.
3. Putting a daily bag limit on the number of birds shot.
4. Introducing special measures that may not be in general application in the country concerned.

Certain administrative restrictions will usually be necessary also; for instance, regulation of camping locations and camp fires, compulsory reporting at arrival and departure, obligation to be accompanied by licensed professional hunter, reporting of wounded animals.

6241   *Brochure and Map*

A map showing where each hunting block is situated does a great deal to advertise the hunting a country has to offer. Along with it, a free brochure should be published containing a detailed sketch of each block; a verbal definition of its boundaries; the periods during which it is open to hunting; the number of hunters allowed in one party; the kinds of animals that may be hunted in it and the number allowed of each species; as well as any special regulations applying to the block including the fees charged for controlled hunting area permits. As regulations for some or all of the blocks will vary from year to year, all such brochures should be dated and not too many of them should be printed at one time.

6242 *Booking*

It has been found advantageous not to allow the booking of blocks more than about 6 months in advance, and to require a cash deposit with each booking. Otherwise, the best blocks may be taken up by speculative bookers who may not use them and thus cause financial loss to government or safari organizations and frustration for other applicants.

6243 *Hunting Area Fees*

Payment of controlled hunting area permit fees should be quite a separate matter from payment for game licences. It is an almost universal principle of law that game belongs to the state and not to the owner of the land on which it is found, although there are exceptions, for in some countries the landowner also owns the game. Hunting rights, on the other hand, are an interest in the land and belong to the landowner. He may retain them for his own use, lease them out, or otherwise dispose of them as he likes. The fees for game licences (i.e. licences to kill game on private or public land) pay for the right to take animals. The fees paid for use of a controlled hunting area pay for a temporary lease of hunting rights in the block concerned, and are similar to the fees that farmers charge for permission to hunt on their land. In Europe, the lease of hunting rights is vastly more expensive than the cost of game licences. Thus, a hunter's controlled hunting area fees could justifiably cost him much more than his game licence fees. So far, however, that is rarely the case in developing countries.

Where, as in the cases we have been considering so far, a controlled hunting area block lies in uninhabited or sparsely occupied state-owned land, the controlled hunting area fees should properly go to government. In many developing countries areas of state land have been designated for use by certain tribes. In such instances, even when there is no question of individual tenure of land, a strong case exists for assignment of fees to representatives of the local people for their common use. This is only proper since it is at the expense of the local people in damage to crops and livestock, and often through their forbearance in not hunting, that game continues to exist. Such a payment provides a way of enlisting local sympathy and cooperation in maintaining game stocks. It also helps break the usual antipathy in local people towards controlled hunting areas, and thus helps to maintain the area as one from which sustained yield of game animals can be expected. Special crop-damage payments should be limited to the inhabitants of recognized settlements, and no payments made to the owners of distant outlying plots which are sometimes locally cleared and planted to invite destruction. The latter is a common practice particularly when elephants are involved.

Payment of controlled hunting area fees can be required in two ways, neither of which is wholly free from objection by the public. In some countries a flat fee is charged for the use of a block, at so much per week or fortnight irrespective of the number of days it is actually occupied. The objection to this is that the hunter pays as much for the worst block in the country as for the best. In other countries the hunter pays a separate fee in arrears for each animal he has killed in the block, the amount varying for each species in the same way that game licence fees differ for different species. This system appears eminently fair, but in certain African countries it has caused ill feeling among overseas visitors who like to know in advance exactly how much their hunting trip is going to cost them.

It is a general practice for game licence fees to be higher for non-residents than for residents of a country. In some cases, non-residents are charged a higher rate for controlled hunting area fees as well. Generally speaking, hunters do not appear to regard this as unreasonable.

6250   Choice of Hunting System

Of the many different approaches to hunting policy by different countries two contrasting systems seem worth mentioning in terms of determining optimum yield.

The first of these is where only a few licences are issued, consequently only a few hunters have exclusive rights to hunting in large areas well stocked with game. These hunters pay the highest fee for licences, trophies and accommodation and I think of this as the Christian Dior system.

This system seems particularly suitable in countries in an early stage of developing its wildlife resource, where numbers of animals in different areas can only be described in such terms as few or many, where access roads are bad and accommodation away from the major cities is either inadequate or non-existent. Under these circumstances hunting is handled mainly by safari companies who provide acceptable transport, elaborate tent accommodation, food, drink and guides for the wealthy hunter.

Under such circumstances optimum yield is in terms of foreign exchange coming into the country; revenue to the government in terms of concession fees from the safari companies hunting licence and trophy fees for each animal shot. So few animals are taken that the question of permissible yield from a population is irrelevant. The main factors determining the numbers of hunting parties accepted are the numbers that can be accommodated by facilities and trained servicing personnel available to maintain the high standard of accommodation, and exclusiveness of hunting rights that combine to provide the quality of hunting experience that this system provides.

At the other extreme maximum numbers of hunters occupy carefully managed hunting blocks at closely spaced successive periods for as much of the year as possible. The numbers of animals removed are of course much greater and the cost to each hunter much lower. I term this second approach the Woolworth system.

Such a system requires a well-organized and well-trained game department and a good number of controlled hunting areas, as well as efficient booking and licensing systems, good roads and accommodation sufficient to handle comparatively large numbers of hunters. Dependable indices to numbers should be available for each of the hunting blocks so that annual permissible harvests may be calculated (see Chapter 7000). Safari companies are usually involved but with a range of accommodation much greater than in the Christian Dior system. Ideally, in addition to tenting accommodation, strategically located hunting lodges and a range of suitable guest houses and hotels will be available in many parts of the country.

Under these conditions 'optimum yield' would take into account both the numbers of animals available for harvesting in each block and the numbers of hunters that can be accommodated.

The Christian Dior system is certainly worth considering in developing countries where the management of the wildlife resource is in its infancy, for it provides a maximum of income with a minimum of government expenditure. The Woolworth system is worth considering in developing countries where a maximum amount of tourist revenue is desired and where it is possible to accommodate satisfactorily a much greater number of visitors.

Normally a greater amount of foreign exchange is generated by increasing the intensity of management and rates of harvests. For example, for some species of sheep hunted in the Middle East, trophy fees are as high as $5000 each. In one year five hunters, if successful, would produce a revenue of $25,000 from trophy fees alone. With a lower fee of $500 per trophy to achieve the same revenue would require fifty animals to be taken; but in the latter case the extra money spent in the country from hunting licences, payments to safari companies and other expenditures might well be ten times as great since fifty rather than five visitors would be involved. Those countries making the greatest income from tourist hunting tend to favour the Woolworth system.

These are by no means the only choices, however. In most countries where hunting for sport is practised some blend of the two systems is used and the systems assume many strange forms depending on differences in emphasis in blending, prestige, exclusiveness, politics and economics. In several European countries special government-controlled hunting reserves are almost exclusively used by high-ranking politicians and, inevitably, their friends. In Scotland, in several large red deer estates the trophy shooting is done by the wealthy landowner and his friends or business associates. Shooting of females and culling males is normally done by employees of the estate, and the carcasses sold. In other estates, where a day on the hill is charged for and the hunter

332

is successful, the meat and skin is still the property of the landowner and is sold to increase the estate revenue. In the United States, licences to hunt and tags to be placed on animals killed are issued by the administering organization and the trophy, skin and meat belong to the hunter. In some European countries the quality of the trophy (size and shape) determines the trophy fee.

There can be no such thing as a generally recommended hunting system for, like so many other aspects of management, the choice of system will depend on the over-all government policy and specific management objectives.

6260 Land-use Problems

Management is comparatively simple under circumstances where controlled hunting areas have been established (a) on uninhabited state lands or (b) on state lands over which a non-hunting tribe has a general lien but where lands have not been divided into individual ownerships. All too often, however, controlled hunting areas have been established on state lands forming the reserve of a tribe of traditional hunters and on which, in fairness, the people must retain their right to hunt. This creates a problem to which, as far as I know, no wholly satisfactory or workable solution has yet been found. The difficulties of such a position are often compounded when the tribe concerned is composed of agriculturists as well as pastoralists and hunters, and livestock and crop protection must also be considered.

Ordinarily it is necessary to control hunting on reserves set aside for hunting tribes only for the well-being of the people involved. Growing human populations, the addition of firearms, steel traps and wire snares to traditional methods of hunting, the increased hunting range of individuals afforded by bicycles and other vehicles, and the breakdown of tribal discipline and hunting traditions[1], all work to reduce game stocks. If game is to be made available for controlled hunting area permit-holders as well, it becomes necessary to restrict the offtake of local hunters even further. Local people normally regard all the game of the area as their own. Although the total of animals taken by controlled hunting area permit-holders is usually an insignificant fraction of the over-all kill, the authorities are posed with the difficult question from locals: 'Why do you stop us from killing our game, but let foreigners come in and take it away from us?'

From the permit-holder's point of view, a block in this type of controlled hunting area is always less satisfactory. He loses the sense of exclusiveness for which he pays. Game is scarcer and wilder and very often his stalk is spoilt by a shot being fired in the neighbourhood at a critical moment.

Although no universal solution to such a situation is known, it is, nevertheless, useful to consider several generalizations based on successful experiences in several African countries.

If tribal discipline is still fairly strong, much can be done by seeking the

[1]In fairness, it should be recognized that many local hunting traditions include effective conservation practices.

understanding and sympathy of the chiefs or elders. There is no better way to do this than assigning the fees earned by the block to the tribe. If the game taken away by 'foreigners' provides money to help build schools, dispensaries, water points and the like, the regulations may become more palatable to the local people. It may be possible, also, to get elders to agree to restrict hunting by their people to certain periods and to open the block to permit-holders during other periods, or else to restrict hunting by local hunters to certain parts of the block. In other cases, permit-holders may be welcome, and obtain considerable cooperation, if they limit their quarry to carnivorous animals such as lions and leopards. Some tribes may agree to harvesting of large and dangerous animals, such as elephant and rhinoceros, which local hunters are ill-equipped to tackle. Under favourable circumstances, elders can be helpful in providing beaters, or in keeping woodcutters out of a certain area, or sheep off a hillside, provided such arrangements are made in advance of time of need.

6270   General Administration

To manage a controlled hunting area system along the lines already discussed requires a game department with a sufficiently large field staff to patrol all controlled hunting areas and adjacent lands. Illegal hunting must be suppressed. Checks must be made on game populations and trends. The activities of permit-holders must be policed to ensure that the conditions of permits are being observed. The headquarters of a game department that runs its own controlled hunting areas must also operate a booking office where the blocks can be booked, permits issued, and fees collected. This will be in addition to the regular administrative sections and those concerned with the issue of game licences, the control of trophies, prosecutions and other routine work.

Some African countries, once they have divided controlled hunting areas into blocks, lease them out for a variable period of years to concessionaires (hunting safari companies). This system may or may not afford a saving in field staff although field staff will always be necessary to suppress illegal hunting and to enforce public law. It may require a full-time officer in close contact with the hunting safari camps to ensure that the concessionaires adhere to the law.

In this system hunters normally deal with safari companies directly and the game department is thus spared the expense of booking.

In principle, a concessionaire, dependent for his livelihood on the hunting within his concession, will spare no pains to maintain game stocks at a maximum. It is thought the longer the period of concession, the more the likelihood of sound management by the concessionaire, However, there is always the danger that the wrong type of individual will obtain a long concession, and manage it with no intention of conserving the wildlife resources but only to get rich quickly by 'mining' the resources without scruple as to how he or his clients do it.

With this possibility in mind it is common practice for a game department, when leasing a concession, to lay down an annual maximum permissible harvest for each species of game animals. Such other general conditions regarding hunting as appear needed are included in the contract.

Once the concession is signed, unless these requirements are observed the contract is not renewed; management is much less flexible than when the game department alone manages the controlled hunting area. If unforeseen circumstances require changes in regulations it must usually be left to the concessionaire to make them or not, as he sees fit.

Finally, regardless of the type of hunting system, careful records must be maintained of the species, sex and age class, and location of each animal shot. This should apply not only to animals shot as trophies but to animals shot for camp meat as well. In several African countries it is not unusual for a hunter to hunt in five or six different blocks during a 3-week hunt. In such a case, he normally employs the same professional hunter for the whole period. Thus he learns the man's hunting habits and learns to rely on him as he moves camp from one block to another. He sees much of the country in the process. Hunting tends to be done from more permanent camps in concessionaire-run blocks and a move to another block usually means fresh negotiations with a different concessionaire and hunting with a new professional hunter or guide unless a given safari company is assigned several blocks.

The processing and preservation of a hunter's trophies is normally not part of game department's duties for such field preparations are usually undertaken by professional hunters. It is important that the chief game warden satisfy himself as to the professional hunter's proficiency in such work, as well as in hunting, before issuing him a licence. The game department should also be in a position to advise on reliable taxidermists to whom field-dried trophies can be entrusted for final preparation and onward despatch. It is essential for game departments to be ready at all times to mark, register and document trophies brought in by hunters. This, of course, must be in compliance with whatever, the country's regulations may be regarding the possession, export, transfer, or sale, of wild animal products.

6300 MANAGEMENT OF LARGE MAMMALS IN NATIONAL PARKS

6310 Policy

In wildlife management, normally habitat and populations of animals are managed together as integral parts of a single management plan since, in practice, it is impossible to separate the two. This is particularly relevant in planning for the management of national parks containing large mammals.

In some of the oldest national parks, overpopulations of some species and the decline of others can be ascribed to past management practices which developed on a more or less *ad hoc* basis. There is throughout most continents an urgent need for plans to include management of animals in parks;

although needs differ, management plans are as important for the well-established as well as for the newly developing parks and reserves.

Management plans are practical tools designed to further the major objectives of a given park or reserve. As these objectives, in turn, relate to an overall national park policy, it is of little use to waste time preparing management plans unless there is a stated national park policy and unless the objectives of the park are clearly understood. It is impossible to over-emphasize the importance of a clear understanding of national policy and specific park objectives since otherwise the management plan may become irrelevant or, at worst, create new problems for the park.

One suggested legal definition of national parks has been included in 'Guidelines in Establishing Legislation on Wildlife and National Parks' (Appendix 4). While this appendix describes various characteristics of national parks, care has been taken to avoid recommending any one definition. After all, in the long run a national park is whatever a given nation chooses to call it. Various organizations, at various international meetings, have tried to form internationally accepted definitions and indeed have achieved agreement at certain meetings. But the definitions keep changing as meeting follows meeting, different combinations of countries being represented at each meeting. In terms consistent with the aims of this handbook it seems more important to emphasize an approach described by W. J. Hart (1966) who, in discussing a systems approach to park planning, presented a spectrum of possible different types of parks or reserves based on the range of social restraints on human manipulation of the natural environment (see Figure 123).

The types of areas he considered ranged from city amenity and recreation areas inside big cities to strictly protected natural areas in remote areas. In considering national policy, regardless of the terms a country finally adopts for the various areas, it seems important to consider the long-term national needs over an entire range of protected areas to help place the development

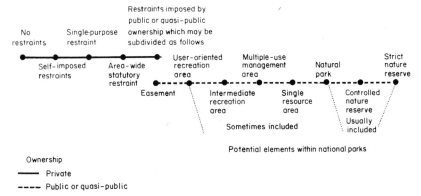

Figure 123 This figure emphasizes a range of various social restraints on human manipulation of natural environments (modified from Hart, W. J., 1966)

and management of any one type of area in perspective. Although Hart has avoided the term national park, a modification of one of his figures seems useful. While various countries have included one or more of the last seven categories shown in Figure 123 within their national parks, emphasis is normally placed on some combination of the last three types of areas Hart lists (i.e. natural parks, controlled nature reserves and/or strict nature reserves).

With the above in mind, and without attempting a definition, there seems a growing international consensus on what national parks are all about. National parks are areas set aside where man can enjoy, as a privileged visitor, the unique natural features of the area and the plants and/animals that are indigenous to that environment under conditions as little affected by his presence as possible; privilege should be sustained for the benefit of future generations as well as for the present.

Some alternative aims have from time to time been proposed in various developing countries as desirable features of national parks, and examples of a few of these, with comments, are listed below.

Alternative or qualifying aims:
1. To keep the environment in the exact form present when the park was created.

    This is difficult without some form of active management because the environment constantly changes in response to climate, occurrence or absence of fire, animals and other factors.
2. To remove every possible influence man might have on the environment and to allow nature to take its course without human interference.

    In most countries this is impossible because the influence of man outside the park will affect the environment within. Also, if parks are managed as strict nature reserves, the desirable features may be damaged as, for instance, by fire or disease or insect plagues, or by a reduction in fire or disease which may have been normal for centuries.
3. To manage the park as a natural zoo, and if animal populations increase to a level at which they damage the habitat, to feed the animals if necessary.

    This is impractical within even the broadest definition of national park. If the area is managed as a zoo this should be reflected in the name: call it a zoo or special wildlife reserve.
4. To manage as a ranch for the production of animals for human consumption.

    This will be inconsistent with tourism and animal-watching in natural surroundings and with the maintenance of sample ecosystems in a natural state, a major objective of national parks (see Table 11).

Since the IUCN publication of W. J. Hart in 1966, several international meetings have further elaborated Hart's basic approach of trying to view national parks in a perspective alongside other legitimate uses of land. Particularly useful summaries of recent developments in national park planning have been made by K. R. Miller, who sees the development of national parks as one aspect of ecodevelopment, whose objective he defines as:

> Fundamentally, the objective of ecodevelopment is to utilize resources to meet human requirements and to improve and maintain the quality of human life for this and future generations. To face this challenge, development must take place in a manner which integrates biological considerations together with economic, social and political factors to address both human welfare and the human environment. (Miller, K. R., 1980.)

In an earlier paper, Miller (1974) prepared a useful guide to making decisions concerning alternative categories for the management of wild-lands. The Miller guide (Table 11) should prove useful for wildlife managers involved not only with management of large mammals in national parks but in several of the other categories as well. It provides an excellent way of comparing different objectives associated with different kinds of land-use (Miller's 'management categories'): in other words, the framework for basing plans for managing wildlife lies within these categories.

6320 Planning

Once we know *what* needs to be done (management objectives), and *why* (within an overall policy), attention should then turn to *how* it shall be done. Planning is simply common-sense thinking which involves the definition of objectives for the park, together with an outline of the steps to their attainment. A plan is essential to performance of effective, efficient, coordinated work.

A description and discussion of management plans, also relevant to national park planning, has been included in Section 6100 above. However, several ideas seem worth reviewing in the interests of improving the effectiveness of management in national parks.

A management plan will differ with circumstances, but always states answers to these five questions. What has to be done? Why? Where? How? . . . together with the cost. When all the jobs have been listed (e.g. establishment of boundaries, protection from trespass, construction of roads and accommodations, development of water supply, ecological surveys, fire control and the like), an order of priority or sequence of operations must be established and specifications set up for each type of operation. The latter should include manpower required, man-hours of work required, equipment and materials needed, data to start work and the date to complete it.

The plan should be revised periodically in accordance with new knowledge of the area, changing budgetary conditions, or charges in public use or the effect of public use. In spite of the prospect of possible future revisions, even the first plan should nevertheless be sound, complete and well-prepared. Without a viable plan, the development of a park will proceed according to

Table 11 Decision-making guide to the alternative categories for the management of wildlands to support eco-development*

ALTERNATIVE MANAGEMENT CATEGORIES

Objectives for conservation and development	National park	Natural monument	Scientific or biological reserve	Wildlife sanctuary	Resource reserve	National forest	Game reserves, farms, and ranches	Protection zones	Recreation areas	Scenic easements and rights-of-way	Cultural monuments	Watershed Programme or River Valley Corporation†
Maintain sample ecosystems in natural state	(1)	(1)	2	(1)	—	2	4	4	4	4	4	4
Maintain ecological diversity and environmental regulation	(1)	(1)	(3)	(1)	(1)	(1)	(3)	(3)	(3)	(3)	(3)	(3)
Conserve genetic resources	(1)	(1)	3	(1)	—	3	3	3	3	3	3	2
Provide education, research and environmental monitoring	(2)	(2)	(1)	(2)	—	2	4	4	2	4	2	2
Conserve watershed production	3	3	3	3	—	(2)	3	(1)	3	3	4	(1)
Control erosion, sediment and protect downstream investments	3	3	3	3	—	(1)	3	(1)	3	3	4	(1)
Produce protein from wildlife: sport hunting and fishing	—	—	—	—	—	(2)	(1)	—	—	—	—	2

Provide for recreation and tourism	(2)	4	4	—	(2)	2	—	(1)	3	4	2
Produce timber on sustained yield basis	—	—	—	—	(2)	—	4	—	—	—	2
Project sites and objects of cultural, historical, archaeological heritage	(1)	4	—	—	4	—	4	(1)	—	(1)	2
Protect scenic beauty and green areas	(1)	(1)	3	—	3	3	(1)	(1)	3	4	3
Maintain open options through multipurpose management	—	—	—	(1)	(1)	—	3	3	3	—	(1)
Support rural development through rational use of marginal lands and provision of stable employment opportunities	(3)	(3)	3	(4)	(1)	(1)	(3)	(1)	(3)	(3)	(1)

( ) Major purposes for employing management systems.
1 Objective dominates management of entire area.
2 Objective dominates management of portions of area through "zoning".
3 Objective is accomplished throughout portions or all of area in association with other management objectives.
4 Objective may or may not be applicable depending upon treatment of other management objectives, and upon characteristics of the resources.
— Not applicable.

* From Miller, K. R., 1980, pp. 94–5.
† In the case of Watershed Programmes or River Valley Corporations, the areas normally include towns, agriculture and other land-uses.

whim, too quickly formed opinions, or outside pressure, and results may prove to be not only costly but unsatisfactory in terms of meeting park objectives.

K. R. Miller (1980) has included a wealth of practical guidelines for planning and managing national parks in his publication on national park planning, and four of these sets of guidelines are included in Appendix 8.

6330 Protection

When a suitable area has been chosen for a park, the policy established, the boundaries well marked in the field and on the map, this is usually followed by allocation of protective staff. Since hunting, farming and livestock grazing are normally prohibited in national parks, a staff of guards has to police the area to stop these activities. One useful procedure is to post guards at strategic points on the periphery of the park, and in addition use roving squads to patrol inside the area. The guards should be informed as to park purpose and regulations. They should be trained in law enforcement and given the proper equipment—at least as good as the poachers—and the legal authority to act against illegal operators.

6340 Ecological Surveys and Studies

Obviously, part of nearly every management plan prepared within the early development years may include plans for obtaining more information to fill critical gaps in local knowledge. Research so stimulated is more apt to contribute to the solution of a critical park problem than the common practice of engaging biologists for the general purpose of learning more about animals in the hope that, somehow, this will help park management.

The particular importance of ecological surveys, at an early stage of development, needs emphasis. Such surveys should be regarded not as separate from other activities but rather as an integral part of management and development.

It is important to develop a sound body of knowledge of the park as a basis for determining refinements in actual management procedure and problems. Much of this knowledge can be obtained through the use of the techniques described in Chapters 3000 and 8000 of this handbook. Some of the information may need be gathered only once, but most of the data will have to be collected or reassessed on a continuing or periodic basis as vegetation and animal populations inevitably change with time.

Without such data and the remedial recommendations which they can provide, it is likely that serious management problems may develop either from forces within or outside the reserve. Over-populations of animals, for example, can adversely affect not only land within the park but land downstream from the park; alternatively, over-populations within the park can be brought about by destructive types of land-use upstream. Such problems may seriously interfere with the major park objectives.

An important source of motivation for developing certain surveys or studies may be based on the assumption that trials of detailed management techniques and assessment of their possible consequences are essential before they are continuously applied throughout the park.

The staff to collect these basic data does not have to be highly educated or specialized, though it is important that a specialist train, direct and monitor the work. Ordinarily, there will be several junior members of a national park staff who are ready and willing to collect data. To be successful, they must be keen observers, be able to read and write and be trustworthy. Given these qualities, and adequate basic training, staff personnel can be most useful in gathering much of the quantitative information essential for the management of the park.

The following are *examples* of specific types of survey or study useful in the early stages of the development of a national park containing large mammals. To assist with planning as much information from such studies as possible should be entered on maps.

1. Species of animals present, their relative abundance and distribution in the park on a seasonal or monthly basis; movements of animals, whether sedentary or migratory; determination of seasonal ranges outside the park boundary, if such exist;
2. Description of habitat components: climate, soils, water distribution, topography and vegetation; principal plant formations; vegetation map of the area; map of fire distribution on an annual basis; major habitat types or existing forms of land-use in areas surrounding the park should also be mapped.
3. Requirements of animal species for food, water, shelter, breeding; distribution of animals in relation to vegetation types and other factors.
4. Identification of key areas for different animal species; extent of grazing and browsing; measurements of grazing and browsing pressures; especially of most critical areas at most critical seasons.
5. Trends in the animal populations: increasing, decreasing or stable; cow–calf ratios; physical condition of animals.
6. Trends in the vegetation: pastures improving, degrading or stable; appearance of hedge forms; deterioration of vegetation classes (trees, perennial grasses); bush encroachment; appearance of accelerated erosion.
7. Historical information on past land-use practices (agriculture, livestock grazing, fire) and on fluctuations in animal populations, as far back as possible.

6350 Management Practices for Animals and Plants

Management of the environment in a national park should be kept to a minimum, but is nevertheless sometimes necessary because biotic communities continuously change following protection. Some habitats, needed by desirable animal species, may be destroyed by another species; or certain herbivores, because of excessive numbers, may degrade their own food supply through over-grazing. In such cases, some form of management may be desirable so long as it is consistent with the policy and objectives set for the park. The tools of management for altering ecological conditions are in the main limited: the principal tools are the rifle and the match; others include waterholes, salt licks, planting or seeding, fencing, moat-digging, mowing, brush-crushing. In principle never use these or other tools *unless* you have good evidence indicating they are *required* in order *to achieve specific park objectives*.

(a) Ideally the variety of animals and habitats found in national parks should be allowed to maintain themselves naturally in some kind of steady state. However, many parks are fast becoming game concentration areas because of the pressure of human settlement outside park boundaries. Game herds which used to move in and out of parks are remaining inside, and more animals (especially elephants) previously living outside are taking refuge inside parks. When efficient protection against poaching is added to this trend, the balance may be disrupted between herbivores and their habitat. Some types of vegetation may be endangered by over-population of one or several game species and a reduction of these populations may become necessary. However, *the removal of animals should be used as a last resort, and only after it has been shown that habitat management measures, such as burning or protection from burning, have failed to stop the destruction of habitat.*

When reduction has to be undertaken, it should be done only after full explanation of the situation to the interested public and to conservation groups. Measures must also be taken to ensure that visitors are not personally aware of, or inconvenienced by, the shooting activity. Population management appears to many uninformed people as contradictory to the concept of conservation and national parks. Actually, it is sometimes essential to the maintenance of both animals and vegetation.

As mentioned elsewhere, if an uncontrolled population is increasing, as is commonly found in recently established national parks, it will eventually adjust to its environment without man's intervention. This applies also to areas in which animals have recently been introduced or reintroduced. However, this normally means that parts of the environment will have deteriorated in a striking way and of course numbers of animals will eventually be much lower than when at the peak of an eruptive oscillation. If these radical changes are consistent with park objectives there is no problem. But if the animal, or habitat, concerned is one of the main features of the park then special management measures should be initiated to achieve a stable environment; for example, with a certain desirable animal species remaining at a higher density than would have been the case if a different form of management had been applied.

Before such decisions are made, administrators should seek the advice of an ecologist skilled and experienced in integrating data from both plant and animal parts of the park ecosystem.

(b) Fire is an effective tool and its use is widespread, especially in the tropics. The use of fire as a management tool in national parks is unfortunately often a controversial subject. However, in many cases, resolving such controversies can be helped by recognizing the following two aspects of the problem.

First fire in itself is not necessarily a bad thing. After all it is a natural phenomenon and even in large unpopulated areas fires have occurred for thousands of years without having eliminated species of animals. Under tribal systems with low human populations, annual fires were deliberately set for hunting or other purposes but because both pastoralists and agriculturalists were normally shifting the frequency with which a given area was fired was comparatively low. As populations became larger and more sedentary and, consequently, as the use of the land became more intensive, the incidence of fires greatly increased. It is this too-frequent use of fire that presents the main concern over fire as a destroyer of certain wildlife habitats.

The second aspect requiring emphasis is that different intensities of fire have different effects under different ecological conditions; different vegetation types; differing rainfall; different parts of the same watershed, for example, low-lying moist areas compared with dry hillslopes. Experiments are needed in each region and in different vegetation types to find the most beneficial use of fire under the local conditions.

Fire has special importance in some national parks where it is used extensively to facilitate visitors seeing game. Some form of rotation system is usually essential to the wise use of fire. Burning can result in major changes if the same areas are burnt too frequently, and these changes are often detrimental. For example, in one of the oldest national parks in eastern Africa, fire was started annually in a tall grassland to allow better viewing by tourists. Twenty years later the grassland had been replaced by encroaching bush which by then was high enough to be equally effective in reducing visibility (see Figure 69(a)). At the same time the habitat had radically changed with consequent changes in the dependent animal populations (Riney, T., 1963a).

With the above in mind it is useful to have recorded on an annual basis the dates and extent of each fire occurring in the park. Annual fire maps can be most useful, both to park managers and planners.

Firebreaks are necessary for confining fires in chosen blocks, as well as to protect riverine forests and other fragile habitats, or to protect man-made structures. There is no fixed rule for standard width, for this is determined by local conditions. However, a road is rarely effective as a firebreak. In some areas a break of 25 yd will be effective; in other areas 150 yd on either side of the road may need to be burned or mowed to reduce the amount of inflammable vegetation. In forests, cuttings from road clearing or from thinings should be burned during periods when fire will not spread.

As mentioned above some kind of rotational system or burning policy is usually appropriate in a given national park at a certain stage of its development. A range of fire management possibilities that may apply either to all or part of a park are listed as:

1. *Accidental.* No deliberate burning; either
    (a) fires left to burn, or
    (b) fires put out as soon as possible after discovery.
2. *Deliberate burning;* either
    (a) early in dry season;
    (b) late in dry season before new growth starts;
    (c) at 2-, 3- or 4-year (or more) intervals;
    (d) annually;
    (e) as often as possible.

Each of these different approaches to burning will inevitably modify habitats in a different way and at a different rate than the others, and even the accidental fires result in a burning pattern which can be mapped and which may then be evaluated in terms of overall park objectives.

(c) Because many parks are becoming heavily populated with game and many are too small to contain all habitat elements required by animals, it is sometimes advisable to develop watering places or salt licks. Such developments can concentrate animals to provide better game viewing but they may create additional problems if concentrations of animals, especially at waterholes, are allowed to degrade the vegetation in surrounding areas. One solution to such problems is to create enough salt licks or waterholes to allow their rotational use, for example once every 3 or 4 years. Water development presupposes a general survey of the soil permeability throughout the area and of the underground water available.

(d) Once the land has been disturbed, the natural process of plant succession normally tends to take place towards the original climax vegetation type. If the area has been used for agriculture, natural recovery may take many decades or even centuries to reach a new climax. In such a situation it may be desirable to speed up the natural processes by planting indigenous plant species which would be expected to make their appearance much later.

(e) The introduction of any species of plant or animal into a national park should be completely prohibited, unless the species is known to have occurred previously in the area and to have been eliminated by human activity. In such a case, where the habitat is considered suitable, and the species is prevented from natural immigration by some environmental barrier, it is acceptable to reintroduce it.

Planting of exotic species for whatever reason is normally inconsistent with national park policies and objectives. For example, if it is desired to stabilize patches of accelerated erosion with colonizing plants, use the native colonizers rather than exotics.

(f) One practice in park management, which has been applied in many countries and especially in West Africa, consists of surrounding the park with controlled hunting areas. This practice serves three main purposes:

1. The controlled hunting areas act as a buffer zone between the completely protected park and the intensive land-use areas. While some land-use activities are allowed within buffer zones, these are regulated and hunting is strictly controlled.
2. The buffer zones allow animals to move more easily in and out of the park; this may facilitate the harvest, outside the park, of migratory game populations that need reduction for habitat protection and thus alleviate the necessity of making the population reductions within the park.
3. The buffer zones if controlled may also help prevent the frequent spread of fire from outside the park, which is common whenever intensive human activities are permitted right up to park boundaries.

When moving from the policy-making to the action phase of park management, conflict may develop between the goal of allowing natural interactions between animals and vegetation without human interference, on the one hand, and the need for active management to prevent major and irreversible changes of the environment, on the other. It is wise policy then to take active measures to change or influence biological habitats only after obtaining firm scientific evidence of their need and effectiveness, and after proper briefing of the interested public.

6360   Management Practices for Tourism

Apart from their role as conservation areas, the national parks are increasingly viewed, especially in developing countries, as an economic asset because of the welcome influx of foreign currency brought in by the growing international tourist industry. This creates another management problem. Tourists and visitors have to be introduced to the park in a way that will have minimum deleterious effect on the animals, the scenery and the natural habitats. The solutions to this problem will vary considerably from one park to another but certain accepted principles should be generally applied.

1. Since it is generally desirable to control the access and movement of visitors within a park, planning may include zoning by which certain areas are kept as natural reserves, others for intensive use (lodges, camping sites), and still others for game-viewing and sightseeing along motor or

foot trials. This makes it possible to assure the preservation of extensive tracts in an undisturbed state. The presence of dangerous animals may also compel park authorities to forbid visitors hiking, or even getting out of vehicles, for safety reasons.

2. The correct placement of roads and tracks is an important part of park planning. The planner should take into consideration the topography, the firmness of the soil, the aesthetic advantages of the area, the distibution of animals and other tourist attractions. The roads are best arranged in a circuit, passing through game feeding areas, reaching game watering points here and there but not following continuously along water courses where frequent disturbances may affect animals adversely. The roads should be good and safe, but not built to high-speed standards. A good system of direction signs is an essential.

3. Living accommodation must also be built for the visitors. This may be inside the park or immediately outside. It will consist of more less sophisticated lodges, cabins, restaurants or camping grounds. These structures should be as unobtrusive as possible and should blend with their surroundings. It is desirable to build with local materials or at least to create a local atmosphere by proper siting, external and internal design, and appropriate landscaping. People generally visit other countries to experience an environment which is different from their homelands, but good standards of comfort and food are usually expected. A high standard of cleanliness and hygiene is essential. Special care must be given to the maintenance of such standards. Where the park infrastructure (roads, hotels) is the responsibility of a separate government department (Tourism, Public Works), the location, maximum density and designs of structures should be controlled and approved by park authorities. Care must also be given to the housing of staff, which has often disrupted an otherwise well-planned ensemble.

4. Many government agencies do not wish to get involved in the administration of lodges and restaurants, and prefer to concede them to private concerns. The concessioner should earn a reasonable profit from his operation but his business must always be in harmony with what is considered appropriate for the area. He must work in close cooperation with the park authority. Rates for accommodation, facilities and services must be approved by the authority. Leases should be renewable subject to conforming to conditions clearly set out in the leasing contract. The sale of souvenirs is a common practice but should be limited to items of good taste and high intrinsic value, mostly handicraft merchandise.

5. Tree lookouts or blinds may be set up in choice locations, usually at game watering points, where visitors may be allowed to lie in wait for viewing or photographing wild animals without being seen.

6. Finally, countries sincerely interested in large-scale development of tourism should review their laws and regulations concerning customs and immigration. Custom and immigration requirements must be reduced to the minimum compatible with the national security of the country.

6370   Education in Parks

National parks have an important role to play in educating visitors in regard to the meanings and relationships of what they see. Hopefully, a visit to a park should heighten a visitor's interest and enjoyment in the natural environment; it should result in awareness of park purposes and policies, arouse a concern for conservation and, if wildlife is an important park feature, acquaint visitors with the needs and problems of wildlife management. This educational role is not easy to achieve but developing countries should make at least a start towards developing in their citizens a knowledge of the concepts and importance of conservation.

1. A first step would be to train game scouts and guides sufficiently to allow them to give intelligent and accurate answers to questions asked by visitors. All park employees should be capable of answering questions from persons who live in areas surrounding the park in so far as those questions may relate to the purpose of the park, the kinds of activities taking place there and the benefits derived from the park that accrue to the local community and to the nation.
2. A good map, and some literature pointing out the main features of the park, should be made available to visitors.
3. A striking and sad fact in the national parks of many developing countries is the absence of visitors from the local populations. The people may well be admitted free and guided tours organized to help them appreciate their national heritage; a heritage which they too often consider as exclusive for privileged foreigners. Examples of important population segments to appreciate and profit from this kind of activity would be schoolteachers, schoolchildren and local villagers. These should be systematically taken to the parks for guided tours. It is an obvious advantage to involve local people to serve as guides for such groups. In no circumstances should translators have to be used in interpreting 'their own park' to the people who live in or near the park.
4. As a park develops, there may be need for a more comprehensive interpretation programme consisting of: guided lecture tours, movies and slide talks for visitors in the evening, and an educational centre consisting of a small museum, conference and projection room, animal orphanage and the like. A good education programme in a national park cannot be developed overnight. However, even in the early stages of park management it should be kept in mind for parks can fulfil an essential role in public education, and the programmes must grow and become diversified as the developing needs arise.
5. One of the best ways of educating the public is simply the conduct of a competent, well-informed staff in a well-run national park. As K. R. Miller (1980) observes: 'The challenge for park departments is to identify innate managerial talent in staff members (and job candidates), and to provide

opportunities for training and educating all personnel in the skills necessary to fulfil park functions.'

6380 International Problems

International problems often arise when national parks meet across an international boundary or when the boundary of a national park coincides with a national frontier. Many problems can arise but they may solved by cooperation and coordination through regular consultations between neighbouring park authorities, or by the creation of a permanent commission for mutual collaboration. International agreements may take various forms, depending on circumstances, for example:

1. *For protection:* coordination of patrols along the common frontier; recognition of the right of guards to pursue poachers into a neighbouring country; use of customs officials to help the guards.
2. *For management:* collaboration for management measures including control of fire, or of plant and animal populations; enactment of common regulations; exchange of information on animal counts, statistics, research projects or exchange of publications concerning adjoining park.
3. *For tourism:* pooling of publicity campaigns, joint organization of tourist circuits, arrangements for a joint policy regarding maintenance of roads, simplification of frontier formalities.

Coordination will be much easier of both countries have agreed on common principles such as elaborated in Appendix 4.

6400 MANAGEMENT TO PRESERVE RARE SPECIES

6410 Introduction

First we need an attitude to rare and endangered animals. They certainly give prestige to the country that has them and looks after them. They are part of a long history of evolution and they add to the variety of wildlife that we have today. In a conservationist's view the complete range of wildlife species is as much a part of the resources of a nation as the cattle, the forests, the iron ore or the soil and water.

Since any stable ecosystem includes many interrelated and inter-dependent functioning parts, the loss of any of these may be in itself an indication of loss of stability. The extinction of a species of wildlife thus has somewhat broader implications than the fact of disappearance.

While these statements might carry little weight with a hungry man, nevertheless it is a hallmark of a civilized country that is can divert some of its resources to improving the quality and variety of people's lives. Variety of wildlife is a small but, to many, an important part of the variety and quality of human life.

One obvious conservation maxim is that no species should be allowed to become extinct without a serious attempt to save it.

6420 Causes of Rarity

There are three main ways by which animals can become rare and endangered: first by losing some essential component of the habitat, second by having too great a pressure placed upon them by predators or disease, and third as a result of widespread climatic change.

Climatic change usually works together with habitat change to make food, shelter or water scarce. These changes are slow, often in terms of centuries, in contrast to the short-term variations in climate resulting in several dry or several wet years in succession.

Wild predators rarely exterminate a prey species, for the simple reason that when the prey is scarce they find it inefficient to hunt and turn to another prey species. The exceptions to this generalization are few and usually exist only over a limited area and under most unusual circumstances.

Man, however, as a predator can exert an increasing pressure even on a declining population as a result of his intelligence and his tools for capture. Thus poaching and uncontrolled hunting for meat or other animal products can make animals rare.

Disease and parasites are unlikely to reduce a species to a remnant since most diseases eventually strike a balance with the host animals. There are exceptions: an introduced disease like rinderpest can seriously reduce the numbers of a species but since the proportion of animals killed falls as the animal become less frequent such diseases are unlikely to keep a large mammal species rare except on a local or temporary basis. However, when a rare animal shares a disease with a common animal species so that the disease is contracted around, say, waterholes, the rare species could be kept rare or become even scarcer; but this is speculation.

The widespread use of insecticides has made certain kinds of animals rare although not, so far as is known, large mammals. Certain of the chemicals are persistent in temperate zones though less so in the Tropics (e.g. DDT and dieldrin); they are passed from seed to herbivore and thence to the predator. Predatory birds are most affected and with the present heavy use of dieldrin in developing countries we may expect eventually to see more large predatory birds affected in developing countries. The difficulty with this cause of rarity is that it can only be detected and demonstrated by chemical tests.

When the effects of pesticides are suspected, government laboratories, universities or museums should be consulted and arrangements made for them to make appropriate tests. Meanwhile, to fully interpret these tests detailed information on past land-use in the area in question should be obtained and in particular the exact form of application and quantities of insecticide applied whenever poisoning is suspected.

On a worldwide scale by far the most usual cause of rarity is man's direct and indirect destruction of the habitat. Sometimes this is obvious, as when forests disappear; at other times it is more subtle and needs research to establish the basic cause.

The significance of the kaleidoscopic changes resulting from the impact of man on large African mammals has been summarized (Riney, T., 1963b) as follows:

1. As large-scale habitat changes take place, the habitat becomes less suitable for certain species and more suitable for others.
2. The habitat changes, therefore, normally result in trends toward certain species increasing their numbers while other species diminish.
3. Some species have become locally extinct as the habitat has completely disappeared.
4. Some species, having decreased in numbers because of diminishing habitat, are subsequently eliminated locally by hunting.
5. Certain species can be eliminated locally by hunting alone.
6. The modern way of eliminating animals is to combine habitat destruction with shooting, for this results in the greatest acceleration of the trend toward extinction.
7. Certain animals greatly increase locally following these large-scale environmental changes, and in Africa examples of such species are spring hare, spring buck, impala, blue wildebeest, buffalo, and, where hunting is not a significant factor, the elephant and black rhinoceros.
8. For most of these species and others not mentioned large increases such as have become apparent in recent years are largely in a transitory stage of adjustment and their present favourable numerical status is by no means a guarantee of future survival.

6430   Remedial Measures

In developing countries a common situation is to know that a species is rare. At the same time the reasons for its rarity may remain unknown, or unproven although they may be suspected. Some remedial action is needed and something must be done quickly.

Under these circumstances there is one obvious assumption that can be made: regardless of what the ideal habitat requirements are for this species, appropriate habitat necessary for its survival already exists in the area where it still remains, otherwise it would not be there. This assumption is important for it provides the basis for the first and second convincingly sound remedial measures. First, to set aside an area or areas where the animal is living, and second rigorously to protect it with enforceable legislation.

In framing the legislation, one needs to use all the scraps of information about the animal that one can get. Historical records may show a coincidence between decline of the species and a change in land-use. Thus one may include not only legislation against poaching, live capture for zoos, but also to control fire and/or grazing.

These emergency actions may give time for further study to pinpoint more precisely what the limiting factors are for this rare species. It would usually be wise to start with research on habitat needs. In this connection it is important to avoid a common misconception. Because a certain area is the only place in the world that holds a certain species this does not necessarily imply that there conditions are best for that species. One may be dealing with a relic species in an out-of-the-way place, already eliminated by man from its most favourable habitats. Because of this misconception it seems convenient to think of habitat requirements in two ways: minimum requirements for existence (although not defined, always present in areas where the species still occurs) and more optimum habitat requirements to provide an environment within which rare populations can develop a considerably higher rate of increase.

When some knowledge is in hand, more refined remedial measures might take the form of, for example:

1. The manipulation of the habitat in order to make it as productive for the species as possible.
2. The installation of dry-season water points, salt licks or both to reduce local movement.
3. The reduction of competitors for food.
4. The control of predators shown to be harmful to the population.

Wildlife knows no political boundaries and frequently the seasonal movements of a species take it from one country to another. International cooperation may then be vital to the survival of the species.

At the same time as over-hunting and habitat destruction are accelerating the process of extinction, another set of human activities is, for various species in widely separated parts of the world, resulting in a slowing-down of the extinction process, and trends are already under way which, when encouraged and further developed, should result in ensuring the safety of at least the major assemblages of large mammals as they are known today. Thus another way of discovering possible remedial measures is to review present favourable trends to see which of these, alone or in combination, is most relevant to the species in question (Riney, T., 1963b). Six trends already gaining momentum are:

1. The tendency for each country to establish wildlife organizations and to develop a system of parks and reserves in which the large mammals characteristic of the particular local environment are protected.
2. The development of relevant local research programmes.
3. The development of the commercial utilization of the more plentiful species of game as a form of land-use eminently suitable to local conditions and as a means of making additional protein and money available, particularly in areas marginal to present agricultural or pastoral practices.
4. The reintroduction of large mammals into appropriate habitat from which they have become locally extinct and in which they are protected and can increase.

5. A recent trend for members of associations of zoological parks to refuse to purchase wild-caught species which are removed illegally from the country of origin and subsequently offered for sale on the animal market.
6. The capture and transport of mammals to zoos and to restricted, carefully selected habitats outside of the country or continent of their occurrence as a final desperate measure for the prevention of extinction.

As one or more of these trends become locally effective the status of a rare or threatened species can be expected to improve. It is thus important to keep in touch with these developments, and to encourage them, to maintain an up-to-date view of the status of rare species in the country of your concern and, as new conditions dictate, to modify the extent to which we designate certain species as critical.

6440   Translocation of Rare Species

This subject is borderline to the present handbook and requires special expertise and close cooperation between wildlife field men and veterinarians.

Translocation of rare species has been used in two main ways: (1) to move individuals from one natural habitat to another, and (2) to extract animals from a natural population and place them under artificial conditions such as a zoo or special quarantine area with the intention of increasing the size of a nucleus population and with the eventual aim of returning such animals to suitable natural conditions.

Regardless of which way is chosen, and particularly in cases where little is known of the response of the species is question to tranquillizing drugs, it is important to have a qualified veterinarian involved at the earliest stages in developing an acceptable technique. An embarrassingly large number of wild animals have been killed by biologists who read the early descriptions of the technique in scientific publications and who then obtained drugs and equipment and blundered ahead without the adequate technical background. In the interest of reducing unnecessary mortality of rare species it is also strongly recommended that biologists with little or no experience with immobilizing drugs develop their expertise under a qualified veterinarian with sucessful experience in immobilizing animals.

The localities to which the breeding nuclei are taken must be selected with care to prevent losses. If possible there should be more than one site in order to provide a margin of safety in case of accident or disease. Many zoos are willing to assist with the establishment of breeding nuclei, and zoos often cooperate in this. Only reputable zoos should be considered.

Once a sufficiently stable and viable nucleus of breeding animals has been established, it may be possible to rehabilitate a number of individuals into their natural habitat (see also Section 8332 dealing with the introduction of animals). Animals reared in captivity do not always adapt easily to natural conditions. Breeding-up in captivity should therefore be regarded as a last resort in the protection of a rare species.

6500   MANAGEMENT FOR THE PRODUCTION OF MEAT

6510   Introduction

Meat production may be based on a short-term policy of removing over-populations of animals, for example a population in an advanced stage of a population eruption. In harvesting such populations some countries have been concerned not so much with making a profit as in reducing the population before large-scale natural mortality could take place and selling the meat locally at a price sufficient to cover the cost of the harvesting operations.

Normally, however, a cropping scheme involves harvesting animals on a sustained-yield basis. This may take place in several ways: by traditional local hunting for meat, through sport hunting or by means of some kind of planned cropping scheme.

Clearly, the details of management will vary greatly with the nature of the cropping scheme. However, the general scheme that follows is easily modified to meet all situations. Harvesting techniques are discussed in Section 7400.

6520   Feasibility Survey

A list of questions and information needed in making such surveys follows.

Question	*Information needed*
How much meat can be cropped?	(a) Species suitable for cropping based on numbers, availability, palatability, etc.
	(b) Estimate of numbers of each suitable species (Section 3421 and Chapter 7000).
	(c) An estimate of the annual production of each suitable species (Chapter 7000).
	(d) Practical cropping methods available (Section 7500).
What technical staff needed?	Biologist(s), field officer, field technician (see Chapters 7000 and 9000).
What equipment and facilities are needed for cropping?	Depending on size and type of wildlife, guns, ammunition, transport (see Section 7500), equipment for handling and dressing carcasses; abattoire; access roads.
Nature of the market for meat?	Local retailer; to railhead; to canning factory; local storage

	facilities; limitations imposed by regulations on meat inspection.
Size of the market and possibility of expansion?	Depends on roads and distances. Other uses of wildlife—horns, skins, etc.
Administration of scheme	See Chapter 9000.
Profit	Costs of administration, all salaries, use of equipment and transport to be placed against likely return for meat (and other commodities). If profit too low, try to cut costs or improve market.
Kinds of problems associated with cropping schemes	Parasites or diseases: (a) affecting human consumption (b) spreading to or from domestic animals; Conflicts with other land-users. Political pressures, vested interests. Meat-processing problems. Marketing and transportation problems.

6530   Organization of a Cropping Scheme

A number of different kinds of people must cooperate closely together to achieve a cropping scheme successful in producing a hygienic product to a suitable market and at a reasonable profit. Successful schemes have been operated by governments alone or by subcontracting to specially licensed private operators with the government maintaining supervising control of numbers and quality of the finished product. Appendix 5 reviews one set of conditions laid down by a Public Health Department in Africa and may serve as a checklist of points to consider in developing national requirements for preparation of dried or fresh meat taken from wildlife.

6600   ANIMAL REDUCTION OR ELIMINATION CAMPAIGNS

6610   Introduction

A special kind of management is required for areas where the aim is to reduce or eliminate animal numbers. If the area is small this is comparatively simple. However, if the area is large, in terms of hundreds or thousands of square kilometres, some aspects of management seem worth special consideration.

Generally three kinds of situation have provided motivation for developing control or elimination programmes. The first situation has involved large tracts or developing lands which form improving habitat for large animals which are consequently increasing and which are incompatible with the developing form of land-use. Deer in planted forest sometimes come in this category. A variation of this first situation is when large mammals have been introduced into an area and come in conflict with existing or developing forms of land-use, for example, following the introduction of deer into New Zealand between the mid-nineteenth and early twentieth centuries. The second type of situation is where land is being cleared of large mammals because they are incompatible with an intended form of land-use, as when large mammals are shot in forested areas in Africa to assist in reducing the incidence of disease borne by tsetse fly. The third situation involves simple predation, as when large predators take excessive numbers of domestic stock or become man-eaters. The first two situations are discussed below as another kind of management, one in which a government department employs shooters for the purpose of eliminating animals.

In each of these situations, before starting operations, priority questions are: (1) to what extent is the problem real, and (2) is the solution proposed appropriate?

Regardless of whether the solution is valid[1] (e.g. if shooting is required) or not, it seems important to briefly discuss control activities because it requires a reverse kind of thinking to normal wildlife management procedures, as when the objectives and management practices are designed to produce optimum yield. In seeking control or elimination we look for weak spots in habitat or population so we can make them weaker. We do everything we can to encourage a trend whereby population numbers become smaller and smaller. We apply every art at our command either to bring the animal toward local extinction or to hold populations at low levels.

Another reason for including a discussion of control operations and of data required to understand their impact on the animal populations is that some of the largest-scale and most expensive control operations ever developed have not only failed to control the populations concerned but have produced some of the highest sustained-yield figures available from wild populations of large mammals. Some control campaigns have thus provided good evidence of the tremendous resiliency of natural populations to harassment by man.

6620 Basic Elements in Management

The basic elements in management for reduction or elimination are the shooter, his techniques and the area shot. These elements are discussed below with data that are required for management purposes. Also mentioned below

[1]See also Section 8400, dealing with politics and wildlife.

are additional data which may be useful under various changing objectives or circumstances.

6621 *The Shooter and his Techniques*

In all occupations men can be very good or very poor at their job, and hunters are no exception. The kind of hunting system used, the results a given hunter gets by applying this system and the area over which he hunts combine to give an idea of what may be recognized as the quality of hunting pressure.

The essential job of a hunter is to find an animal within its home range and eliminate it, then find another and so on. Converted to terms of this handbook it is the 'animal–occurrence' part of the ecosystem with which hunters are involved. A knowledge of a normal size of home range within the shooting area will obviously be an advantage to the hunter; also useful would be knowledge of daily and seasonal movements, breeding behaviour, habitat preferences (including preferences for feeding, resting sheltering, hiding) and reactions to other species of animal also occurring in the block. This knowledge, along with his own experience and with instructions from the control organization, will combine and result in his own individual method and pattern of hunting.

Even if shooting alone were concerned there are so many variations in specific methods and patterns as to make it necessary to precisely exemplify the actual procedure involved: for example, consistently hunting either up rather than down a hillslope; hunting towards the headwaters, or, alternatively, down the drainage; moving fairly rapidly along rather open ridges, or covering carefully all slopes and exposures, even thickets; times of hunting, either in the day or at night; hunting from vehicle, horse or on foot; hunting alone, in pairs, or as part of a team effort, as through drives; the calibre of rifle and whether a telescopic sight or a silencer was used. The above are mentioned as examples of variations with which I have personally observed resulting differences, either in terms of greater or lesser numbers or in different age and sex ratios being shot, or both.

Changes in method and pattern of hunting will affect two basically different kinds of results: the *number* of animals shot and the *proportion* of the existing animals that are eliminated. Failure to appreciate the reality of these differences can be expected to lead to failure in effective control. In terms more understandable to the hunter, there is a temptation to feel successful when after much work one's weekly results drop by half, or lower to a tenth, or so low that it is no longer profitable to hunt an area. No-one would question that the numbers of animals killed are right, but it would be wrong to conclude from this that the hunting had therefore reduced the population by half or 90 per cent. Another kind of cross-checking evidence would normally be required (for example, see Section 3423D). The usual reason for such a discrepancy is the operation of the threshold of visibility as described in Section 3426B.

As noted elsewhere wariness is one factor in determining the level of the threshold of visibility and in explaining changes in threshold levels following hunting it is a major factor. Fortunately, there is an easily obtained index to wariness, *the flushing distance:* the distance at which an animal no longer stands and watches a human intruder but turns and moves away. Although not perfect the flushing distance may serve as a convenient index to the threshold of visiblity in hunted areas for, other things being equal, the further the flushing distance the higher the threshold of visibility; the shorter the flushing distance the lower the threshold.

In recording flushing distances one must keep separate records for observations made from each type of vehicle while travelling at a standard speed (e.g. truck, Land Rover, private car, motor bike) and from those taken on foot, on horseback, elephant back, etc.

Wariness, however, is but one factor determining the threshold of visibility. In managing hunting schemes the practical relevance of this discussion is to learn if such a threshold is in any way interfering with developing maximum efficiency of operations. Demonstrating that such a threshold is strongly in operation provides one useful basis for improving the effectiveness of control operations and these two activities may be combined. A simple way of making such a demonstration is as follows:

1. Carefully and precisely describe the hunting method.
2. Within a given hunting block, record the average number of animals shot per man-week for the previous 3 or 4 weeks.
3. Change the hunting pattern. For example if night hunting has been from dusk to 9 p.m., try midnight to 3 or 4 a.m., change the pattern of hunting routes, the directions of travel, etc.
4. If the change in hunting pattern results in an increase in number of animals taken in the same area, continue until numbers per man-week drop to the level of (2) above, after which even furthur changes may be indicated.

If numbers increase after the initial change in hunting pattern and technique this in itself demonstrates not only that improvements can be made but also that a threshold of visibility has been in operation: the animals having become more wary and having adapted their daily movements and responses to the earlier method of hunting.

6621A   *Measurement and Recording of Hunting Effort*

Hunting effort should be recorded as man-days or man-weeks spent in poisoning, shooting, snaring or trapping or some combination of these.

As it is difficult to separate time spent in hunting from time on the area, the total time on the area should be recorded in man-days, for man-days in the catchment are conveniently checked with past records where such records are available.

Field supervisors as well as shooters should be included in the total man-days when they are in the field, for they are in fact part of the total man effort. Track-cutters, hut-builders (or days spent track-cutting or hut-building) should not be included in the manpower figures representing hunting effort as this would not only give a misleading idea of effort spent but would increase the difficulty of comparisons in future years. Although packers' time should be included in the total cost of operations, it should not be included as part of the hunting effort.

6622   *The Shooting Block*

The second major element in control operations is the 'shooting block'. The shooting block constitutes the basic unit area within which shooters hunt and kill animals. This is the basic division of past and present records of control operations.

In terms of control operations shooting blocks have in several countries become counterproductive because they have been periodically modified in size or because of the way in which they have been used. Shooting blocks can only be regarded as unsatisfactory from an administrative point of view if their boundaries are not fixed but keep changing. This is often a natural consequence following from a misunderstanding of the concept of 'shooting block' and of the way in which shooting should be conducted and records of shooting must be maintained.

A review of some of the main factors which tend to modify the boundaries of these blocks may be useful as one means of profiting from other's past experience.

6622A   *Influence of Bounty Systems and other Factors on 'Shooting Block' Boundaries*

It is appropriate to mention bounty systems as a common past and present way of controlling animals, not as a means of evaluating bounty systems as a method of control but to point out their influence on the relation between the shooter and the area alloted to him to shoot (the 'shooting block').

'Bounty' may be defined as 'a reward, premium, or subsidy, especially one offered or given by a government specifically as a recompense for the destruction of noxious animals' (Merriam-Webster, A., 1946). Bounty systems may include systems known as 'contract shooting', 'skin shooting by government hunters', 'bonus shooting', 'payment for tokens', and 'high-tally system' (that is any payment which depends on numbers of animals shot).

It is useful to recognize all these systems of shooting as bounty systems for each operates on a population according to the same pattern: when numbers of animals taken decrease, the hunting effort generally drops due to a re-channelling of shooting effort to higher-tally-producing areas.

The automatic easing-off of hunting effort, when the tallies become increasingly hard to get, can have, and has had, an effect on the organization of shooting activities in a given area. Thus the problem of the maintenance of a minimum number of tallies can effect the size of the 'block' and has in the past affected even the meaning of the term. 'Shooting block' in some areas has, in the mind of the shooter, come to mean 'that area assigned to me for my (or our) exclusive use in which we are free to shoot as many animals (as high tallies) as we possibly can'. If tallies drop off the shooter's demand is for a return to the higher rate of pay that he enjoyed before the tallies dropped. This is accomplished by raising the bonus modifying his block boundary, or by shifting to another block, or for an increase in the size of the block to enable high tallies to continue. The poaching of animals outside an assigned block is a natural and inevitable symptom of this urge on the part of the average shooter to produce high tallies.

It can thus be seen that the bounty system in all its forms sets the stage for pressure to be exerted from time to time to change block boundaries. This general factor has been elaborated only to serve as an example of one of many factors which have appeared or which may appear to change the opinion of the administrative personnel whose responsibility is the drawing of the most appropriate 'shooting block' boundary lines.

Examples of other factors which may effect 'shooting block' boundaries are: changes in shooting techniques; changes in administration of control activities from one government department to another, which often involves new control policies or practices and either some shifting or major re-organization of 'shooting block' boundaries[1]; and changes due to shortage of shooters by consolidation of two or more blocks.

New blocks have also been created when, for example, pressure outside the game departments requires new blocks to be established where none existed before; when an increased financial grant is made to a shooting district; when inspection or surveys within the operational department have suggested the desirability of creating a new block; by the introduction of superior equipment which sometimes facilitates the creation of new 'shooting blocks' (e.g. air drops of supplies); or by an increase in available manpower.

The process of shifting emphasis between one factor or another as described above helps in building up a dynamic and ever-changing view of the inter-relations existing between (1) the current climate of opinion toward animal problems, (2) the existing shooting effort applied on problem lands, (3) results in terms of numbers of animals killed, and (4) other considerations arising from differences in the ownership, control, and/or use of the land. There is an obvious need for records of shooting activites to be so designed

---

[1]Shifting a responsibility from one organization to another is normally accompanied by strong feelings being aroused, and a common reaction of a newly responsible organization is to sweep away the practices of the former organization and start afresh. Re-designating shooting blocks is an obvious and easy change to make, and this is often done, thus making it impossible to compare new with old results.

that they will stand valid in the face of changing administrative or political decisions (Riney, T., 1957c).

6622B *The Area Shot*

As with the size of controlled hunting areas (Section 6200) in most countries there can be no standard or uniform size; but a brief mention of principle and two examples may help emphasize the need for boundaries of shooting blocks to remain unchanged.

A suggested principle to follow is that the main point of the records is to get as specific a location of each kill as possible. Desired are some unit areas or graded series of areas which will be convenient for recording shooting activities and which can continue unchanged for years to come, to facilitate comparison of today's records with future records. In consideration for the practical demands of administration, use may be made of larger categories but the more specific information should be available when needed.

6622B-1 *Two examples: plantation forestry and wild land*

Plantation forestry

In man-made forests the location of animals shot should pose no problem for the entire forest will be laid out into well-defined blocks and compartments. Care must of course be taken to ensure that the designation of blocks remains the same from year to year although, for forestry or other purposes, the names of the same blocks may have changed. Shooters must be aware of the need for, and trained in the practice of, accurately locating and recording all kills.

Wild lands

In indigenous forests where selective felling is practised, in national parks, tsetse control areas, or unfenced open-grazing areas the fixing of shooting blocks can still be simple, as when the control area has small, well-defined watersheds. In such cases the watershed and shooting block boundaries might well be the same.

In other areas, however, where drainages are large and ill-defined, devising appropriate boundaries can be much more complex and is often difficult. Under these circumstances, wherever possible use natural features (e.g. a river, a lake's edge, a ridge) for one or more boundaries. Some man-made features (e.g. roads, railway lines, power lines) may be used. In selecting boundaries the main guideline should be the extent to which the feature or landmark considered is permanent and easily recognized by shooters.

Sometimes compromises will have to be made and an understanding of the quality of the hunting pressure defined in a rather different way. For example, in one tsetse control area in southern Africa, a hunting zone of 1700 square

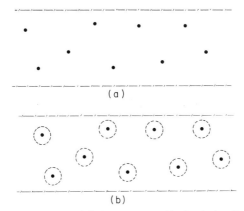

Figure 124   Diagrams (a) and (b) illustrate a kind of detail often necessary to improve understanding of the 'quality' of hunting pressure. The perspective of the annual kill based on an understanding represented by (a) may associate a total number of animals killed with a number of fixed hunting camps (black circles) scattered throughout the total control area. The perspective shown in (b) includes additionally the knowledge that hunters only travelled about 2 miles from their camps, which gave each camp a potential influence over but 12.5 square miles. Thus most of the control area remained entirely without control

miles was clearly defined by 100-mile-long parallel fences spaced 17 miles apart. The control zone was unfenced at either end. Within the area were many hunter's camps, the camps being clearly defined and fixed during the period of the campaign. In this example numbers of hunters and numbers of kills were available for the area and during a 2-year study period detailed figures were available for each camp, whose locations are shown in Figure 124 (a). But this did not give the required degree of understanding of hunting pressure. This was finally achieved only after visiting each camp and observing the precise methods and patterns of hunting. Of particular interest was the size of the area actually covered. By travelling with the hunters on daily hunts, and by transporting them away from their camps into areas unfamiliar to them, it was estimated that the area effectively covered by one camp extended only about 2 miles from each camp. This gives an entirely different basis for interpreting the annual records of kills, as shown in Figure 124 (b).

6623   Minimum Data Required from Government-Organized Shooting Operations

In principle such facts as come in from the field should reveal what has happened in the way of operational work and this information should be of such a nature as to permit study by those interested in evaluating the effect of control operations on the animal population. Specifically, this means that records for each shooter should include: (1) how many rounds of ammunition were used, (2) to kill how many animals, (3) of what species, (4) in which clearly defined shooting blocks, (5) over what period of time. Clear records covering these five points should facilitate the assessment of the cost of operations. These records constitute the minimum requirements for recording control activities.

Such records will produce the following information on a block by block basis, for a given species and season or year: (1) number of animals killed, (2) number of man-days hunting, (3) number of animals killed per man-day or man-week. The most important result is to have records of the number of kills per man-day or week of effort and within a clearly defined area which remains consistently defined from year to year. Only on this basis can useful comparisons be made.

The kill figure constitues the official record of animals killed in the shooting blocks. If extra allowances are permitted, as for those animals shot but not found, or wounded but which got away, then these assumed tallies should not be counted into the official records for the area. If maximum dependability of results is desired, these categories should be kept separate and under no circumstances should they be included in total kill figures. Likewise some precaution should be taken to preclude the possibility of counting the same animals twice, as in the case of supervising field officers splitting their own kills with the men on whose block they are shooting and the same time entering the same tallies under their own name.

Separate sets of records should be kept for each different kind of technique used for removing animals; for example, shooting, snaring, poison, experimental shooting, or some combination of these or others. Within these general categories, if a different method or pattern of shooting is employed or a different poison or method of poisoning used, for example, these records must also be separated: otherwise the changes in the numbers of animals killed would be difficult if not impossible to interpret.

The above section has concerned itself with obtaining reliable records of shooting activities, which is only one part of the information required to assess the extent that control operations are being sucessful. The second kind of information is related to the basic problem which originally stimulated the decision to control or eliminate animals.

6624 *Minimum Data Required to Evaluate the Effectiveness of Control Operations*

There is always a reason for control operations. Sometimes the reason may be more political than biological (see Section 8400), or there may be a valid ecological basis for control. If the reason is valid this should be the basis for selecting the most relevant kind of data required for assessing the effectiveness of control operations. By way of example:

1. In a forest, if it has been demonstrated that trees will not regenerate due to the existence of a certain density of large mammal, then regeneration must

be demonstrably associated with reduction in animal numbers before the operation can be regarded as successful. Figures 137 and 138 on pp. 446 and 450–451 illustrate how the elements constituting such a problem and elements of its solution may be related diagrammatically.

2. If concern is over damage to tree forms then the damage must be clearly defined, criteria should be set for evaluating the damage and acceptable levels of damage should be defined and determined. The operations can then be judged as successful when lower animal density lowers the occurrence of damaged trees to levels accepted by the foresters concerned.

3. If large mammals are being shot to eliminate tsetse fly and the incidence of the fly-borne Trypanosomiasis in cattle, then success is measured in terms of lowering the incidence of the disease in cattle down to acceptable levels.

4. If a species of mammal is being reduced because its present density is considered to be responsible for decreasing vegetative cover and/or increasing accelerated erosion, then, for the operation to be successful, a reversal of these trends in vegetation and soil must be demonstrated.

Although implied in the above four examples it should again be emphasized that in control operations success cannot be measured entirely by the numbers of animals killed or even by a proportional drop in numbers.

Chapter 7000

Techniques of harvesting large mammals

7001 INTRODUCTION

This chapter examines the concepts of carrying capacity, sustained yields and annual harvests and indicates several methods by which harvests may be estimated. Then various techniques of harvesting are presented.

Only a few techniques for determining harvesting levels have been described. Their selection has been based on both the need to reduce field data required to a minimum, and on the need to use evidence that can at present most reliably or easily be obtained under field conditions in developing countries. Techniques for ageing animals beyond their initial period of growth are not emphasized, for example, and indices of various kinds are favoured. For more complex ways of assessing allowable sustained yield harvests, see G. Caughley (1976 and 1977).

7002 SUSTAINED YIELD

Methods used, in the study of sustained yield, can be highly mathematical. Here, however, we are concerned with only a few basic ideas and their application in the field.

Wildlife can be used to provide meat, hides, trophies and sport. There is no difference in principle between the cropping of a population as a commercial undertaking and the cropping of a population by sportsmen. The aim in each case is to harvest some proportion of animals each year without driving the population into decline; this number is a *sustained yield*.

Sustained yield also implies a balance of the population with its habitat.

7003   POPULATION GROWTH AND STABILITY

Carrying capacity

A small population of animals introduced into a habitat that will support more, increases at first slowly, then with gathering momentum and then more slowly as checks to increase make themselves felt (Figure 125). Finally the population levels off, so that the annual production of young is balanced by the combination of mortality and emigration. This level (K in Figure 125) is often referred to as the potential *carrying capacity* or *ecological carrying capacity* of the habitat.

Figure 125 is, of course much oversimplified. It does not indicate the inevitable decrease in those parts of the habitat which are used by the growing population. While it may accurately reflect a trend in a population of rapidly breeding animals, with naturally high mortality like insects, or small mammals, I know of no species of large mammal whose increase follows a similar pattern. Figure 125 is simply a diagrammatic illustration of the idea of a population adjusting to some unknown ecological carrying capacity (K).

The main reason for the difference in the pattern of increase between small mammals (or insects) and large mammals is that the latter have lower mortality, comparatively few offspring per female per year and greater time is needed for the young animals to grow to breeding age. Once started, these factors combine to produce an upsurge of population that soon becomes heavily biased in favour of younger more productive animals. Quick breeders with naturally heavy mortality commonly stabilize their numbers, as soon as the ecological carrying capacity of the environment is reached, as in Figure 125. However, in large mammals productivity is maintained well beyond the level where it can be sustained and eventually a crash in population occurs (Figure 126). This lag in time between cause and effect is so much more pronounced in large mammals it seems worth separate discussion.

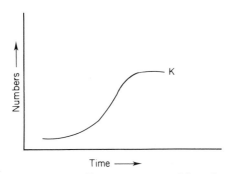

Figure 125   A diagrammatic and (for large mammals) a theoretical picture of a population adjusting to a potential or ecological carrying capacity (K)

368

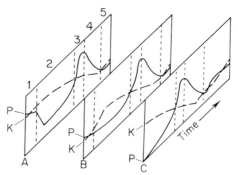

Figure 126 The two main ways in which eruptions start in established ungulate populations: (A) following a sudden decrease in a population, as from a serious disease, following the termination of unsuccessful extermination campaigns or sudden cessation of heavy private hunting pressure; and (B) an increase in the potential (ecological) carrying capacity of the environment, as for example following logging operations, over-grazing by cattle, changes wrought by fire, etc. For comparison (C) represents a typical eruptive curve for an introduced large mammal species. Vertical dashed lines indicate the potential carrying capacity of the environment; solid lines, the animal population; vertical dotted lines, phases comparable to those numbered in (A) (modified from Riney, T., 1964)

For large mammals, to put carrying capacity into a more useful perspective, it is thus useful to consider a series of potentially changing relations between an animal population and its habitat, known as an eruptive oscillation.

7100   ERUPTIVE OSCILLATIONS

Introduced populations of large herbivores, if undisturbed, normally follow a pattern of adjustment to the new environment which consists of a single major eruptive oscillation, the early parts of which parallel Figure 125. A knowledge of this series of sequential relationships between animal and habitat and a recognition of the phase that characterizes one's study population will assist in properly evaluating its status and in planning for appropriate management. An understanding of this potential sequence of events is an important pre-requisite to either the design of utilization or control schemes, or to an unbiased assessment of the extent to which the animals have or may become problem animals on a permanent basis. Likewise, an understanding of the extent to which indigenous ungulate populations are either stable, in one of the phases of an eruptive oscillation, or have stabilized following such an oscillation, is important for understanding the full significance of their present status.

'Balances' between animals and their environment rarely remain still; they are in fact quite normally changing. Even those large mammal populations which are considered comparatively stable are in fact responding to annual changes in the character of the physical and biotic environment. In the literature this uneasy balance may be referred to as an approximately 'steady state' or as a state of 'dynamic equilibrium'.

Such normal yearly fluctuations in numbers may involve 5–10 per cent of the population or more. But if there is a large enough discrepancy between animals and habitat, in favour of the habitat, then populations of large mammals, in the absence of appropriate management, can be expected to proceed through an eruptive oscillation as shown in Figure 126.

In the case of already established populations, this discrepancy can arise in two ways. First, an environment may rapidly become suitable for an eruptive oscillation following any kind of manipulation by man which results in a rapid and massive increase in suitable habitat for the species in question. Typical examples are logging operations, over-grazing, or changes wrought by fire (Figure 126B).

A second way of stimulating an oscillation is when numbers of animals are quickly and significantly reduced, as through serious disease such as rinderpest. The release of a heavy cause of attrition is equally effective, as following cessation of heavy private hunting by prohibiting all hunting, or following unsuccessful extermination campaigns (Figure 126A).

A third situation involving such a discrepancy is when animals are introduced or reintroduced into a suitable environment. Thus introductions of large mammals are normally followed by an initial eruptive oscillation (Figure 126C). Likewise natural dispersal of large mammals into newly created habitat also produces oscillations. These in turn can lead to serious problems as when large-scale forest planting takes place over grassland or moorland, and the new forest areas become over-populated starting from a nucleus of naturally dispersing deer (see also Section 7110).

Once an eruptive oscillation starts it runs a course that, if undisturbed, is predictable in general terms, as shown in Figure 126.

In principle, the greater the suitability of the empty habitat for the species in question, the higher the initial rate of population increase and the more violent the initial eruptive oscillation. At the other extreme, there are rare examples of introductions of ungulates where, without significant hunting pressure, introductions have not led to eruptions. One such example was the introduction of moose (*Alces americana*) into the south-west part of the South Island of New Zealand. This was such poor moose habitat (the potential carrying capacity so low) that the discrepancy between animals and environment, even on the date of introduction, was too slight to start an eruptive oscillation. The very sparse remaining population is fortunate to have survived at all.

The numbers above graph (A) in Figure 126 indicate five phases of an eruptive oscillation. These five possible relationships between large mammals and their habitat are described as follows: Paragraph numbers indicate stages shown in Figure 126.

(1) A situation preceding an eruptive oscillation is indicated by a popu-

lation in some kind of stable relation with its environment. Environment is meant to include the sum of all the forces acting on a population at a given time, and this may include hunting at a constant yearly rate. No large-scale recent change will have occurred in the numbers of animals removed yearly from the population and the main habitat elements observed in the key areas at the most critical season of the year will show no distinct recent upward or downward trends. Obviously, if a newly colonizing population of large mammals is considered this pre-eruptive relationship will be missing (Figure 126C).

(2) This represents the initial phase of population growth in a favourable environment, which can support many more animals than are present. The rate of increase is comparatively high, partly because of higher productivity but mainly because mortality is lower (Caughley, G., 1970) and, as new generations reproduce, the population curve moves steeply upwards. The trends in vegetation in critical parts of the habitat decline, gradually at first and then at an increasing rate: in terms of this handbook the status of the relevant habitats is becoming lower as downgrading trends continue. The physical condition of the population varies between 'fair' and 'poor', during the most critical seasons (see Section 3864).

(3)   This is the period of most rapid population increase. As thresholds of visibility are passed, increases in numbers of animals seen become obvious to most people associated with the area. This is also the period of most rapid decline in habitat, the status of which becomes lower and continues to downgrade. Although the mortality rate increases in this period, the total population continues to rise because of the high proportion of young breeding animals carried forward from phase (2). The physical condition of individuals drops noticeably and in the latter part of phase (3), as the population approaches its crash point, most animals are in 'fair' to 'poor' classes, even in the most favourable seasons.

(4) The start of this phase is characterized by one or more years of exceptionally heavy mortality. Population crashes occur particularly in years when some element of the habitat becomes critical, as in drought years or, in temperate regions, in unusually severe winters. The pattern of mortality shown in Figure 126 indicates a marked decline over a few years; however, the pattern of decline may vary. The crash may involve only one severe season, may decrease by several drastic steps in successive years, or the declining steps may be interrupted by one or two years of favourable climatic conditions. Regardless of the pattern involved, this crash period is usually easily identified by historical evidence: starting in the year when an unusually high number of dead animals were observed. If population indices are available large reductions in numbers will easily be demonstrated for mortality commonly exceeds 30 per cent. Both historical evidence and population estimates may be missing if the crash took place in a remote area or before the memory of present residents. In such cases various observations of the status and trends of key plant species can prove useful. For example, recent marked

drops in the percentage of plant utilization may be observed. Or, for older trees, a sample showing the numbers of years ago terminal shoots of trees or shrubs broke away from the apex of a distinct hedge form (see Figure 53) may assist in dating a sudden past release in animal use. The physical condition of the population will be at its lowest ebb during the early part of the crash period but improving in the latter part of this phase. A typical syndrome (collection of different kinds of evidence) in the latter part of this phase may include, for example: historical evidence of a large die-off a few years previously and some sign of both the physical condition of animals and key species of vegetation starting to recover. If comparative observations are available, some evidence of recovery may be demonstrated within the first year following a crash.

(5) The final phase occurs when the population reaches some new level of adjustment with the modified vegetation. The habitat will still show signs of past heavy use (i.e., its 'status' may be low) but will exhibit recent trends toward recovery. Mortality and physical condition will be better than in phase (3) and (4) and similar to (1), or the early part of (2). The population density is usually lower following an eruptive oscillation than in any previous phase, except in the early phase of an animal introduction. So drastic and memorable are the die-offs which introduce phase (4) that this historical evidence is *usually* easily obtained 10 or more years later.

Once the complete adjustment has taken place, ecologically the introduced population is the same as any other ungulate population in that further major eruptive oscillations can take place only by the creation, rather quickly, of a big discrepancy between the existing population and the potential carrying capacity of the environment.

7110   Initial Oscillations and the Spread of Large Introduced Species

Figure 127 illustrates how a wave of subsequent eruptive oscillations may spread away from the location of an introduction. The graph (A) in Figure 127 indicates an initial eruptive oscillation for an introduced ungulate and extends into a period of post-eruptive stability. The graph shows a situation described for introduced red deer (*Cervus elaphus*) and several other species introduced into New Zealand (Riney, T. *et al.*, 1959 and Riney, T., 1964). Graphs (B), (C) and (D) represent exactly the same process occurring in similar habitat, but at later dates as the nucleus population disperses into areas progressively more remote from the liberation site, and (E) represents a still more remote area not yet reached by the dispersing population but equally capable of supporting the same type of adjustment as it also represents (A) before the initial liberation. Thus different stages of an eruptive oscillation may simultaneously exist at a given time at different distances from the site of an introduction.

Although these above-mentioned studies involved several different species

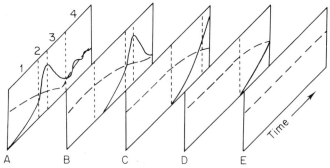

Figure 127 A typical pattern of population responses P number of years following the original liberation at A. The graph (A) indicates an initial eruptive oscillation and extends into a period of post-eruptive stability. Graphs (B), (C) and (D) represent exactly the same process occurring in similar habitat but at later dates as the nucleus population disperses into areas progressively more remote from the liberation site, and (E) represents a still more remote area not yet reached by the dispersing population but equally capable of supporting the same type of adjustment as it also represents (A) before the initial liberation (modified from Riney, T., 1964). (Dashed line indicates carrying capacity, solid line shows population responses)

and different research objectives, in no instance was I able to observe a broad range of animal responses in a single concentrated study. Fortunately, this has been done by G. Caughley (1970) who made a study of Himalayan thar in New Zealand. By selecting a series of different populations of introduced thar, each in a different phase of an initial eruptive oscillation, Caughley was able to piece together a convincing picture of changes in population response following the initial liberation.

Figure 128 represents changes in Caughley's thar populations after liberation. In these areas the most critical time of year is in winter and comparison of the decline in winter food supply as the density of the population increased gives a perspective similar to that shown in Figure 126. As one might predict, rate of increase, fat reserves, juvenile fecundity and the median age of adult deaths also declined. Both juvenile and over-all mortality rates increased. In this study population birth rates remained about the same. Caughey concluded that the change from positive to zero rates of increase was brought about mainly by a rise in death rate. By comparison, changes in birth rate seemed of minor importance.

7120   The Eruptive Oscillation and Management

It should be clear that populations can be maintained on a permanent basis at any point at or below its potential (ecological) carrying capacity. This levelling-off is usually achieved by harvesting. When an annual harvest continues in a more or less steady state of adjustment with its habitat it is referred to as a *sustained yield*.

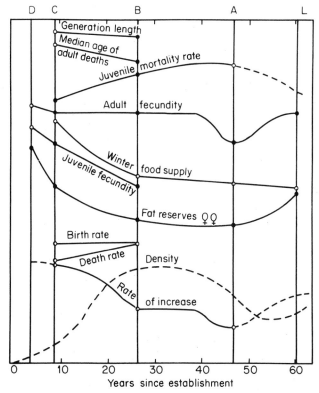

Figure 128 Schematic representation of changes in a thar population and its food supply during an eruptive oscillation following liberation. Samples were taken near the area of liberation 60 years before (L), at varying distances away (A, B, C) and from a most recently colonized area (D), furthest from the site of liberation. Solid circles indicate reasonably accurate estimates, open circles are presented with lesser confidence and broken lines are inferred from the trend of the other statistics (modified after Caughley, G., 1951)

Eruptive oscillations demonstrate the tremendous vitality of the living systems with which we deal. Even with large mammals there is a strong tendency for populations to increase up to the level which the environment can support. As mentioned elsewhere, so strong is this tendency that examples can be found where, during a period when habitat was improving, populations of large mammals increased greatly in spite of considerable government-financed campaigns to exterminate them. On the other hand, if suitable habitat is shrinking, no amount of manipulation of the animal side of the ecological equation will prevent numbers from decreasing. Such natural forces operating within any study or management area are very strong indeed, and if large mammal populations are left to their own devices the ecosystem of which they form a part tends rapidly to become self-regulating and to stabilize at an ecological carrying capacity. Areas with higher ecological carrying capacities will obviously be able to support higher sustained yields.

A practical use of the concept of an eruptive oscillation is to consider some of the implications of recognizing if a population is either stable and independent of, or in some phase of, an eruptive oscillation.

Recognizing an eruptive phase can be a simple matter based on historical evidence, as when recent large-scale increases in animal numbers, or years of large-scale mortality, are known. Where such evidence is unknown it may still be possible to identify the phasing of an oscillation by assessing vegetative or other trends in habitat in a critical season, coupled with an assessment of physical condition of animals, the latter in both favourable and unfavourable seasons of the year. Other types of evidence may be combined to form collections of related facts (syndromes) which, in combination, are indicative of a particular phase or relationship.

This syndrome approach needs emphasis here as well as in defining animal problems, for a single type of evidence can be misleading. For example, evidence of a high proportion of young in a population cannot be used alone to indicate a healthy relationship between a population and its environment. It is true that, as in phase (2) of Figure 126, reproductive success will be good and if the population is held below the level of the potential (ecological) carrying capacity this situation may continue. However, it is also true that a very high proportion of a population in the second and most spectacular stage of an eruptive oscillation consists of young animals. Obviously one must return to a syndrome approach, for some measure of environmental response would also be required in this example.

Since questioning the usefulness of age structure alone may raise a few eyebrows, may I quickly explain that there is nothing wrong with this practice in theory. Far from it. A complete cross-section of a population giving a good perspective of age structure would of course be valuable and can be used in ways explained in other publications (e.g. Mosbey, H. S. *et al.*, 1963 and Caughley, G., 1977).

However, at present, especially in most developing countries, information and capacity for such assessments are largely unavailable. At this point there seems little value in stressing this kind of sophistication when the use of indices will serve management equally well; and in using age structures, much more is required than making assessments of age. For example, great care must be taken not to transfer standards worked out for one species to another species, even one closely related. Perhaps even more important is the need for caution in applying standards in one area to another, for different habitats can produce not only different age structures but different sex ratios as well, even with unharvested stable populations.

In the Himalayan thar study mentioned above (Caughley, G., 1970) emphasis was placed on the difficulty in measuring death rates and rates of increase, and this emphasis is by an experienced field biologist with rather good facilities compared with those normally existing in developing countries.

Caughley felt that female fat reserves in summer provided the most promising index used in his study and, combined with other kinds of evidence, this can indeed prove most useful (see also Section 3860).

In my view a safe procedure to follow in countries in the early stages of developing management of large mammals is to base early management decisions on the most reliable, relevant and easily obtained indices. The selection of such indices will vary with the particular animals and habitats concerned and with the existing level of expertise of the field observers or biologists who gather the data. This approach seems equally valid for more industrialized countries in the process of developing their expertise in wildlife management. Once such a simple basis for management has started, and as the level of expertise increases, more sophisticated approaches may be considered.

The most useful next step forward in sophistication is to obtain information on rates of increase. Rates of increase are good collective indices to population responses as they automatically incorporate other values, such as productivity, age structure and mortality, for which data are usually not readily available. Their determination should be given priority for, as will be appreciated later, they have many uses and can be one of the most practical foundation stones for later building of more sophisticated management programmes.

The above considerations have been used as a guideline for the discussion of harvesting that follows.

7200 DETERMINING ALLOWABLE HARVESTS

In this section simple approaches to determining harvests are described, both for populations being harvested for the first time and for those approximately stable under existing levels of harvesting.

Basic relations

When the density of an unharvested population exists at the level of the carrying capacity of the environment the rate of population increases (r) would be zero by definition; therefore no sustained yield harvesting could take place. Sustained yields can only take place when a population density is below the ecological carrying capacity. Initial harvesting must necessarily lower the density, thus changing a population into a state where it can increase. When this is done, that is when an arbitrary harvest is taken, the population then tends to recover towards the existing ecological carrying capacity.

Figure 129 shows relations between the rate of increase, density of population and carrying capacity (cc).[1] When the population is at maximum carry-

---

[1] cc = the capacity of the environment to sustain a certain number of animals under a given set of conditions.

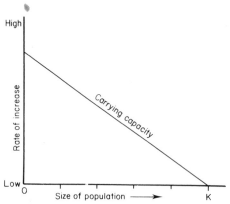

Figure 129 The relation between carrying capacity, within a given habitat, and the two factors which determine its level: rate of increase and size of population. The maximum (potential or ecological) carrying capacity (K) by definition cannot be increasing

ing capacity the rate of increase is obviously zero. The maximum rate of increase occurs when the population is very low, that is when the greatest discrepancy exists between the numbers of animals present and the capacity of the environment to carry many more.

G. Caughley (1977) suggests that, in practical terms, one cannot really tell how many animals to recommend harvesting before harvesting starts. However, in circumstances where harvesting is just starting as a form of planned management it is not so much precision that is required as a safe, conservative way of proceeding. With this in mind do not despair, for the following simple procedures may be used with confidence.

7210 Information Needed and Methods for Determining Sustained Yield

In real life, cropping programmes have to be started with less than complete data. However, provided the operation is monitored and regardless of the simplicity of the system for assessing harvests, field operations can always be adjusted and improved in the light of experience. Harvesting for the first time and improving existing harvests reflect situations different enough to be considered separately.

7211 *Harvesting Populations for the First Time*

7211A *Method 1: 10 per cent Initial Crop*

One African country has adopted an arbitrary 10 per cent for cropping all species of ungulate. This, of course, is very crude. It does not prevent many species from increasing. It can hardly be called efficient management and the effect on the population is often small. It is, however, a safe crop since most ungulates reproduce at a faster rate. Countries without adequate data for better estimates may thus wish to start with 10 per cent harvests. This should only be considered as a rough rule of thumb at the start of management and not as a permanent management system. The procedure includes:

1. obtaining a population estimate (Section 3400);
2. harvesting 10 per cent;
3. one year later obtaining another population estimate in the same season and in the same way as (1);
4. compare (1) and (3).

If 10 per cent harvesting is excessive it may soon be discovered through trial and error as described below (e.g. methods 4 and 5).

7211B   Method 2: Harvest and Monitor Population

Following this procedure the minimum information needed to begin a cropping programme where the population size is not known is an index of population size. The procedure then is to take an arbitrary crop and check the effect of this cropping on the size of the population. One continues to increase the crop until the population as shown by the index begins to decrease. The sustainable crop is a little less than the last crop taken. This method, too, is crude in that it takes time to discover the correct value and it does not distinguish between production and immigration. Its success also depends on the sensitivity of the index of population size. But it is one way to get started and with slightly more refinement than by the 10 per cent method.

7211C   Method 3: Harvest Population and Monitor Habitat

In this system the minimum information is taken from some part of the habitat. Usually only one kind of evidence from the habitat is absolutely required, and this will depend on the objectives of management. If the concern is to harvest until some kind of stability is reached between animals and habitat then the main habitat elements will be assessed at the most critical season of the year for animals. If the concern is to reduce damage by the animals then minimum data required might concentrate on the plants being damaged. In either case, arbitrary annual harvests are then taken and increased until observations taken from the habitat indicate that management objectives are accomplished, under the level of harvesting which will by then be known.

The above three methods may be justified as a way of getting started but if information is taken only from one side of the animal–habitat ecosystem it will be impossible to understand interrelated causes and effects. One can always guess, but for competent management guesses are not good enough. One should understand enough of causes and effects to be able to predict them. This requires that information must be taken from both animals and habitat, even if the approach is simply one of profiting from experience through trial and error.

One way of describing a population at equilibrium with its environment is

378

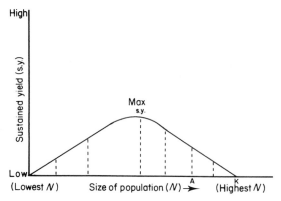

Figure 130   Relation between sustained yield (sv), numbers of animals (N) and different levels of harvesting (indicated by vertical dashed lines). K indicates the maximum carrying capacity at equilibrium with the environment; that is, completely unharvested. Note that, in these terms, sustained yields are always below the environmental carrying capacity (modified from Caughley, G., 1977)

to consider for example stages 1 and 5 in Figure 126A. Another way is to recognize various possible levels of sustained yield as shown in Figure 130. This figure shows the relations between sustained yield (SY), numbers of animals (N) and different levels of harvesting (indicated by vertical dashed lines). K indicates the maximum carrying capacity at equilibrium (environmental carrying capacity) that is, in this case, completely unharvested and stable.

Once approximate stability has been established under a given level of hunting, in terms of Figure 130 the existing population ('A' is used as an example) would by definition have to be lower than K, the maximum population the habitat could carry if no harvesting took place. Sustained yields are thus always taken from populations below the environmental carrying capacity (K).

Caughley (personal communication) suggests that, under the special conditions, where one is dealing with a previously unhunted population of large mammals, initially one could probably safely aim at reducing to 70 per cent of its original density (e.g. A in Figure 130).

The following simple procedure obtains information from both animals and habitat.

7211D   *Method 4: Harvest, then Monitor both Population and Habitat*

Using this method:

1. Obtain an index to the present status and trends for the major habitat elements important for the study animal in question (see Sections 3500 and 3900) as well as an index to animal numbers taken immediately before the harvest (Section 3400).

2. Harvest the population and record the number of animals removed from the study area.
3. After the harvest obtain another index to animal numbers in the same way, using the same sampling areas (or transects) to estimate the percentage by which the harvest reduced the population.
4. As noted elsewhere the indices to animal density before (B) and after (A) hunting can, when combined with the numbers of animals killed (H) be used to estimate the size of the population both before and after the hunt, using

$$P = \frac{H}{(B - A)/B}.$$

5. Continue harvesting at the same level until annual or biannual cross-check observations (usually more indices of the same kind) of both animals and habitat indicate that a new level of stability has been reached.

7212 *Improving Yield from Populations Stable Under Existing Levels of Harvesting*

For populations approximately stable under existing levels of harvesting[1] a check on the suitability of harvesting levels can be made in several ways (Caughley, G., 1977) but two simple approaches seem most appropriate in terms of the kinds of information that can now be readily obtained in the field: either (1) a new level may be established in the same way as described above for a previously unharvested population (e.g. Method 4 above), or (2) harvesting levels may be set by determining the rate of increase of the population.

7212A *Method 5: Modified Method 4*

If the first approach is used, Method (4) may be modified as follows:

1. An index to animal numbers will be recorded annually before hunting starts (if the population has not been estimated see Section 3400 or Method 4 above).
2. The harvest is increased by an arbitrary amount, say 5 or 10 per cent.
3. Annual measurements of the key habitat elements will be made in key areas at the critical season.
4. Under the new level of hunting, observations of both animal density and

---

[1]First, to be safe, a quick cross-check on one's assumption of stability would be useful. The fact that approximately the same number of animals have been removed annually for a few years does not by itself indicate a stable animal–habitat relationship, for under these conditions habitats may have been either improving or deteriorating. A quick cross-check on the recent trends in the main habitat elements in key areas and in the most critical seasons should be used to confirm or challenge one's assumption about initial approximate stability of populations.

habitat will continue annually until some new level of stability has been reached.

5. If greater harvests are desired increase the harvest more gradually, say in steps of 3–5 per cent and at each level of harvesting repeat (3) and (4) above.

7212B   Method 6, based on Rate of Increase

A second approach for use with populations already being harvested depends on a knowledge of the rate of increase or, in a more refined form, this rate combined with mortality and an estimate of the number of young entering the population each year, after the period of calving.

Rate of increase defined

Calculating the level of harvesting using the rate of increase utilizes an important guideline for harvesting large mammals, as described by G. Caughley (1977) and others, that is: set the annual harvest at *the rate at which a population would have been increasing if no harvesting occurred*. This rate *is known as the rate of increase* (*r*). This rate, when converted into appropriate terms, indicates a harvest that will reduce the rate of increase to zero, that is neither increasing nor decreasing the population. Harvest levels calculated in this way are thus sustained-yield harvests. Such harvests are referred to below as sustained yields (sy) and are normally calculated on an annual basis.

It is important to re-emphasize, regardless of which system of management is used, that there are an indefinite number of points—or density levels— at which populations of large mammals may be stabilized with (in equilibrium with) their environment. Equilibrium under a harvesting programme may be achieved with the population at near maximum numbers (e.g. approaching K in Figure 130) or with less than a dozen individuals remaining. So thinking of *the* sustained yield as if it were one value will not contribute much to one's understanding, and will almost certainly detract from effective management. G. Caughley (1977) has elaborated some of these past fallacies. Although a repetition of his discussion seems outside the scope of the present handbook, it is of interest to recognize that fisheries biologists were well aware of this approach to harvesting as early as the 1920s and 1930s (Baranov, F. I., 1926; Hjort, J. G. Jahn and Ottestad, P., 1933; Graham, M., 1935).

How to determine rate of increase (*r*)

First, here is a simple procedure for determining the rate of increase in a way that can be used to set the actual numbers of animals that may be taken annually on a sustained yield basis.

1. Obtain an index to animal density before harvesting.

2. Harvest, recording all kills.
3. Obtain the same kind of index to animal density in the same way after the harvest.
4. Using the harvest total (Ha) and the before (B) and after (A) indices, estimate the population using

$$P = \frac{Ha}{(B - A)/B}$$

5. Before harvesting, at the same season and places in the following year, obtain an index similar to (1) above.
6. The rate of increase can then be calculated by subtracting (3) from (5) and converting to common (natural) log. Thus, if (3) was 50 and (5) was 57 the increase of seven would be 14 per cent, the population having multiplied by 1.14. The natural log of 1.14 is 0.13 and 0.13 is therefore the rate of increase as defined by G. Caughley (1977).

Using this method steps one to three and step five are essential for calculating a rate of increase. Step four uses the same information obtained in the first three steps to estimate the population, so a specific target figure for next year's harvest may be set.

Note that this way of determining rate of increase applies both to species that breed throughout the year and to seasonal breeders.

Following through with the above example, if the harvest was 300 animals ($Ha = 300$) the population (P) would be:

$$P = \frac{Ha}{(B - A)/B}$$

$$P = \frac{300}{(57 - 50)/57}$$

$$P = 2442 \text{ before the hunt, and}$$

$$P = 2142 \text{ after the hunt.}$$

If the population is in approximate equilibrium with its habitat, the rate of increase equals the annual harvest that can be allowed on a sustained-yield basis, when the harvest is distributed throughout the year. This is named by G. Caughley (1977) the *instantaneous rate of harvesting* and designated as H in the formulae. Since, in the above example, the rate of increase ($r = 0.13$) is the same as the instantaneous harvesting rate ($H = 0.13$) this indicates that $0.13 \times 2442 = 317$ animals can be taken annually on a sustained-yield basis, slightly more than the 300 that were harvested. Three hundred would therefore be a fairly safe sustained yield when harvesting occurs throughout the year.

If, on the other hand, the indices before and after harvesting were 57 and 45, and a year later a similar index before harvesting was 50, then:

$$P = \frac{Ha}{(B - A)/B}$$

$$P = \frac{300}{(57 - 45)/57}$$

$$P = 1425 \text{ before the hunt, and}$$

$$P = 1125 \text{ after the hunt.}$$

Similarly the rate of increase would be calculated using the indices 45 to 50 (see above Nos. (3) and (5) respectively). Since this is an increase of 5 above 45, $5/45 = 0.11$, the population had multiplied by 1.11. The natural log of 1.11 is 0.104 which is the rate of increase and instantaneous rate of harvesting. In this case the population of 1425 could support a sustained yield of only 148 animals ($1425 \times 0.104 = 148$), the experimental harvest of 300 being much more than the population of 1425 could sustain each year.

Instantaneous *vs.* isolated rates of harvesting

The above is a discussion of populations on paper as it were, but fortunately it is also usable in the field. There are only two extra bits of information which may require more calculations and may be needed to give the confidence that normally accompanies greater precision. First, there is the matter of the timing of the harvest; and secondly, the extent of natural mortality.

The rate of harvesting discussed thus far has referred to harvesting animals continuously throughout the year. G. Caughley (1977) has called this the *instantaneous rate of harvesting*, which he symbolizes as H. In the above two examples, the harvesting rates of 0.13 (317 animals) and 0.104 (148 animals) were instantaneous rates of harvesting: $H = 0.13$ and $H = 0.104$ respectively.

But most harvests are made during shorter periods of time. Rates of hunting for these shorter hunting seasons (e.g. for 1 month or less) are called *isolated rates of hunting* by G. Caughley (1977) and symbolized by a small h. Fortunately H can be easily converted to h by using the formula, $h = 1 - e^{-H}$, that is, subtract the exponent of negative H from 1 to get h, for example:

$$h = 1 - e^{-H}$$
$$h = 1 - e^{-0.13}$$
$$h = 1 - 0.878$$
$$h = 0.12 = \text{the isolated rate of harvesting.}[1]$$

---

[1] Although this calculation could hardly be simpler, take a non-mathematician's advice and watch your negative signs.

If hunting takes place in 2 months, the rate of harvesting is calculated by $2h = 1 - e^{-H/2}$. Using the example of Table 14 where the instantaneous rate of harvesting is $H = 0.182$, then

$2h = 1 - e^{-H/2}$
$2h = 1 - e^{-0.182/2}$
$2h = 1 - e^{-0.091}$
$2h = 1 - 0.913$
$2h = 0.087 =$ the isolated rate of harvesting for each month.

If hunting takes place over a period of 4 months then the harvest for each month would be calculated by $4h = 1 - e^{-H/4}$ and so on for any number of months. If a population is to be harvested throughout the year, instead of calculating monthly harvests using $12h = 1 - e^{-H/12}$, it is simpler to use the instantaneous rate of harvesting (in this case 0.182) applied to the average yearly population, for it results in the same harvest.

Rate of increase is the most relevant single index for use in calculating sustained-yield harvests. Since this index incorporates both gains (births) and losses in numbers (mortality) it is, in this sense, also a collective index. Table 12 illustrates how different rates of increase may affect both instantaneous and isolated harvesting rates at different population levels. On the environment side of the system, maintaining a favourable state of imbalance means not only continued survival of the key habitat elements but the discrepancy between animals and habitat (which in a sense is an annual unused reserve), offers a measure of environmental insurance against unusually poor climatic conditions. The production of higher sustained yields from lower populations with higher rates of increase, as shown in Table 12, is thus a good general approach to improving management practice and is especially appropriate in areas of variable climatic conditions.

Table 12 Different rates of increase (r) and their effects on both instantaneous (H) and isolated (h) harvesting rates at different population levels

Population	Instantaneous harvesting rates (H)		Isolated harvesting rates (h)	
	($r = 0.14$) $H = 0.14$	($r = 0.18$) $H = 0.18$	($r = 0.14$) $h = 0.131$	($r = 0.18$) $h = 0.165$
1000	140		131	
900		162		148
2000	280		262	
1800		324		297
5000	700		655	
4500		810		743
4000		720		660

Estimating natural mortality

An issue remaining is the timing of the isolated hunting period or periods. The above calculations are straightforward enough in terms of finding the appropriate hunting rates or percentages of the population to be removed. But they are based on evidence taken usually before and after hunting seasons. In fact the actual numbers in the population will be changing throughout the year due to natural mortality.

Annual natural mortality is calculated as the annual loss measured between birth pulses and excluding hunting. For animals characterized by a short period of, say 1 or 2 months, in which the majority of calves are dropped, the greatest numbers of animals will occur immediately after calving period; the lowest number will occur immediately before calving. Differences between indices to total numbers taken after calving and before the following calving season will give an estimate of the rate of annual mortality. If indices are based on animals observed, care must be taken to avoid making the second observations too close to the calving period, for females of several species of ungulates are more difficult to observe within 2 weeks of calving. Using this method will of course include winter or dry season mortality.

To convert annual mortality (M), (for example, 25 per cent), to monthly mortality (m), use $m = M^{1/12}$ or, more simply, $m = 1 - M^{0.0833}$. That is:

1. If throughout the year 0.25 will die, this means that $1 - 0.25 = 0.75$ will live.
2. The number of living each month would then be $0.75^{1/12} = 0.75^{0.0833} = 0.9763$.
3. To convert back to monthly mortality, $1 - 0.9763 = 0.0237$.
4. Multiplying the population by the factor 0.0237 will thus estimate the monthly mortality where annual mortality is 25 per cent.

Examples of appropriate multiplication factors for use in calculating monthly mortality from an estimate of annual mortality are shown in Table 13 for annual mortalities of between 5 and 30 per cent.

Table 13   Calculation of monthly mortality (m)

From annual mortality (M) of	(Using)		Multiply population by
0.30	$m = 1 - 0.70^{0.0833}$	=	0.0293
0.25	$m = 1 - 0.75^{0.0833}$	=	0.0237
0.20	$m = 1 - 0.80^{0.0833}$	=	0.0184
0.15	$m = 1 - 0.85^{0.0833}$	=	0.0134
0.10	$m = 1 - 0.90^{0.0833}$	=	0.0087
0.05	$m = 1 - 0.95^{0.0833}$	=	0.0043

Note: $0.0833 = 1/12$.

This approach to estimating mortality has the disadvantage of being rough-and-ready, and now allowance has been made for a higher mortality rate of young animals in the first few weeks after birth. Other ways of obtaining evidence of mortality exist, some of which are already mentioned in previous sections of this book (Sections 3840 and 3870) but the above procedure should be adequate for most management purposes. More refined procedures requiring field data more difficult to obtain can be found in G. Caughley (1977) and R. H. Giles (1969).

7212C Estimating Isolated Harvesting Rates using Rate of Increase and Estimates of Monthly Mortality

Obviously yields will differ depending on when the hunting season is declared, as between calving periods the natural mortality progressively reduces numbers of animals month by month. The example shown in Table 14 illustrates that a sustained yield (sy) taken soon after calving would be 166 animals, while 4 months later the sy would be only 151 animals. Similar drops in yield can be expected as the year progresses, as illustrated in the last two columns where harvesting is indicated as taking place in the sixth and seventh, and ninth and eleventh, months respectively.

This example involves a population of 1000 whose rate of increase is 0.182 (i.e. $r = 0.182$; $H = 0.182$). An estimate of 25 per cent annual loss through natural causes has been made and each year 0.6 offspring per head is added to the remaining population at the calving season. For comparison, the first column of Table 14 shows the natural mortality month by month in an un-hunted population.

7212D How to Select an Appropriate Method for Estimating Harvests

It should now be clear that there is no one way of setting harvests; there are an almost infinite number of ways and only a few examples have been mentioned above. It should also be clear that various combinations of different methods may be used. The following four practical guidelines may prove useful before deciding either on an initial system or on modification of an existing system.

1. Examine such information as is already available that may be used as an index to population levels, mortality, productivity or level of harvesting. A close look at such existing information may save much time and effort. Before using any figures available over a period of years check very carefully for consistency at the data-gathering stage: same area, same season, same routes, same effort (time and number and kind of observers) and exactly the same method. If these variables have not been the same from year to year see if it is possible to convert to make them comparable. It may be difficult if not impossible to interpret past figures. If the latter is the case, do not use them.

Table 14 Calculating sustained yield for 1- and 2-month harvests

Month[1]	Unharvested numbers	Harvest in first month	Harvest in fifth month	Harvest in sixth and seventh months	Harvest in ninth and eleventh months
1	1000	$1000 \xrightarrow{h = 0.1664} 834$	1000	1000	1000
2	976	814	976	976	976
3	953	795	953	953	953
4	931	776	931	931	931
5	909	757	$909 \xrightarrow{h = 0.1664} 758$	909	909
6	887	739	740	$887 \xrightarrow{2h = 0.087} 810$	887
7	866	722	722	$791 \xrightarrow{2h = 0.087} 722$	866
8	845	705	705	705	845
9	825	688	688	688	$825 \xrightarrow{2h = 0.087} 753$
10	806	672	672	672	735
11	787	656	656	656	$718 \xrightarrow{2h = 0.087} 655$
12	768	640	641	641	640
1	750	625	625	625	625
	750 + 750 bi =	625 + 625 bi =	625 + 625 bi =	625 + 625 bi =	625 + 625 bi =
	750 + 450 = 120	625 + 375 = 1000	625 + 375 = 1000	625 + 375 = 1000	625 + 375 = 1000
	bi(births per adult) = 0.6	bi = 0.6	bi = 0.6	bi = 0.6	bi = 0.6
	r(rate of increase) = 0.182	r = 0	r = 0	r = 0	r = 0
	H = 0.182	h = 0.1664	h = 0.1664	2h = 0.087	2h = 0.087
	sy = 0	sy = 166	sy = 151	sy = 146	sy = 134
		($1000 \times 0.1664 = 166$)	($909 \times 0.1664 = 151$)	($887 \times 0.087 = 77$)	($825 \times 0.087 = 72$)
				($791 \times 0.087 = 69$)	($718 \times 0.087 = 62$)
				Total 146	Total 134

(Annual mortality for all columns is 25%, or 0.0237 per month)

[1] Column shows months after calving.

2. Before applying any method make certain it can easily be handled in the field by existing field personnel. If existing personnel cannot handle the more complex techniques, try an appropriate simpler one. There is much to be said for starting with simple approaches, monitoring the results, developing in-service training courses for field workers and in this way, over the years, gradually evolving more refined techniques. To try a harvesting method for which only unusable data can be obtained from existing field men can only be regarded as a waste of time and effort.

3. Whichever the method for setting harvests, whether it is adopted from others, modified or invented, monitoring (cross-checks) should be applied from time to time to ensure that management practices are assisting in achieving the stated objectives of management.

4. Particularly when harvesting is in its early stages of development, all data on which harvesting quotas are assessed should come from the population and area in question and from nowhere else. This needs extra emphasis because in practice this rather obvious principle is not widely accepted. Ageing criteria developed in one country for one species are cheerfully applied to other countries and even other species. Age structures, sex ratios, mortality rates, rates of increase and any other measure of animal response are not necessarily transferable to other areas. Likewise, neither are criteria based on animal–habitat relations transferable. For example, the percentage utilization of certain species allowable by foresters managing a production forest may vary forest by forest and with different animal and tree species.

There may of course be similarities, close parallels or even identical responses, between certain areas but this must be determined, not assumed. In my opinion the biggest advances in refinement of management technique in the next few years will not come through more precise attention to more and more facts associated with animal–environment interactions. In terms of increased yields it will, I suspect, be more profitable to direct our observations to the particular circumstances existing in the study area in question, to make our decisions, harvest, then monitor the changes, modify our practices accordingly, to monitor again and so on: in other words, so arrange our activities that we can continually profit from ongoing experience in the particular areas for which we are responsible.

7300 MONITORING THE CROPPING PROGRAMME

Most management programmes are designed to sustain the yield of animals, and checks are therefore needed to see what effect the programme has upon the animal population and its sustaining habitat.

There is an almost infinite range of kinds and levels of interaction between animals and environment. Areas can be highly productive or barely produc- tive and two populations with approximately the same levels of productivity may be held at that level through markedly different combinations of ecological factors. Examples of practical implications of this dynamic complexity have been previously mentioned. For example, an eruptive oscillation need never occur, but can be held by management to an early phase; in any area being managed there is not one but many possible sustained yields. Likewise there is no one way of monitoring. In principle:

1. reconsider original objectives of management;
2. review the existing techniques for determining sustained yields;
3. select monitoring techniques after considering (1) and (2) above.

In practical terms, there will be three main variables which can be expected to change and, as they change, monitoring activities may also be modified appropriately. These variables are:

1. *Changes in management objectives*, for whatever reason. This will usually result in selecting different aspects of either animal response, environmental response, or both, as indices to the success of management.
2. *Increasing understanding of the uses and limitations of various techniques in terms of the level of expertise available to apply them.* This may result in more, or less, refined techniques respectively.
3. *Increasing understanding of habitat requirements.* This may result in shifting priority in monitoring towards those habitat elements recognized as critical at the present time. Particularly if habitat management is practised, then as one habitat element becomes less critical another may then assume more importance and so the process may become one of continual refinement.

7301   Suggested Minimum Data for Monitoring

Recording the trend of key habitat elements in key areas and at critical seasons is the simplest way to monitor the adequacy of the crop since it indicates whether the population is increasing, decreasing or stable. Some index to physical condition is the simplest way of monitoring the response of animals to the available habitat. These two kinds of information, combined with an index to animal density, should provide a safe first approach to monitoring. An index to physical condition is preferred to other types of animal response because changes in condition due to different environmental conditions are reflected sooner in fat reserves and in body conformation than in sex or age ratios or in numbers of animals.

7400   TECHNIQUES OF HARVESTING

7401   Harvesting in Fluctuating Environments

Up till now we have been dealing with the simple case of a population living in conditions that remain about the same from year to year. However, many populations, particularly in temperate regions, build up in numbers when conditions are favourable and fall to low levels during periods of unfavourable conditions such as droughts or unusually severe winters.

Cropping the latter kind of population often presents difficulties. A choice must be made between two basic strategies (Caughley, G., 1977).

The first is called a 'mean strategy'. The number to be harvested is estimated for average conditions, and this number is cropped each year irrespective of whether the population is at high or low density.

The second method is called a 'tracking strategy'. As the population rises in numbers the cropping percentage is increased, and when the population starts to decline the rate of harvesting is reduced. The harvesting rate changes as the population changes.

The mean strategy is useful, particularly when the amplitude of fluctuations in numbers is not too great and the average time between fluctuations is not too long. Economically, it is preferable to the tracking strategy because it results in a constant supply of products.

Its disadvantages lie in the danger that the population will be cropped too heavily at low numbers, to the extent that its ability to repopulate is impaired when conditions again become favourable.

The design of a mean strategy requires considerable information. It is not recommended where little is known about the population's dynamics and the natural fluctuations of its environment. Nor is it a strategy for the faint-hearted, as there is always a strong temptation to change from a mean strategy to a tracking strategy when numbers are low.

A tracking strategy is safer, particularly when insufficient information is available on the species and its environment. It is standard practice in several industrialized countries where hunting is controlled by licences, where the annual fixing of seasons is practised and where law enforcement is good. It generally produces a lower average yield over a period of time. In commercial meat production schemes 'tracking' has the disadvantage of resulting in fluctuating market prices and labour requirements. However, where fluctuations in environmental conditions are marked, and have an average periodicity of more than about 5 years, 'tracking' is considered the only workable strategy.

When a cropping programme is in its initial stages, harvesting should first be based on a tracking strategy which can, if necessary, be modified to a mean strategy after more information is accumulated. A mean strategy should not be introduced unless it is based on considerable previous research.

7402 *Selective Harvesting*

So far we have been dealing with the simple case where cropping is random; that is, each sex and age is harvested according to its proportion in the population. Although this method is often the only practical procedure, it may be inefficient. Under many situations the random sustained yield of meat and other products can be improved on considerably by cropping some segments of the population more heavily than others.

7403 *Harvesting of Social Groups*

When the harvested population is composed of social groups, two schemes of harvesting are available:

1. all the animals in a group can be shot, leaving other groups wholly intact; or
2. animals may be shot unselectively from many groups.

Selection on the basis of whole groups can have several ecological implications for those animals remaining. In terms of social organization and habitual patterns of daily movements, shooting a few animals from each group (implying a sudden alteration in group composition) may result in a change in social organization. This in turn can affect daily movements, particularly if the dominant animal in the group was one of those killed. In terms of the wariness of the remaining population, shooting a part of a group is certain to increase the wariness of those remaining which in turn will increase the threshold of visibility, making harvesting more difficult. Wariness in large mammals continues for several weeks and, for some species, much longer. These effects may be averted or minimized by shooting all animals in a group.

In terms of the effect on the habitat, elimination of an entire group can be expected to relieve the pressure on the habitat exerted by that group until it is reoccupied by dispersing individuals. On balance, if group home ranges overlap in key feeding areas, cropping should be directed at whole groups. If group ranges do not overlap and dispersal is slow, a few animals should be cropped from each group. In some species male groups and female groups may be cropped in different ways. For most species this question poses a degree of refinement that makes the question of low priority. However, where wariness and consequent thresholds of visibility are important to management, or where particular over-used areas urgently need relief, then the removal of entire groups can be a worthy objective.

7404 *Selective Harvesting of Sexes*

Most ungulates do not mate for life. The mating system of many large mammals tends either to be promiscuous or characterized by the formation of temporary harems. In such cases the population contains a proportion of males surplus to that needed to maintain adequate reproduction.

This situation allows the population to be biased towards females, thus simulating the farmer's female flocks of sheep served by few males. Production is thus increased. In order to maintain an induced disparity of sexes, males must thereafter be harvested at a greater rate than the females; and inevitably many of the males harvested will be young.

A knowledge of the species mating behaviour is a necessary basis for deciding how great a disparity of sexes should be created by the initial reduction. In monogamous species, a 1 : 1 sex ratio gives the greatest meat production as well as the greatest production of good trophies. For those species whose dominant males have harems during the breeding season, the information can be obtained experimentally by progressively lowering the percentage of males in a population. G. Caughley (personal communication) suggests that the optimum percentage of males, taken from a population managed for meat production, may be at about 20 per cent above the point where fecundity rates begin to drop.

However, to my knowledge, no completely wild species is yet being managed on this basis. A safer procedure would be to accept that, while the above-mentioned optimum ratio may be unknown and difficult to determine, it is possible to recognize existing ratios in selectively harvesting populations that are still on the safe side. In these terms several species of deer have male to female ratios of between 1 : 5 and 1 : 15, with no apparent loss of productivity.

The following aspects of productivity concerning the sexes seem much more important.

There has been much discussion and often heated argument between sportsmen and wildlife managers in different areas over the question of whether one should harvest mature animals of one sex, or of both sexes. Some of these differences of opinion will remain for they may in fact result from differences in objectives of management. Part of the controversy rests on lack of understanding of simple animal–habitat interrelationships.

Resolving this conflict is much more difficult in countries where some form of wildlife management has been practised for many years for there, opinions on both sides of the argument will be firmly fixed. For developing countries, who have no long tradition favouring a certain kind of selective shooting, the following may help.

As a student in the 1940s, when the practice of wildlife management was still gathering momentum in North America, I listened to an interesting lecture on a phenomenon called the 'principle of inversity'. This principle, as described by P. L. Errington (1946) was that 'the rate of population increase in quail is inversely related to population level'. In other words, as has already been discussed, low populations tend to have higher rates of increase. But an unsolved puzzle was presented at the same time and it went like this: Why is it that productivity (in terms of e.g. more eggs per clutch, more young per litter, more young per 100 females) can be increased in certain species by harvesting males alone, while in other species productivity can be increased only by harvesting a proportion of the female population?

The explanation is simple. Improving conditions for farm animals can only be done on a farm-by-farm basis. Likewise, improving conditions for populations of large mammals can only be done on a population-by-population basis. If sexes are geographically segregated during a critical season or are selectively using different parts of the available habitat, then each sexual grouping must be regarded as a separate population and managed accordingly in terms of the relationships between that part of the population and the habitats on which it depends. Put in another way, if conditions for higher productivity are to be improved it *must* involve (either directly or indirectly) the female segment of the population. If the sexes are mixed throughout the year, then shooting males will provide an improvement in habitat elements available to the remaining individuals of the population, including the females who will respond by increasing productivity. If the males are segregated from the females at a critical period of the year, then shooting males may improve the growth rates of the remaining young males and should somewhat improve the survival rates for males, but would have little effect on the female population in a different locality or utilizing different parts of the habitat. Improvement in the productivity of females must involve a more favourable relationship between females and their habitat. For most species of deer the sexes are more or less segregated during at least one season of the year. It is for this reason that, if maximum productivity is desired, some proportion of the female population must be harvested.

In summary, regardless of the management aim, species, or segregation of male and female habitats, if maximum productivity is required a certain proportion of females must be taken to maintain the most favourable rates of increase (mainly through lower mortality of both the remaining females and the younger age classes associated with females). The reasons for harvesting both sexes may differ somewhat but the recommended management practice is inevitably the same in principle.

Thus if we consider areas in which the objective of management is to harvest maximum numbers, then regardless of whether sexes are together throughout the year or are seasonally geographically separated, both males and females would be harvested.

If the aim is to produce male trophies for hunters, harvesting both sexes will still produce best results. For species whose sexes remain together throughout the year, harvesting males only will result in an increase in productivity as mentioned above. However, the additional harvesting of females will be necessary to keep the habitat in even better condition for stimulating the health of the trophy segment as well as the rest of the population.

If the sexes are seasonally separated it is still important to harvest females to maintain the health and productivity of the survivors in the female-occupied areas. This implies reduced mortality rates to ensure a maximum number of young males entering the predominantly male-occupied areas when they grow up and leave the segregated female-dominated family groups.

It is with the above considerations in mind that harvesting of both sexes may be regarded as the common-sense normal practice.

No mention is made of systems of selective shooting designed to produce maximum-size trophies for sporting purposes. Although such systems are used in certain European countries (for example, see de Nahlik, A. J., 1959) this assumes: (1) conditions where individual animals can be closely observed; and (2) objectives of management concentrating on the size and quality of the trophy (e.g. shape, thickness, number of points, length and weight of tusks, etc.). Such objectives are not appropriate to most developing countries.

7405 *Selective Harvesting by Age*

The decision on which age classes should be most heavily cropped requires a knowledge of natural death rates and birth rates at each age. As a general principle, those ages contributing most to the rate of increase of the population should be cropped least. This means those individuals with the highest reproductive value: the juveniles. Reproductive value is calculated for a given age as the average number of offspring produced between that age and death.

In principle selection by age is sound. Biologically it makes sense, but is, nevertheless, considered as clearly outside the scope of the present handbook. This is partly because calculations are complex, partly because of the increased practical difficulties in selective harvesting (e.g. difficulties in recognizing various age groups in the field) but mainly because, even if it were achieved, it would have little effect in increasing the maximum sustained yield (Caughley, G., 1977).

7406 *Selective Harvesting of Species*

When a species to be harvested is one of several species in the same area, the relationship between numbers in one year and numbers in the next may differ from such relationships applied to a single species. A reduction in the density of one species may result in an increase in the density of other species, particularly when diets overlap. Conversely a decline of one or more species may result if the habitat is altered by the reduced pressure of grazing or browsing. These changes, in turn, may affect the harvested species' rate of increase.

Cropping of several species in one area is roughly analogous to selective harvesting of sexes and age classes in one population. The strategy adopted should aim to maximize the production of the total harvestable weight of meat on the hoof.

Although management of more than one domestic animal on the same farm is commonplace, simultaneous management of a number of wild animals for high productivity is a rather new field. Much careful monitoring of different approaches will be required before wildlife managers can feel secure in multiple species management, and presents one of the challenges for future wildlife managers.

A suggested safe approach is: (1) to determine or review habitat requirements for each species, (2) to define clearly the objectives of management, including priority emphasis to be given each species, (3) in monitoring effects of harvesting, to use measures of habitat response as the main indices to the suitability of various harvesting levels. In so far as possible a 'tracking' strategy of harvesting should be used and monitoring should continue on an annual basis for at least the first 5 years.

7407   *The Harvesting Season*

Selecting the season and period of hunting can clearly influence harvesting levels as noted above (Section 7212B). But selection of the season of harvesting can be important to management in several other ways, as the following examples will suggest.

A knowledge of seasonal feeding habits may be important. Setting a season at a time when a high proportion of the population is in easily accessible and easily observed areas may result in an increased harvest, while a season set at a time when the animals were in less accessible or better sheltered areas should produce lowered harvests.

If seasons are set at times when sexes are segregated, and one sex is for any reason more susceptible to shooting, then one could expect a higher proportion of the most accessible sex to be harvested, even if shooting were completely unselective.

Knowledge of the timing and characteristic pattern of dispersal of young individuals away from the home ranges of their parents can influence greatly the decision to assign a particular season for harvesting. Under different conditions and for different species, differences in timing of the hunting season can either accelerate or retard dispersal. The latter effect is particularly desirable in situations where wildlife dispersing away from forests, parks or reserves may constitute a problem by interfering with adjacent forms of land-use.

For example, an introduced species with small home ranges and only an environmental type dispersal (see Section 3242) can have its range of occurrence greatly accelerated by shooting the population during that period when the young are naturally dispersing from their family groups. In this event closing the season during the dispersal period should reduce the rate of spread. Dispersal may be an even more important consideration in assigning shooting seasons for a species whose young of the previous year congregate into large herds and disperse essentially as herds, in both environmental and innate patterns (like the blue wildebeest of Africa). Such dispersing young herds have sometimes been confused with annual migrations. Shooting these yearling herds, just before the start of dispersal, should significantly retard dispersal and thus minimize unwanted conflict with agricultural activities.

The above aspects of a species' natural history should be considered along with a clear understanding of the objectives of hunting in the first place. For example, if animals are hunted commercially for meat, it will obviously be desirable to assign hunting seasons to periods when both the meat and skins are in best condition.

7500   TECHNIQUES OF REMOVAL

7510   Shooting

Shooting with conventional firearms is the most widespread and efficient way of harvesting populations of large mammals. Two considerations are important: minimizing the disturbance caused by shooting and making an instantaneous kill to eliminate the problem of wounded animals. Shooting nearly always causes some disturbance and makes animals more wary, but by using techniques suited to the particular animal, serious disturbance can be much reduced or eliminated. Reducing the number of wounded animals is largely a question of equipment and training and this aspect is discussed first.

7511   *Equipment*

The choice of weapon varies with individuals and circumstances. In Africa a double-barrelled rifle of between .470 and .375 calibre is most suitable for elephant. For antelope, buffalo and hippo a .300 calibre rifle fitted with a medium- or low-power telescopic sight is most satisfactory. A heavy weapon is more easily steadied on the target so that heavy-barrelled target weapons are most suitable for cropping if shooting is from a vehicle. If shooting is on foot, a lighter weapon is obviously best. High-velocity bullets should be avoided as they spoil meat and are easily deflected by twigs and grasses. Telescopic sights if used should be of good quality, with clear graticules (scales that can be seen when looking through the telescope), and with robust mounts.

High-power telescopic sights should not be used at night as the image of the target is too dim for accurate shooting. A 2–4 times magnification is usually enough.

All weapons should be regularly serviced and frequently tested for accuracy.

The small-calibre rifle is in no way suitable for sportsmen, as the close control required to prevent wounding would normally be virtually impossible. However, for commercial harvesting of large numbers of animals, or for certain animal control campaigns, it is often an excellent choice. Small calibres are particularly useful in conditions where visibility is comparatively low, animal wariness is high and stalking is the method. It can be devastatingly effective if certain conditions are rigorously met. In general this means not shooting unless and until certain of a clean kill. Animals up to and including the size of red deer have been successfully harvested in this way. Specifically, the preconditions under which shooting occurred were:

1. Practice prior to hunting to determine the distance beyond which shots would not be taken. This was determined by determining the maximum distance at which a target not larger than 2 in. square (4 × 4 cm) could be consistently hit. In one New Zealand area involving a .22 calibre rifle with

a 2.5 power telescopic sight, a silencer and target ammunition, this distance was 75 yd, with the rifle held, either via a standard sitting or prone position, or against a tree.

2. For the deer only one of two possible shots were taken: the base of the ear, if the animal was standing in profile; or between and slightly below the eyes if the animal was facing the shooter. In both situations, deer dropped in their tracks. These were the only shots risked.

For silencers to be effective the bullet must be travelling below the speed of sound (331.3 m, or 1087 ft, per second). The most readily available commercially sold ammunition meeting this requirement are certain brands of .22 calibre hard-nosed target ammunition. The speed of larger-calibre bullets can be suitably slowed with reduced powder charges but this requires special loading and significantly increases the arc of the trajectory; thus becoming a less practical tool.

The practical advantages of silencers are enormous. I have taken animals from closely feeding family groups, often without disturbing other members of the group. One could stalk up-wind all day and encounter completely undisturbed groups of red deer even at the end of the day, while in the same area a .303 calibre rifle shot disturbed deer at up to 3 and 4 miles distance.

7512   *Training and Testing of Shooters*

Whether harvesting is for control, sport, or meat production, it is important that the quality of shooting be as high as possible, to minimize losses and maximize operational efficiency. If shooting is done by an organization, this normally implies in-service training, and should at least include training: up to a minimum standard, in safety procedures, in estimating distances, in stalking techniques, in dealing with dangerous animals and in maintaining equipment. Still other kinds of training may be appropriate for a particular organization.

7513   *Standards of Shooting and Safety Procedures*

Experience in many countries has shown that the training techniques in marksmanship used by army and police are usually regarded as a basic first step in training shooters for wildlife control or harvesting purposes. Such very basic courses will include instruction in care and maintenance of the rifles, safety precautions, both on the range and in the field, and the attainment of a certain standard of proficiency in marksmanship in prone, sitting, kneeling and standing positions.

7514   *Estimation of Distances*

Estimation of distances in the field is particularly important if precision shooting is required. This ability should be acquired with experience during the training period and can be greatly facilitated through use of a telescopic sight.

Effective control of *dangerous animals* can be more or less routine but it can also demand very quick responses to emergency situations and therefore the addition of a somewhat different kind of training. A course to give practice in snap or reflex shooting at short range is usually arranged with targets suddenly appearing as one walks along the course. Such courses are usually situated along the bottom of small steep-sided gullies so if targets are missed bullets may plough harmlessly into the nearby earth. In addition to this, before shooters are assigned jobs that may require dealing with emergencies involving dangerous animals they should have accompanied experienced men on missions involving the control of such animals and have satisfied the experienced operators that they are competent to handle similar jobs.

7515 *Stalking Technique*

Stalking as used here involves the actions involved in and directed towards placing the shooter in a position where a clean kill can be made.

As mentioned above (Section 3226B) variations in the threshold of visibility will mean that in a given habitat more or less animals will be visible at different distances. The significance of this threshold in stalking is that it will affect the ease with which animals may be located in the first place. As noted above, these thresholds are related both to the wariness of the animals and to the density of cover.

Also related to wariness is the flushing distance, the minimum distance at which an animal being approached by man or vehicle will stand, before running away. It is of course common to find that in areas where animals are wary the flushing distance will be greater than the distance at which a certain kill may be made. Under these circumstances improvement in stalking techniques will have obvious value in making operations more effective.

Knowledge of patterns of daily movement associated with each species is important to include in training courses, as well as an appreciation of the species' social organization, its dependence on various senses and its reaction to disturbance. Such knowledge is useful in planning times and patterns of stalking.

Persons with much experience in stalking become aware of the extent to which their quarry relies on sound, smell or sight to warn them of danger, and they may recognize a range of typical reactions to disturbances of various kinds. Field demonstrations of different kinds of disturbance are usually not difficult to arrange and are particularly useful to potential stalkers in revealing the distances at which wild animals are sensitive to various stimuli.

For example, most species of deer rely more heavily on hearing and smell than on sight, while Himalyan thar and chamois rely much more heavily on sight than smell. I have seen chamois on a high ridge become so disturbed at seeing a man with a rifle walking up a valley over 2000 ft below and well over a mile away that they left the area, moving over the ridge to another drainage. Black rhino have notoriously poor sight but excellent hearing. Each species will rely on its various senses in a different way, a way important for the successful stalker to understand.

Although a man's ability to hear and interpret sounds and smells may considerably improve with experience, his senses are still vastly inferior to those of the animal he stalks. An obvious way of improving the effectiveness of stalking is by using a closely controlled, silent dog who will observe relevant sounds and smells long before the man at the other end of a lead. I have found sheepdogs particularly good as trained stalking companions. They can also be trained to respond to hand signals and can be used to flush animals from thickets of shrub or trees, difficult or impossible for a man to penetrate.

In areas where horses are common, several game species may be approached closer by a man on horseback than by a man on foot. Horses have been successfully used in long stalks in open country by approaching the prey with the stalker screened by the horse. Whether on foot, on horseback, or using a vehicle, if the quarry is aware of the approach, one can normally approach closer by avoiding the use of a direct approach. Movement should be aimed at some point to one side of the animals, aiming at a distance within which a safe shot may be made.

7516   *The Ecological Process of Disturbance and How to Minimize It*

As a basis for improving one's stalking technique, it is useful to understand the process of disturbance in terms of the animal and its surrounding ecosystem. In these terms an animal becomes alerted to or disturbed by some stimulus within the surrounding ecosystem and reacts by increasing alertness, or by responding in some other way. Specifically this reaction may involve the animal standing still, or moving into screening shelter and standing still and looking in the direction of the disturbance, or moving slowly or rapidly out of sight and hearing of the intruding influence. For certain unusually aggressive animals, like the black rhinoceros of Africa, or the grizzly bear in North America, too close a sudden disturbance by man may precipitate a sudden charge.

But the process of disturbance needs elaboration. Any ecosystem will have smells, sounds, movements by various species, all perfectly normal and all normally changing by day and by night. T. Riney (1951) has suggested that large mammals (deer, in this example) are perfectly aware of and kept at ease by these normal stimuli and respond quickly to changes of any kind that may be caused by a disturbance within the range of their senses. The changes may be obvious and direct, as when a man suddenly appears, or is heard or smelt. But much more important for the stalker are the equally important indirect changes, as when a man disturbs a bird, which gives an alarm call, which in turn disturbs the deer. Alarm calls or alarm behaviour by other animals thus provide a common source of disturbance to large mammals. In the same study it was observed that, following disturbance of birds, as by a predator moving through an area, this area was characterized by a 'zone of silence' the presence of which would alert and maintain alertness in the deer. Deer are thus sensitive to sudden stopping of normal bird sounds, including sounds of scratching made by ground-feeding birds.

For example, in the winter of 1948, spotted towhees (a ground-feeding bird) were seen flushing about 30 ft ahead of a coyote (a North American predator) and flying to the tops of nearby shrubs. There were no alarm calls. A female deer under observation could not see the coyote, but followed its progress by watching the birds. A wave of disturbance (sudden flutter of wings) accompanied the coyote in this instance, followed by a zone of silence, the result of the birds' remaining quiet in the shrubs. In this case the coyote had passed in less than 2 minutes. It was 10 minutes before the birds resumed their normal activities and only then did the deer relax and resume normal feeding behaviour (Riney, T., 1951).

With the above in mind the task of the stalker can be now described in a different way. Essentially this means that the stalker should try to move within an ecosystem with as little disturbance to any part of the ecosystem affecting his study animal as possible.

With respect to *smell*, this means adjusting one's movements with the existing direction of the wind firmly in mind. Smell can also work in reverse to the benefit of the stalker. For example, when deer are disturbed a strong scent is released by various external glands. The odour is characteristic and once learned a stalker can be certain that animals have been disturbed up-wind whenever this scent is encountered in the field. To continue stalking in that direction could lead to much wasted time and effort.

A disturbing *movement* obviously involves a visual contact. Then three main considerations come into play: colour and rate and pattern of movement.

It is essential that a stalker dress in inconspicuous clothing that blends as naturally as possible with the natural environment. White or bright colours should be avoided. Although most large mammals are colour-blind the shades and patterns of contrasting colours are easily detected.

Movement should be slow, and combined with frequent stops. As a general guideline, try to imitate the pattern of moving, stopping, looking and listening, used by the species you are stalking when it is undisturbed. This is particularly important in areas with shrubs and trees. In a training school it is easy to make this point. In a forest, have the students spend at least a full minute looking carefully about them, then, keeping their eye on some point 50 m or more distant, move slowly forward one step, then another. It will readily become apparent that even a very slow walk is too fast to enable the observer to recognize standing or bedded animals except by chance. On the other hand, if the observer remains motionless, recognizing movement such as a flicking tail or a twitching ear is comparatively simple.

A characteristic of many ungulates is that they are rarely alone. They usually occur in groups ranging in size from small family units to large herds. This often means that when you see an animal in forest or woodland there may be others that you do not see, who in turn are watching you. In such cases, even when standing still, rate of movement is important in so far as turning of the head; or arm movements, such as raising binoculars, camera or rifle, must be done very slowly indeed. This requires practice.

Although it may not be true, if real improvement in stalking ability is desired, one should assume that the very best stalkers were not born that way but acquired their skills through training, increasing knowledge and experience. Essentially this means self-training, which can be greatly accelerated by constant attention to colour, shape, movement and location. The latter implies giving priority attention to those specific parts of the habitat most likely to have animals at that time of day. Thus one may be looking at the base of certain-sized trees or the edges of thickets for deer colour, the shape of bedded or standing deer and small movements by ears or tail as they flick at an annoying insect, or, at another time, for movement along well-defined trails leading into a feeding area.

Another form of hunting involves no movement at all. Here the hunter waits along a trail, in a tree, near a watering point or feeding area. A variation of this is when a hunter remains in a well-chosen spot while other hunters move about working gradually in his direction, or where lines of beaters are arranged as in 'drives'.

For animals of forest and woodland the acute *hearing* of the animal stalked often poses the biggest problem for the stalker. There are of course times when this is not so, as in a strong wind or rain. Although there is much variation between species, I have assumed that, in woodland or forest, it is impossible for me to walk quietly enough to move close to a wild deer without it hearing my approach. However, it is possible to closely approach wild animals without disturbing them, even though they hear some or all of the sounds of the approach.

A worthy stalking objective is to try to move through a forest without making more noise than the ungulate stalked, and to make that noise at about the same speed and with about the same pattern of stops and starts as made by the animal. Even the snappings of small twigs can be camouflaged by seeing that they occur in frequency and in pattern characteristic of the environment as it may sound normally without man. In this connection the quality of the sounds associated with movement is also important. Jeans and synthetic waterproof clothing, when brushing against vegetation, make a louder and a different kind of sound than fur, wool or rough cotton and should be avoided where careful stalking is required.

It may also be useful to remember that, for wild animals, normal sounds are not necessarily natural sounds alone. It does imply common, consistently heard sounds or observed movements as, for example, cars moving along a road, trains along a track, aircraft passing high overhead, power saws operating in a forest. If such noises are frequent enough the wildlife become so accustomed to them that such sounds are accepted as a normal part of their environment. This has important significance in designing improved stalking and shooting strategies, as noted below.

Bird calls deserve special mention for, while an environment can easily be disturbed by alarm notes and other alarm behaviour, it can likewise be induced to relax and function normally when familiar sounds again come from the area disturbed. This process can be greatly accelerated by the stalker or observer artificially producing normal bird or other animal (e.g. frog) sounds. For example, territorial calls are normal in season, as are various contact calls, the cheeping of young birds, or the typical scratching sounds of ground birds feeding. Close observation of undisturbed behaviour of the animal stalked may reveal many normal sounds difficult to become acquainted with if one's only experience of animals is at shooting distances. The low call of a cow seeking a calf, a calf call, the low contact calls made between members of feeding groups in thick forests: any of such calls that can be imitated may prove of great value on occasion.

If normal calls are not known they can be learned readily enough by close observation of undisturbed animals in an undisturbed environment. A word of warning: the calls imitated must be calls associated with undisturbed animals. If warning, alarm or distress calls are imitated, this will of course accelerate the disturbing effect to the ecosystem and may direct the attention of the animal stalked to the particular site of the call.

7517 *Stalking Strategies*

These vary enormously. Although by no means complete, the following examples may suggest a range of choices, often used in combination.

7517A *A Time of Day*

In addition to movements associated with disturbances, animals naturally move daily between feeding, bedding and watering areas. Such movement patterns will vary by species and take place at different times of day in different areas. In principle, stalking is best done in woodland or forested areas at times when the animals are moving. In open areas, where visibility is good enough to locate animals at considerable distances, stalking for certain species may be more successful at times when animals are either bedded down or are feeding in restricted areas.

Many species that are wary by day are approachable at night. This means that shooting can be done at close range, resulting in fewer animals wounded or lost. Even the report of a rifle seems to cause less disturbance at night than by day, as the animals tend to move shorter distances and there is a lesser increase in wariness. Two night shooting techniques are common:

1. Shooting from a vehicle equipped with a strong spotlight and using a rifle and telescopic sight. A strong spotlight can also be used by a hunter on foot.
2. For smaller animals, shooting with a headlamp and shotgun has proved effective in some African animal reduction operations associated with control of tsetse fly.

The main disadvantage of night shooting in commercial harvesting for meat is the extra difficulty in the subsequent processing of the carcasses once the animals are killed.

7517B *Drives and Other Forms of Teamwork*

Drives can be an effective form of harvesting. They form one of the oldest strategies of harvesting, and are still used in rural areas of many developing countries. They are not always possible and even when they are, they are not necessarily the most effective method of harvesting. Drives normally involve a line of more or less equally spaced men moving in the direction of shooters who are concealed at advantageous places. Their general purpose is to drift animals into the range of the shooters, although there are many variations. For example, instead of shooting, nets or pitfalls or lines of snares may be used. Instead of men on foot, various types of land vehicle or aircraft may be used to move the animals. Sometimes drives will consist of a number of fairly widely spaced armed men moving forward and shooting the animals flushed. In this event it is still useful to have stationary hunters located well ahead of the line, others along the edges and a few following well behind.

When drives are used they require careful control and thorough briefing or training so that every person involved understands exactly his particular function and responsibilities. Briefing before a drive not only sets out clearly the procedure to be followed but should strongly emphasize local rules laid down to prevent shooting accidents. Communications between different parts of very large drives or those occurring in wooded or forested areas are much facilitated by the use of walkie-talkie radios.

The strategy of driving has been most effective in plains, in tundra areas or in areas of open woodland. Drives seem least effective in forested areas or mixed woodland or areas of dissected topography. The major difficulty in the latter areas is the tendency of animals driven to double back through or around the drivers, refusing to leave their home ranges.

Where this occurs a few shooters moving forward well behind the line of the drive may increase the harvest. In spite of this difficulty in areas with dense cover some form of drive may still be the best means of harvesting. Obviously some idea of the size of the home ranges and the suitability of a given species for driving should be determined before large-scale drives for commercial purposes are attempted. Small-scale localized drives can help in providing such information.

Many kinds of teamwork other than drives have been used, which involve two or more hunters. The only limits to the possible strategies are the imagination, ingenuity and skills of the members of the team. Penetrating an ecosystem in a different way often results in new insights and a better understanding of the cause–effect relationships between man, animal and environment. For this reason I believe any new approach is worth a try, particularly in situations where hunters are being trained or where a deliberate attempt is being made to improve the efficiency of hunting efforts.

In the latter case a useful approach in mountainous or hilly areas is to quietly post observers on good observation points well before a given hunting strategy is applied. The observers make careful note of when and where both animals and hunters are seen, and of responses of the animals to the hunters' presence. Hunters and observers then compare observations at the end of the day. This kind of exercise is usually rewarding in terms of increased understanding which can in turn be used to improve hunting strategies.

One form of two-man strategy is worth mentioning as a good basic starting procedure. In this strategy, two hunters 50–100 m apart alternately move slowly forward into the wind. One man stands quietly observing while the second moves carefully forward to a position where he can be seen by the first hunter. He then remains still while the first hunter in turn moves forward. This is a useful plan for country with a mixture of clearings and shelter, and it lends itself to many variations and elaborations.

7517C *Use of Vehicles*

In North America and Europe, where harvesting large mammals is mainly for sporting purposes, the use of vehicles for shooting is normally prohibited. In developing countries, particularly in open country with few people and where harvesting is on a commercial basis, the shooting strategy is often based on the use of vehicles, for locating animals or for driving, shooting, transporting meat or any combination of these. Many types of vehicles have been incorporated into one or another shooting strategies, including bicycles, motorcycles, cars, snow vehicles, aeroplanes, helicopters or boats.

Animals anywhere are easier to locate and approach if they are less wary. There is thus always a tendency for hunters to favour areas that have been rested for a time. In the more consistently and intensively hunted areas it is comparatively more difficult to maintain acceptable harvests for, as noted above, it is often the relationship between hunting pressure and animal wariness rather than low populations that limits the harvest. Vehicles, by extending a hunter's range of operations, thus make it easier to reach areas that have been less intensively hunted.

The nature of the impact on harvesting, and consequently the implications for management, will differ greatly with differing local conditions, as the following three examples will illustrate.

The flood plains of the Zambesi River in Mozambique consist largely of approximately 9000 km^2 of excellent grassland. The land is low-lying and criss-crossed with drainage patterns which are either permanently or seasonally inundated. On higher ground, between the watercourses and swampy areas, live some of the largest herds of plains game in Africa. Many herds of both buffalo and zebra, for example, number well over 2000 animals per herd. Although poaching has traditionally occurred for years along the edges of this vast area, access has been so difficult that the herds have remained virtually unexploited.

In 1976 the Mozambique Government started a programme for harvesting buffalo herds in this region. The programme was based entirely on the existence of several amphibious vehicles that could operate just as easily on swamps and rivers as on dry land. Shooting was done from these vehicles; without them this particular scheme would be impossible (Riney, T., 1977).

In the cool temperate mountains of the South Island of New Zealand, in the 1960s, another kind of harvesting developed which also depended mainly on a vehicle, in this case a helicopter. With differences in elevation between stream bed and nearby ridge of 3000 ft or more, red deer populations living at higher elevations could be shot but, since the region is largely without roads, it was impossible to exploit these populations commercially. Helicopters provided the basis of a hunting strategy involving landing hunters on the ridges and, later, collecting and transporting carcasses to the nearest accessible ground-vehicle collection point. The helicopter thus made possible the development of large-scale harvesting and a thriving trade with European markets in meat and skins.

An example of how vehicles may be important as a means of providing access and thus concentrating hunting effort comes from California, where deer management of various types has taken place for many years. In one study area, largely forested, it was calculated that in a 1947–48 hunting season over half of the mule deer killed in the most heavily hunted area were shot within 1 mile of the closest road (Leopold, A. S. et al., 1951).

Shortly after this study closed, and in the same area, the increase in four-wheeled vehicles considerably extended hunter penetration into previously un-hunted areas; a few years later motorcycles extended the distance from access roads even further. Eventually special regulations had to be made to cope with the uncontrolled growth in the use of such vehicles.

The key point here seems to be that access roads and the use of various types of vehicles can certainly affect the numbers of animals harvested and should be considered as one of the tools of management that can be used (regulated) to contribute to the goals of a particular management area. Regulation of the use of access roads and vehicles can thus indirectly be as effective in reducing or increasing harvests as the setting of seasons, or limiting the numbers of animals killed per hunter.

Although involving no different principle, the use of vehicles may have a special significance in developing countries where wild animals supply a significant part of the local subsistence diets. For example, in several regions in the 3000-mile-long woodland savannah belts in Africa south of the Sahara, commonly one man from each village was a hunter, who supplied meat for his village. Originally the meat was bartered locally. When bicycles became available in the region for the first time, these hunters quickly modernized. Because doubling the distance easily hunted away from a village increases the area available for hunting about four times, the hunter became more effective in supplying meat to his village. In recent years the sale of motorcycles[1] has further enlarged an individual hunter's scope by many times.

[1]This particularly refers to those motorcycles requiring only a mixture of oil and petrol, and that feature automatic gears and sealed units; consequently they require practically no maintenance.

While the effect of the early hunting strategy on animals is unknown, it was obviously conservative, for the traditional practices operated for hundreds of years without eliminating animals. The effect of the introduction of the bicycle is also unknown but thought to have simply increased the sustained yield. To predict the effect of the coming of the sealed-unit motorcycles is much more complex and difficult. Not only does the motorized transport greatly increase the radius of operations and facilitate the transport of meat, but these two advantages combine to change the character of the hunter's occupation. In addition to bartering he can now move meat into more distant villages for a cash return.

Vehicles, under some form of control, by extending a hunter's choice of daily hunting areas, make it possible for a greater portion of time to be spent with less wary animals and facilitate good sustained yields. If uncontrolled, vehicles can quickly lead to excessive kills and declining populations.

If wildlife is to remain a significant part of the protein diet of rural populations, then the time when a country can first recognize a trend away from simple subsistence hunting towards some form of commercial utilization of wildlife is a good time for wildlife management to start. This is a particularly critical period for those conspicuous or easily hunted species; their local survival may depend on the speed with which sensible policies for managing wildlife can become effective.

7517D   *Time and the Consistency of Hunting Effort*

Particularly if hunting is for sporting purposes, knowledge of the changing hunter/success ratios over time may be useful. Typically these ratios indicate a sharp decrease in success, starting shortly after the opening of a hunting season. As a typical example, in 1949, a study of mule deer in a mountainous area in California revealed that over 70 per cent of the total harvest was taken in the first 3 days of a 1-month season (Leopold, A. S. *et al.*, 1951).

This would suggest that extending the season another month would not have made much difference to the total kill, providing other relevant conditions remained the same.

7517E   *Planning Harvesting to Minimize Conflicts with Other Land-uses*

Game cropping and tourist viewing are usually incompatible because cropping increases the wariness of the wildlife, making them more difficult for tourists to see and photograph—and also because most tourists are not attracted to an area when animals are being harvested. With this in mind it may be necessary to separate the cropping from the tourist viewing, in time or space or both. The different activities may occur in the same area at different seasons, or separate areas may be designated for exclusive use either by tourism or for hunting. If the latter 'zoning' system is used the areas hunted can be changed from year to year as may be required.

Management strategies intended to cope with animal problems are most appropriately based on and designed around the problem and its required solution. Therefore, discussion of other ways of developing hunting strategies to minimize conflict between wildlife and various forms of land-use are deferred to a later discussion of animal problems (Chapter 8000).

7517F *Suggestions for Selecting or Modifying Shooting Strategies or Hunting Techniques*

An initial hunting strategy is usually chosen subjectively, whether designed as a control campaign, for commercial harvesting or for sport hunting. It is recommended that, preceding an initial decision, consideration should be given to the objectives of harvesting or control, the nature of the environments involved, various characteristics of the species and the amount and nature of shooting pressure that can be applied to the area. Shooting then starts, using a selected strategy. Normally, if the annual hunting pressure remains about the same, a more or less consistent annual harvest will result: a 'sustained yield' as discussed earlier in this chapter. The sustained yield is essentially the natural result of having some form of pressure consistently applied in an area year after year, providing the yield does not exceed the carrying capacity of that area.

In situations where the objective of management is to increase yields, one may simply experiment with various kinds of changes in shooting strategy, providing the changes are socially and legally acceptable in the country and region concerned.

However, in my experience, an increased understanding of animal–habitat relations, using different hunting strategies, can be gained much more quickly where the objective of management is to decrease yields: where animals are either to be reduced to lower numbers or eliminated altogether. If such areas do not exist in a given region the creation of a special experimental management area, where excessive control (over-kill) can be attempted on an experimental basis, is worth considering. For example, one easily modified procedure may be to:

1. Harvest, following one consistent strategy. This will involve a defined area and a consistent number of hunters. At the same time record total kills and total hunting effort. Continue until yields become consistent in terms of total kills and kills per man-week or man-day.
2. Apply a different strategy to the same area while maintaining the same records. For example, change the method of hunting or the number of kinds of hunters. Continue until yields become consistent.
3. Change the strategy again, until yields again become consistent. After a few such changes it is often revealing to change back to the original strategy as in (1) above.

Although forested and woodland areas are the most difficult areas to harvest, even there changes in hunting strategy can result in considerably increased harvests.

A normal response to change in hunting strategy is for yields to be temporarily increased, then subside to lower levels. However, when the harvest becomes stabilized at a lower level following a change in strategy, it may mean one of two things. The population may have in fact been reduced and the lower levels may be a result of fewer animals present. On the other hand, the animals may simply have become more wary while their numbers may have remained approximately the same, or have even been increasing.

If some index to population size has been made at appropriate times the situation may be easily assessed. However, if appropriate population indices are not available for an area of particular concern one simple way to find out if there are many more animals present than suspected is to try the following procedure.

1. Learn and record the details of the existing hunting strategy.
2. Assume that the animals concerned will also be aware of these details, not consciously in human terms, but in the sense that once hunting starts the animals become accustomed to being approached at certain times and in certain ways. One way of testing this in mountainous or hilly areas is by placing observers in strategic positions to watch hunters hunt and, at the same time, observe the reactions of the animals to the hunters' presence.
3. Do something different; that is apply a different strategy and observe the results. Similarly, strategically placed observers can significantly contribute to understanding how the animals are reacting.
4. If the first change does not quickly produce a higher yield per man-day, try another strategy.
5. If several changes in strategy do not improve yields even for a short period, while the animals are becoming accustomed to the new strategy, then one may in fact be dealing with comparatively low populations.

In terms of timing, no firm rules can be laid down but in my experience in forest and woodland areas sudden increases in yield associated with a marked change in hunting strategy will be recognized within the first 3 days. Stabilizing to a consistent lower level takes as long as 2 weeks.

The implications of this kind of information for wildlife managers should be clear. If a management objective is to hold a sustained yield at a certain level, managers should encourage a consistency of strategy and effort known to have already produced that yield. If the objective is to achieve higher yields, in addition to methods mentioned elsewhere, a change of shooting strategy alone may improve yields. Where the objective is to achieve maximum reductions in population a series of changes in strategy may be indicated.

7520 Immobilizing Drugs

In recent years, specialized weapons, such as special low-velocity guns, airguns and cross-bows have been developed to shoot hypodermic syringes that inject a drug on impact. These have been mainly used to immobilize animals for marking or for translocation, as mentioned in Section 3300, and have been suggested as a possible means of cropping animals.

At first sight, drugs may be thought to be useful for cropping animals in that they do not have the drawbacks of conventional firearms. In particular, they are practically noiseless; their quietness reducing disturbance. Indeed some successful trials were made by the late J. Uys, in Zambia, in 1963 in which a few complete family groups of elephants were taken (J. Uys, personal communication, 1963). However, drugs have other drawbacks; they are expensive, the guns and cross-bows operate effectively only at very short range, the drugs are slow to act and the risks to operators much greater than with conventional firearms. A small field research team, skilled in the use of immobilization techniques, can safely kill a few animals by injecting over-doses. However, at present, both the expense and added risk to operators would seem to rule out the use of drugs as a practical procedure for large-scale cropping operations.

7530 Snares and Nets

Most large mammals may be caught and killed with snares. However, this method is illegal in most countries because of the consequent numbers of maimed animals and because it is aesthetically repugnant. Large animals are very strong and snares have to be commensurately strong to hold them. Wire or nylon rope snares are most successful but even these break occasionally, allowing the animal to escape with the snare often cutting into its neck or limb. For this reason snares should never be left unattended. They should only be used as a means of holding the animals until shot or captured alive. A variation of the snare is the 'snare net' which can be made by splicing many snares into a line stretched across an area where animals commonly cross and stretched between trees, for instance. Animals are then driven into the snares and shot or captured immediately. Once again, snaring should only be con-sidered as a method of detaining an animal until it can be dispatched, and frequent visits should be made to all set snares.

Animals such as small to medium-sized deer have been successfully harvested by driving them into nets, where they are either dispatched or captured.

7540 Trapping

While trapping is the main method of harvesting fur-bearers and is exten-sively used in the control of problem animals, it is little used in harvesting large mammals. However, wider use could be made of small corral-type traps which may capture up to five or six animals at a time. Such traps are normally portable.

Corral traps are small enclosures with a device for automatically closing the trap with an animal or animals inside. They are easy to construct and use. Traps should be located on trails used frequently by a number of animals, or in well-used feeding areas. They are baited with a preferred food. Scents of various kinds can also be used to assist in attracting the animals to and inside the trap. The traps are normally closed by the action of an animal feeding on the bait which releases an elevated door allowing it to drop in place.

Materials for the sides of the trap may involve wood, or a collapsable metal frame covered with netting and opaque plastic sheets. For example, portable corral traps were improvised in a California study area where the research workers had access to cheap lumber from a nearby timber mill. The circular enclosures varied from 25 to 50 ft in diameter and were constructed of 8×8 ft board panels. The height of the trap walls will depend entirely on the jumping ability of the species concerned. In the California example, each panel consisted of nine 1 in. thick and eight 1 in. wide planks, each 8 ft long and nailed to 2×4 in. uprights. The lower three planks were spaced closer together than the rest (Leopold *et al.*, 1951). One panel was constructed with hinges near its top so the lower part formed a door which was held open by a small cable which crossed to the top of a panel on the opposite side of the trap where it led down through a pulley and ended in a ring which (as the trap was set) was slipped over the protruding end of a smooth $\frac{1}{4}$ in. bolt and held in place by pressure from the weight of the uplifted door. The door fell closed when the ring was pulled from the peg by a string attached to the snap wire of an adjacent rat trap screwed firmly to a panel. This string was strong enough to release the ring, which was adjusted finely at the outer end of the bolt, but was easily broken by the weight of the door. The rat trap was in turn sprung by a thread attached to the treadle of the trap and stretching in front of the bait.

After entering the open trap and approaching close to the bait, the animal presses against the string attached to the rat trap, thus springing the rat trap, which in turn pulls the ring supporting the door off its bolt, allowing the trap door to fall shut. If additional bait is placed at two or three places away from the trip wire it increases the changes of catching more than one animal in the trap at the same time.

Corral traps were originally designed to capture deer in forest conditions in order to mark them, or for transporting to other areas. The use of live traps could, however, easily be extended to small-scale harvesting operations. The advantages are that they are comparatively inexpensive to construct and maintain, are portable, and the animal may be kept alive. When in use the trap causes little or no disturbance to the population and when not in use the trap may remain open and unattended without danger to animals or man. Some disadvantages are that set traps must be checked frequently as with snares; for some species they may have to be frequently moved so that they may be contacted by new family groups, and trapped animals provide an easy prey for large predators.

7541 Corrals or Drive Traps

Some form of driving animals into an enclosed space is probably as old as hunting itself. Certainly, for hundreds of years wild Indian elephants have been captured by driving herds into massive wooden enclosures. In the 1940s and 1950s various game departments in southern Africa have developed the use of corrals (locally called 'bomas') to the extent that their use in some areas is a practical and economic means of harvesting several species of large mammal. These corrals may be infinitely variable but are essentially of two main types: permanent and portable. In both types the operation consists of manoeuvring groups of animals into progressively smaller spaces where they are eventually contained and held, either for moving elsewhere or for harvesting.

7541A *Permanent Corrals*

These corrals were developed in the 1950s by the Department of Nature Conservation in the Republic of South Africa for use in capturing large numbers of several species. Their objective was to capture and transfer animals to private farms or game reserves from which they had been hunted to extinction in previous years. The major species involved in these restocking operations were blesbuck, impala, springbok, zebra and both blue and black wildebeest. The use of corrals was found to be quicker, more economical and more effective than tranquillizers where the aim was to capture large numbers of animals and were for these reasons preferred (Riney, T. and Kettlitz, W. L., 1964). Permanent corrals became a feature of departmental game reserves.

Thus in the Transvaal each game reserve from which game is distributed has a capture pen specially constructed and located to serve the capture needs of that reserve. For example, one reserve built a combination of drift fences, holding pen and chute. There the opening between drift fences was 1 mile wide, with each drift fence 1 mile long. Game is drifted by cordons of men and horses into a stockade-type holding pen with a fence 8 ft high, and from this holding pen the animals move into a long narrow chute from which they are loaded on a truck.

No one type of corral or technique is equally effective on all mammal species. The Transvaal game capture operators have accepted this principle so that the use of corrals in the capture of their main species has developed along several lines. Here are a few examples.

Blesbok

In 1960 the Transvaal Department of Nature Conservation considered it necessary to capture at least 200 blesbok to make the operation economically worthwhile, because experienced game staff had to be assembled from all over the Transvaal. Blesbok are not segregated by sexes and normally entire herds are captured by gently manoeuvring them into the open funnel, then slowly shifting them down to the holding pens from which they are run into the crush pen. In one reserve, for example, the end crush at the end of the long funnel can hold about forty blesbok (see Figure 131). An important

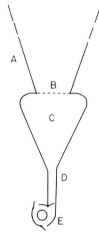

Figure 131   General plan of a typical funnel installation for capturing blesbok or springbok. (A) The outer wings consist of a six-strand wire fence $4\frac{1}{2}$ ft high, varying in length depending on the size of reserve or farm but for large herds extending up to 1 mile. (B) Again depending on herd size the size of the gate varies from an ordinary road gate to a gap 120 ft wide. (C) The keeping pen also varies in size; for example, for 500 blesbok the long sides should extend about 500 yd. (D) There is no gate at the constriction (to about 4 ft wide) the walls of which are still $4\frac{1}{2}$ ft high and consist of $\frac{1}{2}$ in. wire mesh. (E) This $\frac{1}{2}$ in. wire mesh also covers the walls of the circular catch pen, the 3 ft gates opening as indicated (after Riney, T. and Kettlitz, W. L., 1964)

detail of construction is that the width of the final chute preceding the crush pen must be about 1 ft wider than the pen width of 3 ft to allow blesbok to turn around. Without this important modification, the force of the blesbok behind piles up against those near the gate, and casualties occur. After not more than forty blesbok have entered the crush pen, the gate at the end of the chute is closed by means of a long wire leading 20–25 m from the pen to where a man can remain inconspicuous. The sides of the final circular pen are about four feet high and consist of heavy wire screening on a solid frame.

Springbok

Springbok may be driven into holding pens and crushes like blesbok, but with the following differences. They are considerably more difficult to drive than blesbok, and it is recommended that the drive should not commence until they are of their own according grazing well inside the mile-wide mouth of the funnel.

Before handling springbok it is important to alter the sides of the crush pen, making them solid, with iron or plastic roofing sheets, to prevent the horns from going through the wire netting.

412

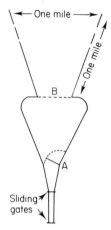

Figure 132 General plan of a typical drive trap used for wildebeest. (A) The 14 ft gate is operated from the outside and the final chute between the two drop-gates is $2\frac{1}{2}$ ft wide. Note the rounded corners on the holding pen to prevent injuries. (B) The opening of about 120 ft can be closed, usually with movable panels. The sides of the holding pen and the final chute are about 5 ft high and these fences are of wood (Riney, T. and Kettlitz, W. L., 1964)

Blue wildebeest

Blue wildebeest may be captured by driving into a holding pen, but it is useless to use a circular pen with low walls such as is suitable for blesbok and springbok, for one man cannot handle one animal. The corral plan is therefore modified as shown in Figure 132. Straight crush pens with sides over 5 ft in height are used, and once in the final chute wildebeest are moved into individual transportation crates. When wildebeest are manoeuvred into individual transportation crates this is usually done by three men handling the animals by the horns and tail.

Cape buffalo

Buffalo have been handled in a strengthened version of the wildebeest trap but tranquillizers must be used before handling and transporting.

7541B *Portable Corrals*

Portable corrals represent the biggest breakthrough in harvesting large mammals populations that has taken place in recent years. Particularly in the 1960s several commercial harvesting operations were successfully mounted in East Africa, but these involved mainly elephant and hippo both of which species were over-populated. Preceding this development by over 15 years, South Africa had been gradually improving the commercial use of wildlife so that, between 1955 and 1960 between 2000 and 3000 private farms and ranches were using wildlife as an increasing contribution to the total individual ranch income (Riney, T., 1962).

Although the South African developments at this period involved large ungulates, the circumstances permitting this development were not generally applicable even elsewhere in Africa. Mainly involved were a few easily harvested species, contained in ranches or game reserves. However, when attempting to harvest most of the large mammal species on extensive wild lands, South Africa was faced with the same main problem found throughout Africa. This was simply one of logistics. Before utilization could be recognized as successful, some method was needed to considerably improve on the existing practice of individual hunters shooting animals, which had then, somehow, to be extracted from the bush with the skin and meat in good condition. Furthermore, animals had to be taken in sufficient numbers to place the harvest or capture–removal exercise on an economically viable basis.

Portable corral traps provide one answer to this dilemma, and have already proven successful in southern Africa. This method has enabled very large numbers of animals to be captured; for example 3500 were taken in 4 months in 1969, with very little damage, losses or even major disturbance (Oelofse, J., 1970). Species thus captured were wildebeest, zebra, nyala, impala, kudu, warthog, waterbuck and Cape buffalo. Entire herds have been captured, transported and liberated as a unit. Eventually, when the technique can be more widely applied, it should prove equally valuable elsewhere.

Basically the large portable corrals are similar in plan to the funnel-type permanent corrals described above, but differ in that they are designed to be quickly erected and removed and are based on nets and dull-coloured, opaque, woven polypropylene plastic sheeting supported by steel cables. Since portable corrals are widely adaptable it is worth considering two variations to provide a better basis for further modifying and improvising.

Plastic capture traps

The first of these portable traps to become well known was the Oelofse plastic capture corral which was developed in Natal (Oelofse, J., 1970). The design of the Oelofse trap is shown in Figure 133 and a Zimbabwean modification is shown in Figure 135. Both designs have such common features as long wings and wide mouths to the trap. They are usually built with more than one enclosure and the final holding enclosure is reinforced with netting behind the plastic walls.

The Oelofse trap requires about 1600 yd (1463 m) of plastic in 100 yd (91 m) lengths, 10 ft (3 m) wide, preferably of a pale grey or beige colour to harmonize with the surroundings. Coloured plastic is used in the wings, as shown in Figure 133, to help shift animals into the mouth of the trap. And coloured plastic strips when tied on the curtains that close behind the animals seem to encourage the animals to move in the opposite direction; that is

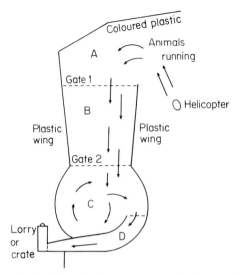

Figure 133 Plan for the Oelofse plastic capture corral, after J. Oelofse (1970). Animals are drifted into the vicinity of Gate 1 either by lines of men or aircraft. The gate at the entrance to the outer compartment (B) is about 200 yd (183 m) wide; the inner compartment (C) is smaller and roughly circular, with an exit chute (D) for leading animals either to crates or lorries for transport, or for killing and processing in the adjacent area

deeper into the trap. About 800 yd (732 m) of nylon netting is required to strengthen the final capture compartment, beyond Gate 2; 3200 yd (2926 m) of $\frac{1}{4}$ in (6 mm) steel cable, in 100 ft (31 m) lengths and fitted with hooks and eyes to make joining easier, are required, along with a cable strainer. Four sides of timber, 14 × 8 ft (4.3 × 2.5 m) are needed for the crush pen; two 14 ft (4.3 m) sides for the loading ramp; one 14 × 4 ft (4.3 × 1.3 m) floorboard strong enough to carry the animals climbing into the lorries and eight 14 ft × 3 in. (4.3 × 7.5 cm) poles to be planted in the ground and to which the loading pen partition and loading ramp are tied to stabilize the loading section of the trap.

The approximate size of the Oelofse trap is about 300 yd (274 m) deep and 200–300 yds (183–274 m) wide at the mouth. Cables are strung 8 ft (2.4 m) above the ground, and a lower one on the ground and immediately below the first and tightened with a cable strainer. The plastic material is then tied to both cables with short pieces of wires, or preferably with hooks as illustrated in Figure 134.

Zimbabwean modifications

The Zimbabwean Department of Wildlife Management and National Parks uses both a modification of the Oelofse design and a funnel design to capture

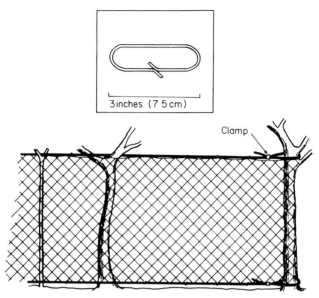

Figure 134   Method of attaching net or plastic to cables at 2 ft (55 cm) intervals with wire or high-tensile steel hooks (insert). Cables are strained by winch before clamping

a wide range of species (Child, G., personal communication). Because of its economic importance and widespread distribution in eastern and southern Africa, the latter, a corral specially designed for impala, has been emphasized but, as noted below, various modifications are easily made for other species.

The impala corral consists of four compartments: an entrance compartment (K) and three progressively smaller working compartments, L, M, and N, the latter ending in a crush leading either to a loading ramp or carcass-processing area (see Figure 135). Enough plastic sheeting is required to extend 1105 yd (1010 m), of which 415 yd (380 m) are required for curtains dividing the compartments. In addition to the plastic, compartment N has walls made of ¼ in. nylon netting 8–10 ft (2.4–3.1 m) high. Walls are supported by ¼ in. steel cable. The entire trapping equipment is carried in two 5-ton trucks. The number of people involved is three supervisors and forty labourers. Three walkie-talkie radios are used, and radio contact is maintained with a helicopter or light plane; these are particularly useful for driving and for giving the signal when to close the openings. Time taken to erect a trap by an experienced team is under 2 hours.

Failure or success depends on several properly executed working practices. For this reason various procedures based on a combination of South African and Zimbabwean experience are reviewed in terms of various stages: finding a suitable site, erecting the corral ('boma') positioning the staff and making the drive. If the animals are to be transported, loading the animals on trucks and transporting becomes the final stage. If animals are to be processed for

416

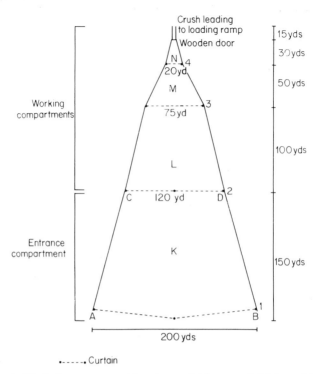

Figure 135 A Zimbabwean trap with a large, fairly open initial catching area and three successively smaller handling compartments into which smaller numbers of animals can be manoeuvred and held pending the final harvesting operations. Numbers 1 to 4 represent the location of curtains. The gates A–B and C–D are closed by two curtains which meet in the middle

meat, the final stage will be killing, removing the intestines and placing the carcasses in a cool or cold truck for removal to a refrigerated storage depot, or, if distances are not too great, directly to market.

Finding a suitable site

Ideally an aircraft must be used to get an idea of animal densities, distribution, etc. Once an area has been found where the impala are concentrated (e.g. near water) the team then looks on the ground in that immediate area for a suitable corral site.

As one can see from Figure 134 the corral is ideally set in a small belt of trees and shrubs with the entrance just tucked in the shrub, leaving the front of the corral fairly open. This will make it easier for the plane or helicopter to drive. In compartment (K), Figure 135, the area is still fairly open with a number of game trails leading further into the corral. An area is sought for compartment (L) which is covered with trees and shrub at its entrance (gate C–D) so the impala have no sight of the plastic sheeting further inside the corral.

In open country camouflage is achieved by positioning the site behind a ridge, using poles instead of trees to keep the cables and plastic material in position. J. Oelofse (1970) suggests that, if possible, the site should be up-wind of the animals to make the driving operation easier.

Erecting the corral

The erection of the trap will involve (a) stretching the cables using as many trees as possible; (b) affixing the netting; (c) either raising the opaque plastic sheeting or placing it, camouflaged, in instant readiness for erection; and (d) arranging for closure of the entrance, and other subdivisions within the trap as required for different species. More specifically, the corral is erected by stretching the cable between trees at a height of 8–10 ft above ground. Another cable is stretched along the ground. Both cables are brought up to high tension by means of a winch. The net walls of compartment (N) are then tied on top and bottom with either bailing wire or hooks at 2 ft intervals.

The plastic is then tied to the bottom cable in compartments (L), (M), and (N), folded up and camouflaged with grass and leaves. The top of the plastic is not hooked onto the top cable until the animals have passed through the gate (C–D).

Once the sides of the corral have been erected all the curtain cables are put up. Pieces of wire may be placed at intervals of about 3 ft (1 m) on these are cables to facilitate attaching the plastic as soon as the animals have entered (Oelofse, J., 1970). The plastic material for the gates is folded at each side of the gates, so that they may easily be drawn across the opening. Note that the first or outer gate (1) and the second gate (2) in Figure 135 each consists of two curtains (A) and (B), and (C) and (D) respectively, while gates (3) and (4) are both single curtains.

The crush and ramp are then erected at the end of the corral. If vegetation permits it should not be seen from the entrance to the corral.

Netting is of obvious importance in the final enclosures, as it is only here that the animals are apt to come in contact with the walls. Also the plastic in these sections should not flap in the wind so that animals are alarmed or can see the way outside through gaps in the walls (Harthoorn, A. M., 1976).

Positioning the staff

Once the corral has been readied and the drive is about to commence the labourers are placed in groups around compartments (L), (M), and (N), ready to lift the plastic. Foremen with walkie-talkie radio and three assistants are concealed behind curtains (A), (B), (C) and (D).

Everyone involved, both on the ground and in the air, must be thoroughly acquainted with the precise plan of capture and particularly with the assign-

418

ment of responsibilities for closing gates, raising plastic sheeting and so on. During the drive it is of course absolutely essential that personnel at the trap remain quiet and out of sight of animals. Vehicles will also be removed from the vicinity of the trap before the drive commences.

Making the drive

For drifting and manoeuvring the animals into the trap, and for signalling the ground crew as to the precise moment for closing the trap curtains, the use of aircraft is an essential part of the capture operations. This requires not only a suitable craft—preferably a helicopter—but a consistent, experienced, interested pilot. The practice of hiring aircraft is normally not satisfactory.

With the ground crew out of sight but in position, the aircraft then takes off with one passenger who maintains radio contact with staff on curtains (A, B, C, and D). Once a herd of about 100 impala has been located the aircraft proceeds to drift and manoeuvre the desired herd or group of animals towards the opening of the corral, the aircraft acting like a sheepdog. This may take up to $1\frac{1}{2}$ hours if the animals are about 2 miles from the corral. However, distances normally are but a few hundred yards. Impala have been driven from 3 miles and buffalo from up to 10 miles into such corrals by helicopter.

Once the animals are inside gate (A–B) and in a position favourable for closing, the signal is given—usually by an observer from the air, curtains (A) and (B) are quickly closed and the bottom cable is then brought to full tension between the curtains as during the drive this cable is buried so as not to distract the impala.

The impala then make their way through gate (C–D) after which another signal is given by radio for personnel operating curtains (C) and (D) to quickly close. As curtains (C) and (D) start to close this is also the signal for the personnel placed around compartments (L), (M), and (N) to hook the plastic on the top cable. At the signal the plastic is immediately held high by labourers, while others complete the hooking operation.

The animals are then further subdivided into smaller compartments and transported, shot, or rested as is appropriate to the species and objectives of capturing. In the case of impala the adult rams are quickly destroyed as they gore the females. Zimbabwean operators suggest that this is through frustration at being confined in a small space. The remainder of the impala herd is then left to rest for a few hours. They have the compartments (L), (M), and (N) in which to roam as curtains (3) and (4) have been left open.

In the final enclosures the animals may be separated into groups or shifted into the final compartments by rows of men carrying a length of plastic sheet before them like a wall (Harthoorn, A. M., 1976).

Modifications easily possible

Flexibility seems the key to the success of the portable game trap. With the same materials, labour and time, the trap can be constructed according to either an Oelofse or a Zimbabwean funnel design and from these other modifications can be made to adapt the trap for capture of most large mammal species (see Figures 133 and 135). For example, the funnel shape seems best for impala, eland, zebra, waterbuck and hartebeest. The round (Oelofse type) trap seems most appropriate for buffalo, kudu and wildebeest.

In capturing impala using the funnel design, cables and nets are erected and the plastic sheeting is camouflaged on the ground but not raised. Once animals are through the entrance the plastic is quickly raised and held in position by tyres on the bottom and hooks on the top. Table 15 shows suggested modifications for other species.

Hartebeest are caught in a trap similar to that used for impala, but with hartebeest the most critical aspect of capture is the selection of the trap site, usually done from the air.

Trap location is also of great importance in capturing waterbuck, for the distance of driving should not exceed a few hundred yards. For sable the plastic should be already up and wings should extend in a funnel shape on either side of the entrance.

Several herd species can be corralled close together with a minimum of fighting, but there are of course exceptions. Zebra provide a special problem in that they become difficult to manage in close quarters as they kick and bite. The practice is to keep them inside the pen for a few days until they take their first water voluntarily and even eat a little of the hay that has been left for them. After this further transport is much easier (Harthoorn, A. M., 1976).

Where buffalo are concerned (Figure 133) the Land Rover is driven inside the corral which is 200–250 yd (183–229 m) wide at the front and with no wings leading out from the opening. The plastic is already up and in position before the buffalo are driven in. The shape of the trap is approximately circular. Once the animals are inside, the compartment is divided into smaller compartments to minimize the disturbance factor.

Table 15   Suggestions for initial positioning of netting, plastic and wings for several species and traps of both designs

Species	Funnel Type			Oelofse Type		
	Netting	Plastic	Wings	Netting	Plastic	Wings
Impala	Up	Down	Absent			
Eland	Up	Down	Absent			
Waterbuck	Up	Down	Absent			
Hartebeest	Up	Down	Absent			
Zebra	Up	Down	Absent			
Sable	Up	Up	Present			
Buffalo				Up	Up	Absebt
Kudu				Up	Up	Absent
Wildebeest				Up	Up	Present

Wildebeest are caught in a trap similar to that used for buffalo but with wings extending out from the entrance.

Roan antelope tend to cause fatal damage to each other unless special precautions are taken. 'High percentages of losses have occurred in some cases ... during roan capture, and no losses in other exercises. A large enclosure with plenty of cover and gentle driving with a reduction of panic to a minimum appear of paramount importance' (Harthoorn, A. M., 1976).

J. Oelofse (1970) notes that the helicopter should fly low and for most of the drive move the animals slowly. But they may be stampeded when the animals come within two or three hundred yards of the gate, or otherwise when at a distance at which it is certain that the animals will enter the corral. The helicopter remains in the air in the vicinity of the outer gate until the outer curtain has been closed securely. Sometimes, when the animals are driven on foot instead of helicopter, they are walked in (Harthoorn, A. M., 1976).

A. M. Harthoorn (1976) rightly notes that a great deal of skill and experience is required in the siting of the funnel and enclosures, and the driving itself. Considering the expensive outlay in materials and helicopter he urges that those planning to use this method first consult experienced persons who have already used it successfully. If experienced persons are not available, correspond with them. Start small, profit by trial and error and gradually develop your own procedures for your own area and species. The results are well worth the effort.

If such traps are scaled down in size and adapted for deer that commonly feed in certain areas during the day, the following procedure may be used:

1. Determine normal feeding times and normal daily movements between bedding and feeding areas.
2. Prepare trap while deer are in bedding areas. Leave open and unattended for 2 or 3 days or until deer movement becomes normal with the trap in place.
3. Operate trap in the usual way, either by waiting until the deer are inside the (now enclosed) feeding area or by driving.

7550 Animal Loading and Transport

The loading of trucks is normally done when the temperature is cool.

The herd is then driven up slowly with a moving wall of plastic sheeting, as mentioned above, in the direction of the crush. The idea is to cut the herd at either curtain (2) or (3) into groups suitable for a truck-load.

Once animals are in compartment (N), they are again moved up slowly with plastic sheeting until they are in the crush with the net and curtain (4) closed behind them. From here, with a little gentle persuasion, they are manoeuvred into the truck which has a sliding door that can be shut quickly as the last animal enters the truck. The truck is then started immediately as the vibration of the engine has a noticeable calming effect.

Standard-sized wooden crates can be developed, satisfactory for animals ranging in size between eland and impala. The major features of such crates are that they are highly planed and smooth on the inside, with rounded edges. The top is slatted but the sides are closed. Various special crates are necessary for exceptional animals, such as the giraffe. In addition to a larger crate, giraffe in southern Africa are transported with a special head-harness covering the eyes, as this not only reduces irritation to the eyes from the wind but has a calming effect.

Dr A. M. Harthoorn (1976) suggests that animals being transported may be divided into two groups: animals moved elsewhere immediately after capture; and those moved only after becoming accustomed to captive conditions, artificial food and living for several weeks in a holding enclosure. He notes that the first group have the advantage of frequently possessing better physical conditions and resistance to stress. The second group has less fear and is more familiar with crates and man's activities. It has the further advantage of having adapted to an artificial diet, which can be an advantage on very long trips. He further warns that if animals are held for only a few days in holding pens and are moved too early they may be in a considerably worse state than those newly captured and transported immediately they are caught.

7551 *Practical Guidelines*

With experience, a list of practical points or guidelines to improve operations will naturally grow. Here are a few such guidelines as described by an experienced Zimbabwean trapping crew.

1. Do not destroy vegetation in outer gate of corral. Leave as natural as possible. Animals will detect anything unusual.
2. Absolute silence must be maintained whilst erecting the corral so that animals do not move out of immediate vicinity.
3. If animals are kept in corral over 36 hours water must be given.
4. Impala must be guarded against leopard at night as they are very susceptible–and the leopard know it!
5. Never get excited (shout, etc.) when driving the animals in the corral with plastic—take it calmly and move slowly, and so will the animals.
6. Vehicles transporting animals to new areas must stop (engine off) at least once every 2 hours for 15 minutes so that the animals can urinate.
7. The crush area must be watered down so as to cut down on dust which may hamper the loading, for example animals may not see entrance to truck.
8. Holding the animals in the crush and leaving them quietly for a few hours before loading has a calming effect—thus there is no chaos up the ramp when loading.
9. The $\frac{1}{4}$ in. (6 mm) cable is easy to handle if rolled up onto small drums in 100 ft (91 m) lengths.

7552 *Uses and Limitations*

Portable corral traps have their limitations. Reedbuck, oribi and bushbuck, for example, are more appropriately caught by manual drives into nets. For very large mammals, such as elephant, giraffe, hippo and both species of rhino, animals are captured individually through the use of tranquillizers.

Mobile traps, as exemplified above, are particularly useful to departments who wish to capture large numbers of ungulates, as in reduction and reintroduction programmes. While the expense of building and operating such a trap would probably be unjustified for research purposes alone, the little extra time and expense required to individually mark animals being transported for release in another area would seem to be well worth the trouble.

Portable traps represent a step forward in recent progress towards commercial harvesting of wildlife for meat, skins and other products. Harvesting over-populations of elephant and hippo were immediately worthwhile without such traps because of the size of the individuals. However, before the portable corral traps it was difficult to get enough ungulate carcasses together to fill a refrigerated truck, for example. Now the concept of multiple pastoralism, utilizing commercially several species of wildlife, becomes practically possible.

The above section on methods of harvesting is by no means complete. It should be clear, particularly from realizing the recent date of the invention of the portable large corrals, that the future will certainly hold other important breakthroughs. However, hopefully, the range of choices discussed above will stimulate the reader to develop modifications and inventions of his own. For example, many variations can be developed which would fall in between the portable, self-tripping corral trap and the large portable Oelofse or the Zimbabwean corrals.

While the large portable traps are not appropriate for most deer, smaller versions can be used for those deer species which herd together in considerable numbers and with comparatively gentle topography. The use of smaller net and plastic traps instead of the wooden panel traps mentioned above can be considerably developed, particularly in areas where a combination of commercial forestry and the commercial utilization of deer is desired.

Many improvements are still needed to improve the effectiveness of harvesting woodland and forest animals. The practical wildlife manager will not wait for these needed techniques to appear in technical journals and handbooks such as this, but will go out and get the job done by inventing his own traps, capture devices or harvesting methods.

Chapter 8000

Animal problems

8001   INTRODUCTION

In this chapter, animal problems are used as a means of summarizing and re-identifying various concepts mentioned earlier, and an attempt is made to put them into some kind of practical perspective. This is done partly through emphasizing some basic attitudes to wildlife management, 'a way of looking', and partly by emphasizing problems as a basic core of reality around which wildlife studies and management and the formation of wildlife policy may be based.

Although it has become a recognized discipline, wildlife management is interdisciplinary by its very nature. Changes in habitat inevitably affect wild populations and since the most common and dominant cause of change is man himself, wildlife management is (whether we would have it so in our ideal world or not) inevitably involved with social attitudes, with economics and with politics. Thus the management of wildlife commonly involves compromises and is almost always a synthesizing process. The job of wildlife surveys and research is one of asking relevant questions, obtaining relevant information from either animal or environment or both, drawing appropriate conclusions and making consequential recommendations, testing the recommendations in practice, monitoring the results and modifying the management practices accordingly. This process goes on and on for management is neither a one-action or a one-year affair. Effective management must always be a continuing process.

In the preceding chapters many techniques have been described for gathering facts. The scheme of classifying techniques as described in Chapter 3000 was designed to help the field worker classify and perhaps be a bit more objective about wildlife techniques, which after all are not important in themselves. They are simply useful or not in terms of achieving long- or short-term research or management objectives; in helping to solve some problem that may represent a difficulty of the moment. Developing a specific working plan (such as described in Chapter 2000) is usually a safe way to plan a practical effective programme of action. This is an important early step towards clarifying objectives, after which relevant techniques can then be considered.

Throughout this book phrases such as 'appropriate evidence' and 'relevant techniques' have been used. One may well ask: appropriate or relevant to what? The easy answer is . . . to the objective or job in hand. This generalization at first glance may seem of little use when one considers the many kinds of animals living under so many vastly different habitats and under different social and economic conditions. The task of describing detailed management possibilities for each species of large mammal is often regarded as a practical impossibility, given our present limited levels of understanding. But if our thoughts, energies and activities are oriented around and directed towards a specific clearly defined problem then we have a 'key' as it were, that will allow us to work within even the most complex ecosystem, in a meaningful way.

With all this in mind, as a kind of synthesizing way of summarizing the book, it is useful to consider some simple suggestions for dealing with a selected series of problems of various kinds.

First, by way of selective review, consider what Dr J. Lockie has referred to as 'a way of looking' which may suggest a useful frame of mind or approach to the 'animal problems' aspect of wildlife management. Then selected problems will be discussed on a simple 'how to clarify' and 'what to do about it' basis. Hopefully this may be one way of fitting together the complex jigsaw puzzle of animal–environment relations in a way that makes a little easier the job of dealing with wildlife problems. The secret lies in selecting only the kinds of information from animal and from environment which are relevant to the problem in hand. Finally, a few examples of problems are included as examples to illustrate how the handbook may be used in practical ways.

8100 A WAY OF LOOKING

It should be clear from preceding chapters that the natural phenomena with which we are concerned involve ever-changing interactions between animals and their environment. Wildlife management consists of using, usually manipulating in some way, a complex of habitats and animals to achieve some purpose. The wildlife manager seeks sufficient understanding of these interrelationships to allow him to organize his programme of action in the best way to achieve his particular goal.

The student of wildlife may take any one of a number of approaches. He may relate the individual animal to certain aspects of the habitat it lives in; he may investigate the relation between a herbivore and its predator; he may study the effect of a single herbivore species on a single plant species. Any such specific study may contribute extremely valuable background data and may be vital to a specific management problem. They will, however, rarely in themselves solve the problems that face the manager of large wild mammals. This is because many animal problems are basically due to an imbalance in the ecological system of which they are a part, and ecological systems can be complicated.

The ecological system or *ecosystem* is a functional unit, the basic ecological unit (Evans, F. C. 1956), and is composed of interrelated parts. It can be any size: the world, or your garden. It can be a national park or game reserve or managed hunting area, or a particular study area within the park or reserve. The parts of an ecosystem (the areas where we work) are, for example: animals, plants, soils, climate and people. A stable ecosystem has all parts working in harmony, but the balance is always in adjustment since the parts of the system are so closely interrelated that any upset to the ecosystem is reflected in changes to one or more of its parts; a significant change today in one part may in time be reflected through the rest of the system. We can easily detect such changes. The real problem, however, is to determine the cause of the imbalance and this usually requires that the ecosystem be looked at as a whole. If one studies part of a problem one gets, at best, only part of the answer.

But an ecosystem is extremely complicated, and a serious problem for the investigator is to know where to break into the system and what to measure in order to answer his questions. There is no shortage of factors to measure; the question is, which are relevant and which are not?

A multitude of techniques are available for making measurements. However, using even precise techniques and measuring many factors does not guarantee solutions to your problems. Techniques are tools and, like other tools, are useful only when they are correct for the job. Likewise, the measurement of factors of the environment is only useful if it contributes to the solution of a specific research or management problem.

The following sections expand the idea that the ecosystem is the unit of study, that one must view it as a whole and that one must measure the parts relevant to the problem as a means of defining as well as solving problems that involve large mammals and their environment.

8110 Asking the Correct Question

Animals in themselves are not pests. They only become so when they conflict with man's interests or activities. Usually the problem situation is not seen as a whole and it may be presented in a misleading way. It then becomes necessary to rephrase the problem so that it may be more clearly seen in terms of the whole environment.

Consider some problems as they are commonly presented to the wildlife manager, how they might initially be viewed and how the questions posed might be rephrased.

A common animal problem as presented to the wildlife manager is: *this animal is too numerous, how can we get rid of it*? However, the mere statement that an animal is too numerous does not constitute a problem. The statement should be rephrased thus: *to what extent is the animal detracting from the major use of the land in question*? In this way the relevance of the complaint about numbers is put into perspective and it is stated in a way that allows useful study.

It may well be that the animals are doing serious damage. In some instances this may require their destruction, although control schemes of this sort rarely achieve eradication and so are continuous activities and a constant drain on finances. In other instances, the damage is done by animals prevented, by some upset in their environment, from fulfilling their normal annual succession of activities, such as seasonal movement.

It is worthwhile then to ask: *why are these animals pests at all*? It may be that fencing, a change in land-use, or increased shooting by poachers has driven the animals to behave differently and so to come into conflict with a particular form of land-use. It might be possible to ease the situation at the source. If so, the solution would be permanent.

Where wildlife is a tourist attraction, there is a temptation for managers to increase the numbers of wildlife in order to provide bigger and better spectacles. The question might be asked: *how can we have more animals at this particular site or how can we have animals all the year round at this site without detracting from its value in other ways*? A knowledge of essential habitat requirements (see Chapter 5000) should facilitate relevant analysis of the study area and eventual improvement of the habitat by strengthening weak links. For example, since water is often one limiting factor in semi-arid areas the provision of more waterholes in the dry season may bring the animals in or hold more animals nearby.

This, however, takes no account of other aspects of the animals' habitat, particularly food. In dry areas the habitat usually suffers by too generous an artificial increase in available water. The resulting damage to the habitat by the resulting concentration of animals may be so bad as to ruin the wildlife spectacle for many years. It is all too easy to make a quick and superficial decision and thus to kill animals with kindness.

The question that should have been asked, given that more animals were wanted, is: *will the total habitat in this locality stand an increase in the numbers of animals*? The answer to this question decides whether we even try.

Taking another example, the statement may be made that certain species of ungulate are carriers of sleeping sickness. Therefore, in order to eliminate this disease these animals must be killed. The question commonly asked of the wildlife manager is: *how best can we kill off the ungulates*?

This question, however, is based upon an assumption which may be a dangerous basis for a control programme. The basic question which should have been asked is: *what effect will the elimination or drastic reduction of ungulates have on tsetse control*? This question may in turn stimulate other questions such as: To what extent is it necessary to eliminate animals to achieve control of tsetse? Are there other alternatives? Which is the cheapest? If ungulates are to be removed, which species?

These questions may give rise to a wholly different and more relevant answer to the problem of the relation of animals to sleeping sickness than the first question posed ('How can we kill off the ungulates?')

Faced with a declining cattle or sheep industry on marginal land, the rancher often tends to blame the most obvious feature of the problem area that is outside his control. Consequently wild herbivores are often in a situation where influential graziers are making statements like this: 'These wild animals are competing with cattle (or destroying my soil and vegetation) and reducing my profits. If something is not done soon the cattle (or sheep) ranchers in this area will be forced to abandon their land.' The implication is that the wild animals must therefore be removed. The statements may be perfectly true and the implication correct.

However, the wildlife manager cannot assume this statement. If he does there is little he can do but organize a campaign to get rid of the wildlife. Instead he would be well advised to ask questions that will help him understand more clearly the cause or causes of imbalance in the ecosystem. He might well ask a general question like: '*What is the relation of both wild and domestic animals to soil and vegetation?*' Or, more specifically he can ask: 'What feature(s) of the habitat are deteriorating?' What is the relation of each animal species to each of these features? What are the habitat requirements of each species and to what extent is each domestic and wild species competing or contributing to deterioration of the habitats? There are usually several ways of asking a question without having the assumed answer built into the question, just as there are several ways of cleaning a fish, skinning an animal or lighting a fire.

In many instances in different parts of the world, it has been shown that domestic animals were too numerous and were themselves destroying the habitat. In the presence of vested interests and in the absence of relevant information wildlife is often unjustly blamed.

These simple examples are given to make the point that questions must often be rephrased both to make them more relevant and also to alllow them to be studied in such a way that increases chances of a solution.

However, in order to ask the correct questions one must have some knowledge of or feeling for the situation. This in itself may require study; it also requires an approach to complexity that on the one hand deals adequately with the animal in the ecosystem and on the other selects what matters in the context of the problem.

8120   Signs and Symptoms (the syndrome approach)

Any disease has a number of symptoms by which it can be recognized. Doctors use combinations of symptoms to make diagnoses. This is possible because the various parts of the body and their functions work in harmony when the body is healthy, but are disrupted during illness. The assembly of symptoms that characterize a disease is termed the *syndrome* of the disease.

An ecosystem can be looked at in the same way. In health, its various parts are in a harmonious relationship. An upset to this harmony produces 'illness' and a pattern of symptoms (a syndrome) peculiar to the 'illness'. A rapid increase of one animal species, the disappearance of an age class of trees, an increase in the speed of erosion, all are examples of signs of 'illness' or upset in nature.

The syndrome approach to problem-solving looks at these signs singly and in combination and assists in learning both the fact and the nature of the disturbance that the ecosystem has suffered. This allows problems to be more accurately defined.

The syndrome approach is infinitely variable and thus not readily defined as a rule-of-thumb approach. However, as part of a 'way of looking' it seems worth emphasizing. A typical example of this approach discussed is the recognition of various stages of an eruptive oscillation (see Section 7100) which can best be done using a syndrome, or combination of different kinds of evidence, as described.

8130   Indicators of Illness and Health

In order to put the syndrome approach to problem-solving into practice, appropriate indicators of change of state must be located. This in itself may require survey or research. Then the indicators must be measured.

Since the state (or present condition) of an indicator is a symptom and since, as we have seen, symptoms may suggest an upset in the ecosystem, it follows that at least some indicators may tell much more than simply the state of the indicator animal population or indicator plant that is measured.

I think of these as 'collective indices' but labelling them is not as important as recognizing that the present status of, and the nature of, change in certain selected indices may reflect the state of health in the ecosystem as a whole.

There are parallels in other fields. Temperature in man is an indicator that gives a clue to the state of the person's health. Peace between nations might be taken as an indicator of social stability. The number of industrial strikes is a measure of the much wider field of work/employee relations. Certain indices may reflect the state of a whole system.

Some examples of useful 'collective indices' are: a cover of annual grasses occurring where perennial grasses should be, means the succession process has had a setback for some reason; the presence in large numbers of rodents or guinea fowl, doves, wildebeest, elephant or any other single species indicates a past upset to the ecosystem (usually as a result of heavy grazing or too frequent fires); a recent increase in the perennial grass cover may mean that a previous upset is now righted and that the animal, plant, soil relation is improving; the presence of a heavy load of silt in a permanently flowing river can mean accelerated erosion is occurring upstream.

Measurements are usually made in the form of an index or *indices* rather than absolute measurements. This is because absolute measurements are time-consuming and therefore expensive, whereas many indices are quick and easy to measure. As a general rule, the measurement one makes should not be any more complicated or accurate than it needs to be for the solution of the problem. Thus for some purposes a total count or estimate of an animal population is necessary; in other instances *relative* differences in density (for example by dropping counts, or numbers of animals systematically counted along roads) may be adequate. It may be necessary sometimes to know the actual rate of revegetation of bare ground while in other instances the demonstration of recent recovery is enough.

8140   Thresholds of Change

It is a matter of common experience that there are levels of intensity above which something works and below which it does not. Anyone who is not colour blind will know (and can check for himself) that colour vision only occurs in daylight. As the intensity of light fades at the end of the day, there comes a point at which colour vision switches off and monochrome night vision begins. We say that there is a *threshold* of light intensity below which colour vision does not work. Likewise, there is a threshold stimulus needed to make a muscle contract. There is also a threshold pressure which will most efficiently operate the trigger of a rifle. Anything less will not operate.

Thresholds are common in everyday experience and are probably common in nature too, but less in known about these.

One kind of threshold, the threshold of visibility, has already been described in Section 3426B. Because of the operation of this threshold a population may be increasing but unseen or rarely seen until the threshold is reached. Once the threshold is exceeded animals may be readily observed and further increases in visual indices may correspond to actual increases in the population. Conversely, a population may be thought to be greatly decreased by shooting, and a standard and careful roadside count might be used to 'prove this' while in fact increased wariness may have simply raised the threshold of visibility. If a census method is based on counting animals in the field, the automatic operation of this threshold can lead to quite false conclusions for even large changes in counts of animals seen or tallies of animals killed, for they are not necessarily related to corresponding changes in actual numbers of animals in the area.

Not only are there thresholds above which something happens or does not, but there are also thresholds at which the rate of change alters. Imagine a habitat that is being too heavily grazed and perennial tussock grasses are slowly disappearing, leaving bare ground between the clumps. Erosion increases steadily through the action of water and rain as more and more ground becomes exposed. But there comes a threshold when so much ground is bare that erosion speeds up and the habitat deteriorates more rapidly. If we remove the cause of the disappearance of perennial grass *before* the threshold is reached, recovery can take place. However, if the cause is removed *after* the threshold is reached, the habitat may continue to deteriorate; or if the cause of deterioration is removed after the threshold is passed it may be very difficult and slow to bring the habitat back to what it originally was.

It is possible, though proof is lacking, that some animal populations cease to be viable below a threshold of size that allows the appropriate social organization. Below this threshold it may be impossible to protect the remnant.

There are doubtless many thresholds of processes in nature that are still not appreciated. They are important in the practice of wildlife management. Perhaps the reader can himself discover other thresholds since they are most likely to be seen by observant field workers.

8150   When Cause and Effect are Separated in Time

It is usually easy to observe the causes for short-term changes between two parts of the ecosytem. When major changes take place this too is often simple to explain if you happen to be there at the time. But if you enter the area 5 or

100 years after even a major environmental disturbance, the interpretation of what you now see may be more difficult.

The changing course of a river, the large-scale clearing of a forest, the introduction of domestic animal grazing for the first time or the sudden emptying of land by people who had intensively occupied it for many years are types of ecological disturbances which may take dozens or hundreds of years for the consequent effects to work their way through an ecosystem.

Being aware that present effects, in the form of existing animal–habitat relations in a problem area, may spring from causes a few or many years ago is an important kind of awareness to cultivate. This may prove a useful attitude in any assessment of relations between animals and their environment. It is particularly useful to keep this time-lag possibility in mind in understanding animal problems.

This concept can lead to an interesting series of questions which may be important with respect to understanding several types of wildlife problems—for example: how many years after the Masai or Fulani leave a village site, or an Australian rancher abandons his over-grazed land, can the surrounding area be recognized as formerly occupied by pastoralists? What changes have taken place in the West African ecosystems since firearms were made available; or following the protection of a national park; or after the introduction of a new road, or following the cessation of active slave trading.

This notion has been emphasized in several earlier sections. For example, the importance of historical information and how to get it was emphasized in Section 5320. And the usefulness to management in learning which of the various phases of an eruptive oscillation characterize your study population of the moment was noted in Section 7100.

8160   Things As They Are and As They Will Be

Earlier mention has been made of status and trend as two ways of looking at animal habitat. There is a third way—Potential—which is not elaborated in this handbook. However, it is now useful to distinguish between these three 'ways of looking' as they provide a starting point for three very different sets of questions. First, the present state of a population or habitat (or the existing relations between the two) is called its *status*. Second, the way in which the population or habitat or interrelationships is changing is called its *trend*. And third, the level of productivity to which the population or habitat can be brought (regardless of what the present situation is) is called its *potential*.

Regarding these three concepts two points have been selected for emphasis in this chapter: one concerns *trend*; the other, *potential*.

The direction in which habitat change takes place (the trend) is called *upgrading* or *improving* when there is improvement of some desirable feature; when there is loss of desirable features, which might be litter and perennial grasses, it is called *downgrading* (deteriorating). Or there may be no change, in which case the habitat is *stable* (maintaining).

It is clearly very important to define the criteria used in deciding what is meant by downgrading and upgrading. This is of course closely linked with the definition of most practical animal problems. Sometimes there can be no doubt of the way to evaluate evidence. The covering by grass of bare ground that is susceptible to accelerated erosion can only be considered as upgrading. However, the change from grassland to woodland may be upgrading if woodland is more desirable than grass; but if grassland is preferred, say for grazing animals, then the encroachment by bush is downgrading. Objectives vary. Moving up the successional stages may be an objective, or maintaining a diversity or a maximum interspersion between two habitats, or holding an area stable but at a lower successional stage. Clearly the criteria used in deciding what is upgrading and what is downgrading must be defined.

The third question one may ask of a habitat is how far it can be improved (what is the *potential*) for this or that species of animal. If the habitat is degraded there are two things one can do in the absence of other information. First, one may compare it with similar habitats that have had different management or are untouched; or second, one may experiment by treating the degraded habitat in different ways.

One false 'way of looking' at *potential* must be mentioned because it is commonly used and its continued use may delay an understanding of *potential*. I refer to the practice of obtaining evidence of status from a number of different sites, some poor and some good, and regarding the best of this series as a potential which could be achieved by the others through proper management. This may be true but it is not necessarily so.

In considering an area as it may be (potential) we must normally have more information than a description of things as they are now (status). A little common-sense reflection should reveal obvious difficulties in using 'present status' to indicate 'potential', for example, the 'best' area observed may still be well below its true potential; the worst area may be so degraded in terms of soil loss that, even with complete protection, it would not improve to its former state.

Potential, however, has another dimension that is dependent on technical innovations and this we cannot foresee. It seems unlikely that any new technique will replace either adjustments of grazing pressure or controlled burning on arid savannahs or both, for many years to come. The possibility, however, must be kept in mind. Likewise, it may be possible to improve management in such a way as to give a higher yield than is now possible (see Chapter 7000).

8170 Proving Yourself Wrong

To effectively deal with an animal problem one may need but one *relevant* kind of information. More commonly measurements of a number of indicators may be required: e.g. an index of the trend of the vegetation; an index of the numbers of wild and domestic animals; an index of the use made of the vegetation by domestic and wild animals; an assessment of its recent trends.

Suppose an interpretation of these measurements indicates that the habitat is deteriorating and bare ground is increasing. A consequence of such deterioration would be active sheet or gully erosion and one might predict an increase in silting in rivers and perhaps the drying up of rivers that previously flowed permanently. These additional observations, taken from other similar areas but with a longer history of abuse, provide cross-checks on the original interpretation.

Historical evidence can also be used as a means of cross-checking. If any evidence suggests the drying up of rivers which once flowed permanently, an independent cross-check would be to ask old men in the locality about the previous occurrence of crocodile and hippo, for example, or other animals associated with permanently available water (see Section 5322).

Again, the start of deterioration may be dated by the age of trees or shrubs that have encroached upon what was once perennial grassland. It might be found from historical records that this coincides with the start of pastoralism or cattle ranching in the area, or a big increase in stock numbers. Conversely the age of trees and shrubs inside what previously was an erosion gully can give an indication of when erosion stopped and repair began. Historical records can then be checked and old men interviewed to provide clues to the cause of the trouble.

It is usually possible to think of consequences of any conclusion and these provide ideas for cross-checking on the conclusion. One may thus try to prove oneself wrong (to test one's own assumptions or early conclusions), thereby strengthening one's confidence (or not) in the interpretation of, and clearer definition of, a problem.

8200   THE PROBLEMS THEMSELVES

A cat scratches a chair; a rat gnaws into and short-circuits an electrical power supply; a predator kills a chicken, a cow, or a man; bats foul attics; a plague of rodents or a band of baboons, or elephants or pigs destroy crops; hare and deer prevent the regeneration of forest trees; kangaroos and wild ungulates in many parts of the world compete for food with cattle or sheep .... The simplest way of defining animal problems is to recognize that *animals are only labelled problem animals when they come in some sort of direct or indirect conflict with man's interests or activities*. Regarded in this way wildlife problems are also human problems. These problems exist because some animal is adjusted in such a way with its environment that it interferes with some human interest. The animals then become known as 'problem' or 'noxious' animals, 'pests' or 'vermin'.

From a purely technical point of view if we are concerned simply with the management of wildlife populations, then as G. Caughley (1977) has elaborated, the problems associated with management are but three:

1. the treatment of a small or declining population to raise its density,

 2. the exploitation of a population to take from it a sustained yield, and
 3. the treatment of a population that is too dense, or which has an unacceptably high rate of increase, to stabilize or to reduce its density.

Caughley calls these respectively conservation, sustained yield harvesting and control. Technically, these are true and useful classifications.

There is an important difference between Caughley's way of classifying problems and the approach used in this chapter. Although the first and third problems listed by Caughley can involve direct or indirect conflict with man's interests, the sensible exploitation of a population is by no means in the same category. Caughley is referring to the technical problems a wildlife manager has in managing a population. The definition as here used (i.e. the extent to which animals are interfering with some human interest) refers to animal problems as they become known to or are referred to wildlife or national park organizations or to various sections of the public. The reason for emphasizing such a small point is that the importance of defining the problem (by whatever criteria) will be emphasized and re-emphasized in the following discussion. Caughley's second problem associated with management has been elaborated in Chapter 7000. The following section follows the definition given at the start of this section and concentrates on ways of approaching various animal problems.

A few kinds of animal problems are already common in many developing countries and other problems are apt to arise as development continues. The problems discussed below are by no means complete, but the different types of problems provide, in practical terms, an extension of the concept of 'a way of looking' and a way of integrating much information presented separately in previous chapters.

8210   Animal Problems and Problem Animals

Problem animals may be indigenous or exotic; wild, domestic or feral. They are usually, but not always, animals of some economic importance.

We may be concerned because members of a particular animal species are rare, as with the bongo in Africa or the banteng in Asia, because the continuing presence of introduced large ungulates threaten the existence of a unique local flora. For example over-populations of goats or deer can devastate vegetation on small islands, and large mammals under certain conditions can become important in preventing forest regeneration and in contributing to accelerated erosion. Likewise, again under certain conditions, native mammals may become grossly over-populated to the extent that they become problem animals, as, for example, the three largest species of kangaroo in Australia; deer in parts of North America; wildebeest in parts of Africa.

Various human and environmental factors combine to modify the status of wildlife problems.

In considering wildlife problems generally, it is important to realize that wide variations in environments exist, even in areas geographically close to one another, and in these different environments different problems exist. This is true of most countries. Wildlife departments are not normally dealing with one generalized wildlife problem; they are concerned with many specific wildlife problems. This is a logical result of concern over several different species of animal in different environments.

One important general set of problems revolves about animals, depleted vegetation and/or accelerated erosion and exemplifies a series of problems that can be extremely complex, for always several and usually many factors contribute to the problem's development and should be considered in defining the problem. An important factor contributing to the complexity is the presence of widely divergent human attitudes. Such problems are variously defined as 'deer' or 'kangaroo' problems, as 'forest' problems or as 'erosion' problems, depending largely on the interests or set of interests against which the problem is defined.

For example, in New Zealand, in the 1950s I encountered two such divergent opinions on what was at that time referred to as 'the deer problem'.

1. Several conservation organizations and at least two Government Departments believed that every small group of deer was a direct threat to the survival of many native plant species and, further, that if the deer were eliminated so, therefore, would be the erosion and forest problems. They favoured complete extermination of all deer from the North Island; they pressed for enlarged Government shooting operations.

2. On the other hand, the New Zealand Deerstalkers Association believed that private deerstalkers alone were capable of keeping most deer herds under control. They did not regard deer as seriously detrimental to regeneration of native forests, except, perhaps, in certain localized areas. They strongly opposed any attempts to exterminate deer from New Zealand. Hunting was for them a preferred form of recreation; one they felt was a legitimate way of using undeveloped forest lands.

The passing of time can allow new factors to develop, old ones to diminish or disappear and this in turn may be reflected in changing attitudes toward animal problems. The above observations from New Zealand were made in the early 1950s but are no longer applicable in 1980. In the intervening 30-year period deer gradually have assumed an economic significance not predicted by either of the opposing sides of the 1980s. Commercial harvesting of deer from the forests for sale to an increasing European market has been followed by a growing number of economically viable deer farms who still obtain forest deer for re-stocking purposes and to build up herd numbers. Deer meat has become a valuable commodity, as have antlers in velvet for which there is a great demand in Asia. There is no longer serious pressure for eliminating all deer from New Zealand even by the most dedicated conservation organizations.

436

436

Since wildlife problems and human interests are, by definition, so closely linked, it is only natural that many private and public organizations have special concerns regarding wildlife. In New Zealand, in the 1950s, 20 organizations were public (12 national and 8 provincial), and 9 were private organizations. Since there are 13 provinces this means about 125 interested organizations in one country, each viewing wildlife problems in a somewhat different way. Several European countries, Canada and the United States, have many more; many developing countries have considerably less.

8220   Relation of Wildlife Problems to Different Forms of Land use

A simple and practical way to obtain a perspective of wildlife problems is simply to review the various problems in terms of the existing forms of land-use and special interests associated with these uses. The following list is in no way complete and refers to no particular country. It is intended merely as an example of a kind of outline or guideline list that may easily be prepared for any country or region.

OUTLINE GIVING EXAMPLES OF RELATIONS OF WILDLIFE
PROBLEMS TO PATTERNS OF LAND AND WATER USE

Agriculture lands
  Grazing lands
    Competition between domestic animals and wildlife
    Stock mortality due to predatory wildlife
    Disease transmitted to stock from wildlife reservoir
    Erosion induced by overpopulations of wild animals, feral animals, or
      one or both of these in combination with domestic animals
  Crop lands
    Wildlife depredations on crops (e.g. rabbits, deer, elephants, kangaroo)
    Problem of balancing farming aims with production of upland game for
      sport or meat (e.g. quail, pheasant, guinea fowl, deer)
Forested lands, woodlands
  Exotic forests (forest plantations, usually of introduced tree species)
    Animals retard or prevent growth of plantings or natural regeneration
    Wild or feral animals associated with accelerated erosion
  Native forests and/or woodlands
    Animals retard or prevent regeneration, or accelerate erosion
    Rare or vanishing species threatened with extinction
Parks, reserves, wilderness areas, unoccupied natural areas
  Protection of native animals and their habitats
  Special management to arrest the decline of and increase number of rare
    species
Ponds, marshes, swamps
  Requirement of preservation as a habitat for:

Rare species, indigenous birds or other animals, waterfowl
While balancing above needs with;
Drainage demands
Inland waters (lakes, rivers and streams)
Requirement of preservation as a habitat for:
sport fisheries (trout, salmon, etc.)
commercial fisheries
local subsistence fishing
waterfowl or associated mammals or reptiles (crocodiles)
While permitting rational use for other purposes:
hydroelectric urban water supplies, etc.
Urban areas
Town and garden pests
Depredations on gardens
Predation on fowls
Disease from wildlife to domestic animals (foxes and rabies)

When wildlife organizations are involved in decisions relating to the assigning of particular forms of land-use to certain areas it is well to remember that the pattern of land-use itself can set the stage for the development of some problems with animals.

For example, in most of the French-speaking countries of West Africa it is normal to have a strict protected game reserve, or national park, surrounded by a zone of limited hunting (usually including an area featuring tourist hunting). Areas of general hunting, forest exploitation, grazing and arable lands are normally outside these two inner zones. Since the wildlife is administered and managed by the Forest Department, many potential problems involving wildlife in conflict with other forms of land-use are thus minimized.

In most of the English-speaking African countries the pattern of land occupancy had been more haphazard, with the result that intensive grazing areas or agricultural lands are not uncommonly located adjacent to national parks or game reserves. It is understandable that problems of crop damage and competition with domestic stock develop more frequently in such circumstances.

It is suggested that at least one of the two most expensive activities associated with wildlife, the practice of fencing (the most costly is the bounty system mentioned later), could be inexpensively solved and in some cases eliminated if some modification of the French system of zoning were adopted. This is increasingly being recognized in several parts of East and southern Africa.

8230 Two 'Ways of Looking' at the Question of Solving Problems

8231 *Short-term Solutions*

A system which is applied extensively to wildlife problems in many countries is to use a research organization, in the same way as a trouble-shooting crew of electricians is used to locate a break in a power line. This kind of study is usually part of a larger problem but, because of its critical local application, the object of the study made is often one of trying to solve a current problem in a specific locality.

It is usually brought to the attention of the wildlife authority by a local private agency or group of individuals and can involve any of the kinds of problem mentioned above. For example the issue may be unusually heavy damage to crops by elephant, baboon or kangaroo. Complaints are lodged either directly with the government organization concerned or with higher authorities up to and including the ministerial level.

At its best this short-term, or *ad hoc*, approach to a local problem involves investigation particularly designed to remedy a specific local situation. The *ad hoc* approach typically deals with the symptoms (the crisis of the moment) and not the underlying causes. Often the remedy is deceptively effective because the action taken temporarily satisfies the organization or interest groups immediately concerned. Such research and such 'solutions' seldom result in a sound management programme or long-range solution, even in dealing with similar future local problems. However, considerations of policy often render some immediate effort necessary.

8232 *The Solution of Problems on a Longer-term Basis*

In dealing with longer-term, more permanent, solutions the game is to look beyond the crisis of the moment and search for underlying causes. Action directed towards correcting such causes has as its general objective the elimination of the problem altogether, or reducing the problem to manageable proportions; that is to a level that will be tolerated by the group of land-users concerned.

It is easy enough to understand that it would be useful to have good information on all the contributing causes and that this information should be relevant to the problem concerned. But this is not always easily done and it is often not possible for a wildlife or park officer to obtain all relevant information without the help of other workers in other disciplines.

For example, imagine a situation where graziers are complaining about wild animals competing for food with their domestic animals. Their overriding concern is with degrading lands and if recent trends continue they realize they must eventually abandon the land.

Even if investigations showed a clear overlap in feeding areas and in kinds of plants eaten the problem would still be unsolved if the domestic stock were too numerous, regardless of the presence of wildlife. If this were the case even complete elimination of the competing wildlife would not halt the downgrading trend, although their elimination or reduction might slow it down. Close collaboration with range management or pasture research officers might be necessary to clarify the long-term problem. Long-term solutions to problems are often superficially regarded as wildlife problems, while, after the problems have been more clearly and ecologically defined, the responsibility may be shown to rest with an organization other than the wildlife organization.

Conflicts between wildlife and crops or cattle, like bush encroachment, are only symptoms of an imbalance in the ecosystem. Local remedies such as fences, or shooting campaigns may remove the symptom but at great cost with serious inroads on the wildlife resource and without solving the basic problem of mismanagement of domestic animals.

For long-term stability, something more basic must be done about the causes of the imbalance. This can only be done adequately through carefully planned regional land-use policies. These policies should aim at the most efficient and appropriate sustained land-use patterns, which implies a rational use of all renewable resource, including wildlife.

No matter how complex the problem or how clearly the main long-term responsibility may lie outside the wildlife organization, there is always one kind of contribution wildlife organizations may make in dealing with any wildlife problem. Especially with any animal of economic importance, or when dealing with rare species, it is important to understand the relation of the species to its environment and to associated species. More specifically, and immediately more practical, but part of the same question, is the query: What are the minimum essentials for the existence of the species in a given area? In still other words: What are the factors (or parts of its lifetime habitat) which limit its existence? This general approach, when used by a trained worker, forces him into the heart of most wildlife problems—of understanding the animal as it relates to its environment.

One or more studies of this type inevitably lead to information of wide practical application because we achieve a more fundamental understanding of the relationships between the animal concerned and the environment concerned, and particularly of those relationships which are relevent to a long-term solution of the problem in question.

The question of determining habitat requirements has been discussed in Chapter 5000.

8240   Wildlife Problems are More than Technical Problems

When we deal with wildlife problems the technical basis for our recommendations and decisions will be our understanding of animals and their habitats. But when it comes to defining the problems and designing and implementing some effective action the technical aspects form only a part of the real-life problem we have to deal with. Often in resolving an 'animal problem' one must be as concerned with various economic and social (including political) factors as with animals and their habitats.

One might question the value of wildlife biologists stepping outside the purely technical boundaries of an animal population–habitat complex. However, if we are to be successful in developing sound long-term trends we must on occasion take some of these many non-biological factors into considera-

440

tion. Since the problems themselves are partly (sometimes entirely) non-technical, so likewise must be the solutions. To keep this discussion from becoming too academic consider the following as an example of commonly encountered situations which may prevent a wildlife manager from effectively dealing with a wildlife problem on a technical basis alone.

The problem as presented to the wildlife biologist to solve may not in fact exist, or it may be part of a much larger or more basic problem. A well-meaning but ill-informed person or organization (local, national or international) may bring political pressure to bear on the responsible government organization to solve a problem which in fact does not exist. These pressures usually involve preservation issues and most commonly take one or two forms: either the pressure is directed towards saving species from extinction (species which are in no danger whatever) or strong public criticism may be made of annual harvests taken from animals so abundant as to be regarded as pests by government authorities (often under carefully documented and controlled plans of management and monitoring). Wildlife organizations often refer to this kind of situation as 'inappropriate or irrelevant pressure on their organization by ill-informed or un-informed special interest groups'. The greater this kind of pressure, the greater is the need for a wildlife organization to explore ways and means of improving the effectiveness of their extension and public relations efforts.

Over-specialization can, and often does, become a factor in slowing down progress in dealing with animal problems. A biologist may be so highly trained in a narrow aspect of animal-environmental relations that he can see a problem only through the eyes of his specialty (e.g. disease studies, food habits studies, forest regeneration). There are of course occasions when a specialized approach is relevant. Where it is not relevant then the findings of an irrelevant study can considerably obscure understanding of the real problem and so waste an organization's funds and postpone a solution. A specialist in another field (agriculture, veterinary medicine, engineering) may take a decision that, when implemented, may actually stimulate the development of a wildlife problem.

Other non-biological factors may on occasion dominate animal–habitat relations and modify, aggravate or even cause animal problems:

Stubbornness can play havoc with attempts to use a common-sense approach on an animal problem. An elderly boss, or a local, national or international organization with a way of thinking and operating as inflexible as granite exemplifies this. These individuals or organizations may not be interested in changing a long-held view, no matter how convincing and relevant new evidence may be in making clear the need for a change in their attitudes. This makes it difficult, particularly for a junior biologist, to define the problem in a more rational way.

Vested interests are perhaps the most common non-biological factor interfering with the long-term solution of animal problems. In this category are individuals or organizations (e.g. involving mining, agriculture, pastoralism, forestry or urban land development) who have strong interests in the outcome of a wildlife policy decision affecting their livelihood or their investment. This sometimes takes the form of conflict between two adjacent or overlapping forms of land-use, with wildlife caught in the middle.

Ignorance of the real problem involved may in turn sometimes be due to lack of men trained in wildlife management or with little experience in defining problems.

This list is by no means complete. The point emphasized is that animals' problems are usually entwined with and obscured by factors that have little to do with animals and habitats, except that indirectly they may exert a major influence, or even dominate the scene. This is considered important enough to elaborate in the last part of this chapter under 'Politics and Wildlife' (Section 8400).

8300   AN APPROACH TO SOLVING PROBLEMS

8310   Tackle the Problem Systematically

When one has a practical problem presented by administrators or land managers the following proposed steps may be of assistance as a guideline for coming to practical terms with the problem.

1. Obtain as clear a statement of the problem as possible in writing. Review available evidence of the existence and nature of the problem. Sometimes a review of published or unpublished material may suffice; often it is necessary and always desirable to see the problem in the field. Even if the visit is but once, brief and superficial, this will be useful later in the preparation of an initial working plan.
2. Analyse the problem as stated and compare and relate the stated problem to the specific aims and objectives of the department responsible for the area or for the species concerned.
3. If this proves a straightforward exercise, with no inconsistencies, irrelevancies or conflicts then the problem can be accepted as stated, a detailed working plan prepared for an appropriate study or management operation (see Section 2030), approval can be obtained for the detailed working plan and work can then be started.
4. If, on the other hand, there are inconsistencies, irrelevancies or conflicts with departmental aims and objectives, re-form the proposal to bring it closer in line with those aims and objectives. Discuss such alternative approaches or studies with the department until a more acceptable orientation for the study has been mutually agreed.
5. Proceed with the formulation of a detailed working plan, final approval by whoever must approve it, then start the work as in (3) above.

The above has been written as for a biologist or wildlife manager working for a government organization responsible for wildlife management. It is strongly suggested that universities, institutes, museums or other research organizations contact the responsible government organizations in the early stages of conducting an investigation on an animal problem. This should in no way compromise a 'scientific' or 'objective' approach. It should ensure that the study is conducted with a better appreciation of the non-biological environment surrounding each problem and mentioned above. Keeping the responsible organization fully informed and encouraging their cooperation in the planning or execution of the project can go a long way towards increasing the chances of the results of the investigation being used and the problem eased, or perhaps solved.

Where there is a genuine desire to solve an animal problem there is little room for the 'ivory tower' type of academic approach. If real life is to be helped, then real life should be involved at the birth of the project.

8320   Examples

8321   *A Deer-forest Problem*

The following simple example came to my attention while writing this chapter. A European forest service arranged for a biologist to undertake a study to answer the following questions: Can deer be reduced by shooting? If so, what effect will this have on forest damage? Although the example is from Europe, this is not an uncommon way for foresters in developing countries to present a problem to a wildlife manager or research worker. It illustrates an attitude by the organization responsible for management that, if accepted unquestioningly by the biologist engaged to answer the question, could lead to irrelevant waste of time and effort, and with no guarantee of helping the service concerned. Following the guideline procedure suggested above:

1. The statement was presented in writing as noted in the preceding paragraph.
2. *Analysis*. Deer can of course be reduced by shooting. If the biologist was young enough not to be able to convince the administrators of this fact from well-known local examples then a simple dip into the literature on deer should do the trick. The second part of the question seems almost as obvious. If deer are doing some damage to the forest, then a reduction in deer numbers should reduce the damage.

   Questioning of the relevant senior forester by the wildlife officer revealed that the major objective of the forest service is to produce trees for timber, at least in the areas they wanted him to investigate. Further questioning revealed that the way the project started in the first place was that certain foresters had noticed that many trees were being used by animals in the forest. Although deer, sheep, hares and rabbits were all in the forest, deer were regarded by foresters as the main animal responsible for damage to trees, and these observations were in fact the basis of the problem as stated.

3. Close comparison of the problem as stated, with the stated objectives of forest management and with foresters' present understanding of the problem, revealed inconsistencies and irrelevancies. There was no evidence that deer were in fact the culprits, it was simply a 'well-known fact'. In conversation it became clear that in some parts of the forest areas deer were frequently seen yet 'damage' was not serious enough to concern the forester.

4. Without elaborating the difficulties rising from this comparison, in a brief visit to the forest a fresh look was taken by the biologist and forester working in close collaboration. A major point raised by the biologist and accepted by the forester was that much of the significance of the results produced by any future study involving deer damage depended on a rather precise definition of damage. Furthermore, since the forest service was responsible not only for the formation of national policy on forest land but for accomplishing specific management objectives, it was therefore considered the responsibility of foresters to define damage in terms of their various classes of forest (e.g. recreation, production, etc.). As a first step it seemed desirable for foresters to set standards and criteria by which damage could be assessed.

A project was therefore devised to assess what percentage loss would be acceptable to foresters for seedlings, saplings or poles in the different types of forest and for each species of tree with which they were concerned.

In this first project a forester was to be in charge with the biologist assisting and devoting particular attention to developing simple ways of assessing such losses in ways acceptable to the forester.

A second stage was visualized whereby the biologist would assess the extent to which the various animals in the forest were associated with and responsible for unacceptable levels of use as defined using the newly developed standards. It was felt that the results of this work would in fact constitute a clearer and more useful definition of the problem, both from the ecological and the forest management points of view. It would also provide a sound basis for developing a third phase or stage of the study.

The third stage would involve testing the uses and limitations of various methods of control, including direct control of animals, as by shooting, and including indirect reduction of animal damage by manipulating the habitat, as, for example, by planting less palatable species of trees.

Although it was agreed that some combination of the second and third stages might be possible, priority would be given to stage one, and the subsequent priorities reviewed once the prerequisite criteria had been established.

Of course those familiar with game or forest departments will recognize that another less efficient approach may have to be used. A wildlife manager may find himself in a situation where the above approach is inadvisable for some non-biological reason. For example, if the administrator (or one's chief) is perfectly clear in his own mind and is satisfied that his way is the way the problem must be defined and solved and no arguments, then a junior wildlife manager, for example, has little choice but to smile and accept. However, it is usually still possible to develop a useful project by accepting the job as stated, but with careful attention to the working plan, ensuring that the 'objectives of study' in combination with the 'methods' section will produce information leading to recognition of some more basic or more relevant problem: in effect, a re-definition of the problem arising from an initial study of the 'problem as given'. For example, taking the question as originally stated the objectives would be:

1. to determine if deer can be reduced by shooting in a national production forest,
2. to learn the effect of different levels of deer reduction on forest damage.

And to these might be added:

3. to develop simple techniques for recognizing acceptable and unacceptable levels of use by each kind of forest animal present (if other browsers and grazers are present they must also be studied to place the information gathered on deer into a perspective);
4. to experiment with both animal and environmental methods of control as a means of reducing forest damage and to demonstrate the comparative effectiveness of these experiments.

In introducing data-gathering techniques (Section 3000) a scheme was presented for classifying techniques based on whether information was gathered from the animal or environmental parts of an ecosystem and from which of three levels—occurrence, utilization or response. Since this framework has been followed in presenting most of the techniques it may also prove useful in thinking about various elements of an animal problem as well as elements required or desired in reaching a solution. Figure 136 shows how an animal problem (solid lines) and one solution (dashed line) may be presented in this way.

Consider the deer-forest problem mentioned above. As originally defined by the forester the problem elements involved were too many deer which prevented regeneration of trees in a production forest. In Figure 137 solid lines are drawn connecting problem elements as they are involved at three different levels of the ecosystem (occurrence, utilization and response).

But deer-forest problems are usually not this simple. This problem may in fact be as stated, but this should not be assumed or completely accepted without some kind of cross-check. Normally where deer occur in forests there are several other species of animal present which also utilize trees. Rodents eat

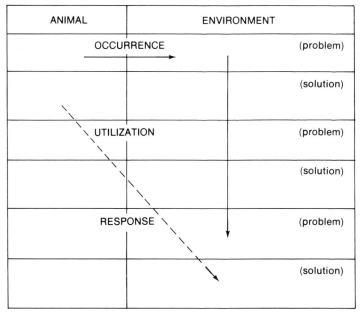

Figure 136 Semi-diagrammatic scheme for planning the gathering of data to define problems, then plan appropriate remedial action. In this example solid lines indicate that animals and environment are interacting at the occurrence level, resulting in some kind of unfavourable environmental response, the latter being regarded as 'the problem'. Dashed lines indicate elements of a proposed solution which involves modifying the occurrence of animals in some way to change the objectionable environmental response. This simple scheme has proved useful both in selecting techniques for use in further defining a problem and for testing alternative strategies for arriving at a solution, as illustrated in Figures 137 and 138

seeds and seedlings, hare and rabbit can strip bark or eat terminal shoots and twigs of seedlings, sheep and cattle also feed on woody plant material and some kinds of game birds, or pigeons, parrots, finches, etc. nip terminal shoots. Each species will relate to the regenerating forest in a somewhat different way. Only when this kind of information is in hand can one place deer-forest relations in perspective. In other words, the 'deer' forest problem should be defined more precisely, even if it has to be done quickly and crudely for a start. Some kind of objective evidence can be obtained at each end of each line shown in Figure 137. To objectively define the problem, data-gathering should be planned to permit close comparison of, and demonstration of relationships between, these different aspects or elements of the problem. In the interests of simplicity let us assume that more detailed investigation reveals the problem to be essentially as shown in Figure 137, with the deer being principally responsible for the lack of forest regeneration.

Figure 138 shows four alternative options for solving the problem shown in Figure 137. Dashed lines connect elements involved in the proposed solu-

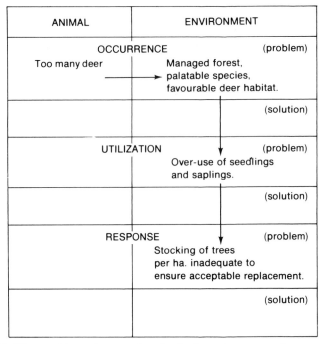

ANIMAL	ENVIRONMENT
OCCURRENCE (problem) Too many deer	Managed forest, palatable species, favourable deer habitat.
	(solution)
UTILIZATION (problem)	Over-use of seedlings and saplings.
	(solution)
RESPONSE (problem)	Stocking of trees per ha. inadequate to ensure acceptable replacement.
	(solution)

Figure 137   The problem elements involved where too many deer are preventing regeneration of trees in a production forest. Some kind of relevant evidence can be obtained at each level as indicated. To objectively define the problem, data-gathering must be planned to permit close comparison of, and demonstration of relationships between, these different aspects or elements of the problem

tions. Solution (a) may take one or two forms, but involves only the reduction of animals to a level that allows enough regeneration to meet the standards set by foresters. Solutions (b) and (c) involve modifying forest management practices to make the forest trees less attractive and less vulnerable to deer. In (c) it was assumed that the forest was essentially a recreation forest. In this case deer are desired as an amenity and maintained as an extra attraction for visitors. The standards set for permissible use by deer in a recreation forest would differ from standards of use in forests managed primarily for timber production. Elements in the fourth optional solution (d), involve changes in management of both animals (deer) and their environment. On the one hand deer would be harvested for meat or sport. On the other hand increasing the proportion of unpalatable species of tree will make the forest less vulnerable to deer.

In any of these options, unless standards are already available and permissible thresholds of use clearly defined, each would require such standards to be determined and set as guidelines for demonstrating success in terms of the adequacy of forest regeneration. Forest regeneration can of course be measured directly but this takes several years to learn the practical result of a reduction in deer numbers. The advantage of evidence taken from the 'utilization level' is that, once the relationship between regenerating trees and percentage of annual growth removed annually has been established, a test on the impact of a reduced deer population can be made within a year after reduction has taken place. Thus in forest management, option (2) has distinct practical advantages over option (1) in Figure 138(a).

Plan (1) would work, but it might take several years to demonstrate the adequacy of forest regeneration. By using plan (2), once the standards of utilization are set the adequacy of the reduction in deer numbers can be assessed in a year's time (see Section 3700 for more details).

8322   *A Problem of Competition*

Several other kinds of problem can be defined and elements of proposed solutions visualized in this same way. Consider a problem presented to a wildlife service by a grazier's organization as a problem of wild animals competing with domestic animals to the detriment of the stock. The case, as presented to the wildlife manager, can be simply put. In terms discussed above the grazier's version of this problem and his proposed solutions may be shown diagrammatically in Figure 139(a) and (b) respectively. But problems of competition are seldom as simple as this. Figure 139(c) lists a few questions, the answers to which would help define the problem more precisely and provide a better basis for proposing a working solution.

This extra information taken at the Occurrence level alone (Figure 139(c)) might well result in a drastic re-statement of the problem as, for example, if the results showed that, except for a small area of overlap, deer and sheep occupied quite separate habitats and that sheep was the only species consistently associated with downgrading areas. In this event, reducing numbers of wild herbivores could not be expected to solve the problem. A solution involving either reduction in sheep numbers or improvement of sheep habitat, or both, would be required to halt the downgrading trend and result in improving grasslands (Figure 139 (d)).

8323   *Problem of Maintaining the Habitat of Animals*

The habitat of a wildlife species may change for good or bad in three ways. First, as a result of heavy use by the wildlife species itself; second, as a result of some outside factor such as fire, scrub clearance, heavy use by other wildlife or by domestic animals or some natural catastrophe such as an earthquake or flood; and third, as a result of natural plant successions. The three may be related, for plant successions may be initiated by either a release of pressure (a reduction in animal use) or some other disturbance (e.g. fire, strong wind, logging).

The maintenance of appropriate habitat can only be done if one knows

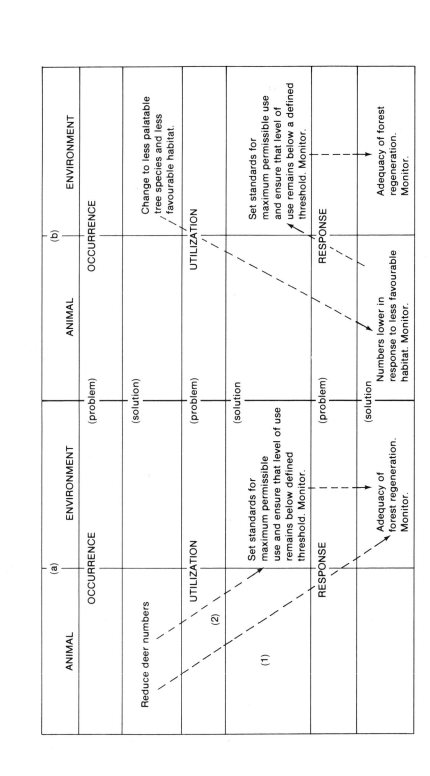

(a)

ANIMAL		ENVIRONMENT
OCCURRENCE	(problem)	
Reduce deer numbers	(solution)	
UTILIZATION	(problem)	
(2)	(solution	Set standards for maximum permissible use and ensure that level of use remains below defined threshold. Monitor.
RESPONSE	(problem)	
(1)		Adequacy of forest regeneration. Monitor.

(b)

ANIMAL		OCCURRENCE	ENVIRONMENT
			Change to less palatable tree species and less favourable habitat.
		UTILIZATION	
			Set standards for maximum permissible use and ensure that level of use remains below a defined threshold. Monitor.
		RESPONSE	
Numbers lower in response to less favourable habitat. Monitor.			Adequacy of forest regeneration. Monitor.

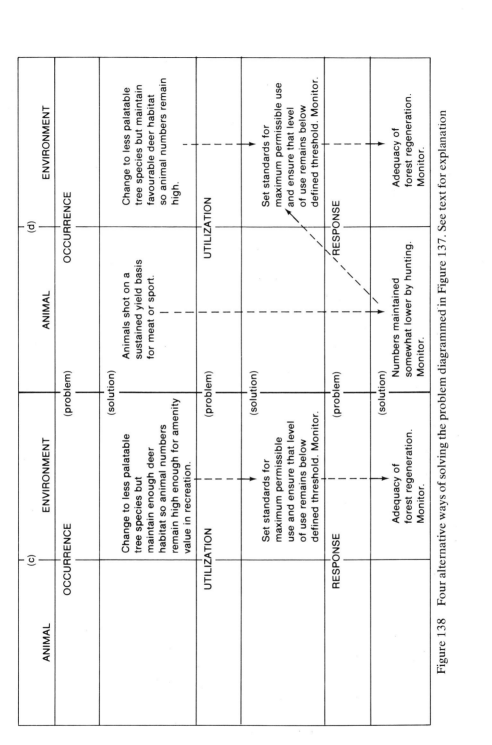

Figure 138   Four alternative ways of solving the problem diagrammed in Figure 137. See text for explanation

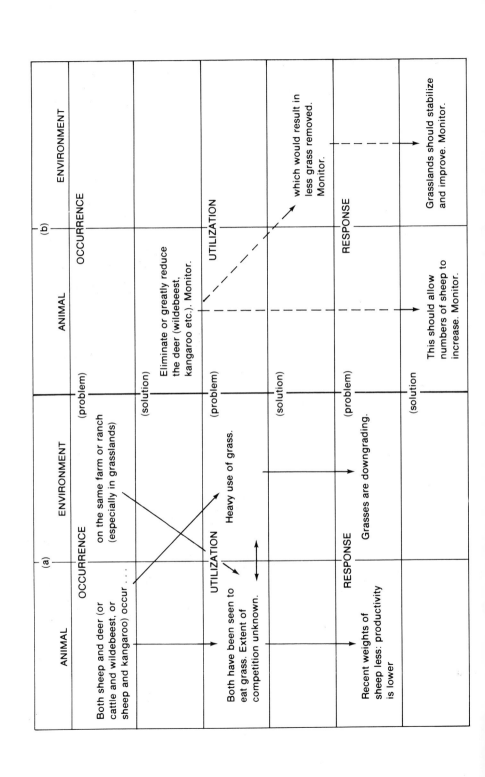

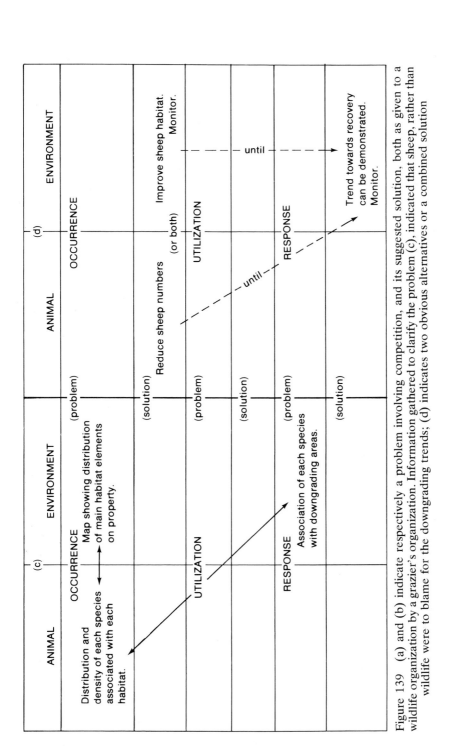

Figure 139 (a) and (b) indicate respectively a problem involving competition, and its suggested solution, both as given to a wildlife organization by a grazier's organization. Information gathered to clarify the problem (c), indicated that sheep, rather than wildlife were to blame for the downgrading trends; (d) indicates two obvious alternatives or a combined solution what is appropriate. Normally we wish to maintain the habitat for a particular species or combination of species and this requires knowledge of the habitat needs of wildlife (Chapter 5000). Examples of questions which may be asked and answered are:

1. Which elements of the habitat are required to maintain the desired numbers and kinds of wildlife species and how are they best interspersed? (see Sections 3400 (Occurrence–Animal), 3500 (Occurrence–Environment) and Chapter 5000 (Determining habitat requirements)). This decided, the next questions are:
2. What is the present state of the vegetation? (Section 3500).
3. In what direction (trend) is the vegetation changing? (Section 3900 Response–Environment).

The status of vegetation is defined by describing objectively what is present on the ground. This tells us how much remains of the habitat necessary for the wildlife species (see also Section 3500).

The trend of the vegetation tells us whether or not the habitat is changing and if so in what direction (upgrading, downgrading or stable)(see Section 3900).

4. Is the habitat changing as a result of an ecological succession? What agency has set this succession in motion?
5. Is the habitat changing as a result of pressure by domestic or wild animals? Which animals?

Habitat management may be adequately based upon indices to the numbers of animals concerned combined with indices of status and trend of the existing habitat elements. Desired changes in the habitat can be effected either through harvesting of selected animal species or by direct manipulation of the habitat.

These generalizations may be more or less true, but habitat management is not usually discussed in such general terms. Normally one manages habitat as a normal part of a solution to (management practice directed towards) more specifically stated problems, for example:

1. the problem of maintaining the existing variety of habitats in a national park to ensure maintaining a number of dependent animal populations;
2. maintaining a particular habitat of sufficient size to ensure the survival of a declining species;
3. improving habitat to permit increased harvesting of an economically valuable species;
4. reducing or eliminating one or more critical elements of the habitat of a pest species.

453

Whatever the specific management practice, if it affects the animals' habitat
(as it inevitably will) the habitat should be monitored to ensure that the
objectives of management are being achieved. And *once a management plan
comes into effect, monitoring becomes an integral part of management.*

8324 *Crop Damage*

To further emphasize the importance of monitoring as a trick to profit from
past experience, consider a simple case of crop damage by baboon, the most
important destroyer of crops in Africa (Riney, T., 1967a).

The problem posed is one of a troop of baboon in a recently established
commercial orange orchard. For some trees as much as 80 per cent of the
oranges are estimated to have been damaged or eaten by baboons. The
orchard owner is desperate, for if losses continue he will have to abandon
his orchard.

The baboon shelter at night in trees near the orchard or in a rocky cliff
$\frac{1}{2}$ mile from the edge of the orchard. They thrive. One troop numbers over 100
individuals.

Figure 140(a) shows the problem, and three trial solutions are presented in
Figure 140(b), (c) and (d). For each type of management tried it is important
to monitor both baboon numbers and loss of oranges. Trials should be on a
small scale at first. As management continues the monitoring may demonstrate the problem is being solved to the orchardist's satisfaction. However, if
the level of damage continues (as well it might) then information produced
through monitoring should be of real help in re-defining the problem. And as
problems become more precisely defined, more precise and relevant solutions
may then be possible.

8330 Two Special Problems for Developing Countries

Most of the problems that concern wildlife officers have been dealt with in
preceding chapters. However two kinds of problems are here given special
emphasis: those associated with rare and/or endangered species and other
problems associated with introduced animals. This is done with the needs of
developing countries in mind for it is well to recognize that some problems
seem to go hand in hand with various stages of a country's development.

8331 *Rare and Threatened Species*

As J. B. Trefethen observed (1964) extinction can come to a species in one of
several ways, with or without human assistance. Man, however, has been the
direct or indirect cause of the extinction of most species that have disappeared
within historic times (Trefethen, J. B., 1964 and Fisher, J. *et al.*, 1969).
J. Fisher and his co-authors, N. Simon and J. Vincent, also noted that man-
induced extinctions of mammals are over two and one-half times as great as

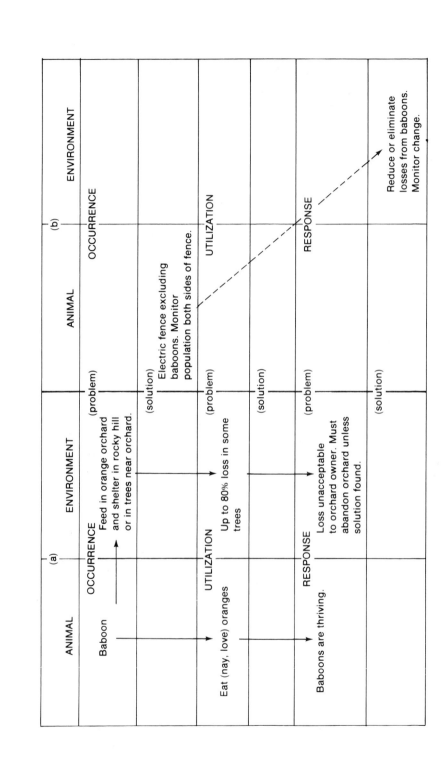

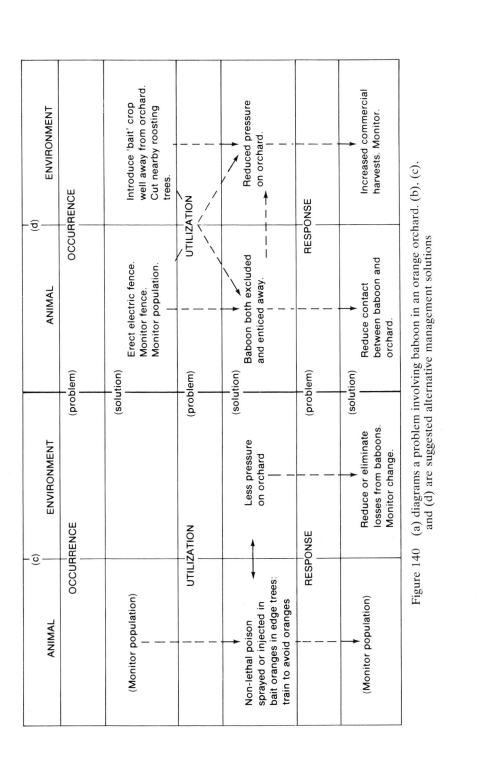

Figure 140 (a) diagrams a problem involving baboon in an orange orchard. (b), (c), and (d) are suggested alternative management solutions

extinctions due to natural causes, and five-sixths of the mammals presently known to be in danger of extinction have come to their present state because of man's activities.

Animals can become extinct through perfectly natural causes. When a species, through evolution, becomes so specialized it can exist only in very restricted or limited habitat, or it becomes dependent on a single restricted food, as a species it becomes very vulnerable. Many such species have become extinct following gradual changes in habitat such as may accompany changes in climate extending over very long periods of time. Nevertheless, since most animals have become extinct through activities of man it follows than man can prevent the extinction of animals by appropriate management practices.

There is no great mystery about the disappearance of animals due to man's activities: as land is cleared of forest, forest species decline; as perennial grass declines, so must those species which require perennial grass as a critical part of their habitat; as permanent streams disappear so do the fish. Such gross changes are of obvious importance but of equal importance are changes in less obvious ways: changes that may affect the interspersion of habitat elements, or their seasonal availability or the fact that other competing species may have changed in numbers with a consequent change in habitat.

The significance of many complex changes resulting from the impact of man may be summarized as follows:

1. As large-scale habitat changes take place the habitat becomes more suitable for some species and less suitable for others.
2. The habitat changes, therefore, normally result in trends whereby certain species increase in number while other species decrease.
3. Some species become locally extinct as the habitat completely disappears, or some species, having decreased in numbers because of diminishing habitat, are subsequently eliminated locally by hunting alone.
4. The modern most effective way of eliminating animals is to combine habitat destruction with shooting, for this results in the greatest acceleration of the trend towards extinction.

In thinking about 'rare' and 'threatened' species in developing countries it is important to distinguish between these two concepts. If a country has signed a treaty regarding endangered species, such as the CITES Convention, then in matters relating to the implementation of the treaty, CITES definitions will be used. However, not all countries have signed such treaties and particularly for such countries, if they are genuinely concerned with maintaining their existing animal species, then it may be useful to consider the two following common-sense definitions and their implications for management.

Rare species may be so regarded because the species exists in one, or only a few, restricted areas, or the present populations are everywhere very sparse. In some instances the classification may be influenced by the fact that the species is rare in the sense that it belongs to a relatively small zoological group, or that it represents some unusually distinctive zoological type. In other cases the classification may be influenced by the fact that the animal in question is commercially attractive and thus in danger of being over-exploited, such as crocodiles have been in many countries. Species that are rare are of course not necessarily threatened (endangered).

To be regarded as a threatened (endangered) species there should be either some evidence of a trend towards extinction in recent years, or developments that seriously threaten, or which, if continued will threaten, the species. The threat may be either direct, as through increased shooting activities, or in-direct, as through destruction of habitat.

A useful aim for the organization concerned with wildlife in each develop-ing country is to have a clear and well-publicized list of rare and threatened species. This is not often easily achieved.

There have been, and continue to be, very real difficulties in preparing such lists. The following list of difficulties can be used as a guideline for an organ-ization which wishes to become more effective in saving and rehabilitating rare species still found within their national boundaries.

1. The species *vs*. subspecies question. There is little doubt that priority emphasis should go to species but the taxonomic status of animals in many developing countries is imperfectly known save for a very few groups. There is an urgent need for the preparation of checklists of mammals. There is an even more important need to reach some measure of unifor-mity between collectors speaking different languages and of different nationalities with respect to standard measurements recorded from mam-mal specimens. Whenever possible, museums, universities and game departments should cooperate with their equivalent institutions in other countries with the aim of obtaining national, international and continental perspectives of the rare species that concern them.
2. There is commonly a scarcity of library facilities or documentation centres that include the relevent literature in the appropriate languages for the country concerned.
3. Since one must largely depend on correspondence and published works, difficulties arise from differences in the local criteria used for including animals on critical lists. Cooperation between institutions and between countries is obviously a good procedure to encourage.

In thinking about rare and threatened species in developing countries, it is important to distinguish between these two concepts. However, these are not the only terms used in considering endangered or threatened species. The IUCN Red Books, which are commonly used as a guideline in implementing the CITES Convention (Convention on International Trade in Endangered Species of Wild Fauna and Flora) present five definitions, ranging from 'endangered' to 'out of danger', these are reproduced in Appendix 6 of this book. In the same Appendix is included a statement of fundamental principles taken from the CITES Convention, the whole of which is used as a guideline for controlling imports and exports of rare and endangered species by customs officials and specially assigned wildlife officers in the many countries which have signed this Convention.

Once identified as threatened the wildlife organization may proceed with appropriate research and management to ensure a species' survival.

In research terms the determination of the species' habitat requirements is always a top priority, for this knowledge is basic to effective management regardless of which form the management may take. Specific management actions may involve one or more of the following:

Complete protection of the species.
Designation of a sanctuary area.
Assigning staff to manage the area and species.
Annual monitoring exercises to keep up to date on the present status of the species, changes in numbers and changes in any element of the occupied habitat.
In certain areas where a species cannot be adequately protected in its native haunts, it may be necessary to consider capturing and removing a number of the animals to a safe place (sometimes, but not always, outside the country) where they can be allowed to increase to such numbers as will permit reintroducing the species into formerly occupied areas.

Two precautions are:

1. Protection alone may eliminate certain species; for example, in situations where the species requires an early successional stage and where protection results in developing a uniform climax vegetation.
   Guideline. An initial interim management goal would be to keep existing habitat elements present in about the same proportion in which they now occur.
2. Areas formerly holding the species are not necessarily now suitable.
   Guideline. Introduce only into habitats known to be suitable. Know the species' habitat requirements.

It is usually in the early largely exploitative phases of development that the greatest loss of wildlife occurs. This is particularly so as land is being cleared for settlement, or development for more intensive use, for this tends to create a more uniform landscape. In the absence of a detailed knowledge of habitat requirements of individual species the safe practice to follow, to maintain the largest number of viable species in an area, is to maintain the greatest possible diversity of habitats.

8332 *Problems associated with Introduced Animals*

Introduced species of plants and animals have played a large part in man's gradual development of more appropriate and effective use of the natural resources of the world. Crops such as wheat and the potato, animals such as cattle and sheep, and trees such as eucalyptus and conifers, are in these days easily introduced and, as an incidental result, intensively used fields, pastures or plantations look much the same from one country to another.

Most of the problems arising from the introduction of animals have resulted from the introduction of domestic animals, either into a vulnerable environment or with an unsuitable system of management, mainly over-stocking. The two main kinds of problems concerning wildlife officers are loss of vegetation and destruction of habitats which sustain wild animals, and the introduction of parasites or diseases which may spread to either domestic or wild populations.

The question of introductions of large wild mammals is assuming increasing importance in developing countries. Reasons for introduction of animals have included reintroduction of animals in national parks where they had become extinct, or the introduction of certain species into parks or reserves to provide a supply of meat and skins, or as a tourist attraction, either to hunt or to photograph.

With the increasing ease of transport and the improved methods of capture and transport developed in the last two decades, we now have a power to move animals about. This facility did not previously exist, and is now on a scale not previously known in history. Before the last European war, the largest number of introductions in any one region was in New Zealand where, between 1860 and 1914, 175 to 200 introductions involving fifteen species of large mammals were made. However, between 1946 and 1959, many thousands of large mammals have been reintroduced into South Africa, and in the Transvaal Province alone introductions of over 1000 animals a year took place for over a decade and involved over twenty large mammal species (Riney, T., 1962; Riney, T. and Kettlitz W. L., 1964).

It is thus possible to introduce and reintroduce large mammals, but because it is possible, this is not *reason* to do so any more than jumping into a pool with large crocodiles is desirable because it is easy to jump. Introducing the wrong animal or combination of animals can have damaging effects on the environment or on other desirable species, and can destroy the same values the introduction was intended to enhance.

Consequences of introduction should be predictable. As long as we see introductions of large mammals as a hit-and-miss affair, we betray the inadequacy of our ecological understanding of the animal in question. The confident prediction of consequences following an introduction is a natural and worthwhile aim of any competent wildlife organization. We can for many large mammals already predict certain types of consequence; for others we need more knowledge and this knowledge can best be gained through study of the introductions of the species in question or similar species that have been made in recent years.

There is need to clearly define problems resulting from or thought to result from introductions, for introducing an animal does not necessarily imply trouble. Most ungulate introductions have in fact been successful and have not caused serious problems. Even in countries where problems have resulted from introductions in other parts of the same country introduction of the same species may have resulted in little or no problem. This applies even to the deer in New Zealand, judging from a study made in 1956 which showed that the areas of highest deer density were not necessarily the areas that were troublesome from the standpoint of accelerated erosion, or regeneration of native forest (Riney, T., 1956(a) and 1956(b)). And even the notorious red deer is not necessarily bad in all New Zealand areas (Riney, T. *et al.*, 1959).

The field of investigation within disturbed ecosystems is only just developing, and exploitation of opportunities arising from introductions can make an important contribution to the understanding of such disturbed systems (Riney, T., 1967b).

Introduced populations of large herbivores, if undisturbed, normally follow a pattern of adjustment to the new environment which consists of a single eruptive oscillation (Riney T., 1964 and Chapter 7000). It is normally not until stages following this initial phase are reached that species can be permanently evaluated in terms of the extent to which they: are successful or unsuccessful; become problem animals; are considered useful for basically the same reasons as indigenous animals (Riney, T., 1964).

The initial oscillation is important to understand before assessing the extent to which the introduced ungulate is to be a temporary or a permanent problem in the area of its introduction. The most important question in this connection is the extent to which the introduced ungulate interferes with the land-use policy during the various phases of its initial adjustment and after (see Figure 126). If the interference is slight or nil, then, in this particular context, the introduction can be considered a success. If there is evidence that trends are continuing to the detriment of the land-use policy in question, the introduction must be regarded as unsuccessful and there may be reason for the reduction or elimination of the introduced species. This criterion should, but unfortunately does not normally, apply to introduced domestic animals as well as wild species.

Introductions should never be undertaken lightly and when they are made should be only with the approval of the responsible government organization.

With the above considerations in mind there is good motivation for research to analyse and more clearly define problems associated with introduced animals. The field of problem analysis within disturbed ecosystems is only just developing, and exploitation of opportunities arising from introductions can make an important catalytic contribution.

Several useful questions may be asked when considering the possibility of introducing a species of large mammal.

1. What are its habitat requirements, even in general terms? (How adaptable to a wide range of conditions, how restricted to a narrow range?)
2. Would over-population of the species interfere with the major form of land-use to which the area is now or will be devoted?
3. What is its potential eruptive curve under ideal conditions? (see Chapter

7000). This will provide a basis for estimating the time it will take for the species to make its initial adjustment (that is in the area where it is first introduced). If an initial oscillation is not desired this knowledge will assist the timing of control operations to allow the initial adjustment to take place without an initial oscillation.
4. Is the population controllable?
5. What is the pattern of dispersal, and at what time of year does dispersal take place? This may be important in answering (4) above.

Because of the inherent dangers involved in introductions, many countries and several international organizations have been concerned. The following has been modified and expanded from a statement of conclusions reached during an IUCN meeting held in Lucerne, Switzerland, and lists steps which should be taken to avert or reduce dangers which may be associated with introductions.

1. Accidental introductions should be minimized and, if they do occur, should be promptly investigated to determine if practical remedial measures are necessary.
2. Species should be introduced into national parks and equivalent reserves only if they formerly existed in the area.
3. No species should be deliberately introduced into a new habitat unless the presence of the species conforms with, and can be expected to contribute to, the type of land-use assigned to the area and unless the operation has been carefully and comprehensively investigated and planned.

Examples of important considerations in (or guidelines for) such planning are that:

1. No species should be considered for introduction to a new habitat until the factors which limit its distribution and abundance have been thoroughly studied by competent biologists and an estimate of the timing of the initial oscillation and an appraisal of its dispersal pattern has been made.
2. No species should be considered for introduction into a new area, if stock of any comparable native species is present in the area and can be built up, or is no longer available but can be rehabilitated by reintroduction, or extension from a nearby range.
3. The species to be introduced should be reasonably expected to enhance the economic and/or aesthetic value of the area and provide a minimal foreseeable risk of conflict with land-use policies in the area of introduction or in adjacent areas to which the species might spread.
4. The individuals to be introduced should be, to the satisfaction of competent authorities, free of unwanted communicable diseases, viruses and parasites.

5. The species, if introduced, should not threaten the continued presence or stability of any native species or distinctive population, other than those which are pests, whether as a predator, competitor for food, cover or breeding sites, or in any other way arising from its characteristics and ecological requirements.

6. The knowledge of how the species could be eliminated, or at least kept under control in its new range, should be available; and the means of doing so by a socially and economically acceptable method.

7. Whenever controllable experimental areas are available which contain habitats typical of those into which it is proposed to introduce the species, these should be used to test locally the suitability of the introduction before it is made generally.

8. Animals should never be introduced into areas where one or more elements of their required habitat is downgrading.

9. Both the techniques of capture and of transport should be carefully planned with a view to eliminating mortality and wounding; minimizing disturbance and with full awareness of the costs involved.

10. Immediately following an introduction, continuing studies should be undertaken of the introduced species in its new habitat and especially of its rate and pattern of spread and of its impact on the habitat. The importance of initial close monitoring cannot be over-emphasized.

8333 *Problems involving Dispersing Animals*

The question of naturally dispersing animals may pose a serious problem in some situations. For example, consider a situation where it is planned to convert large areas of a comparatively open land into forest and there is a species of wild ungulate suited to forest or forest edges living nearby. If the pattern of dispersal of the species will result in natural colonization of these newly planted forest areas then eruptive oscillations will inevitably occur, following exactly the pattern described for deliberate introductions (Figure 126C).

As with newly introduced or reintroduced animals the definition of a problem involving self-colonizing animals, at a given time, will involve both the present extent of animal use and the extent to which the population is stable. If the population is managed it will be either stable, increasing or decreasing and the latter two circumstances may be indicative of some phase of an eruptive oscillation. If the population is unmanaged it may be either in some phase of an eruptive oscillation, or stabilized following an oscillation (see also Chapter 7000).

8340 The Problem of Comparing Different Kinds of Data taken from a Study Area

There is a constant interplay between animals and their environment and whether we wish to simply understand the processes involved, or we wish to define a problem or indentify its causes, we must find some some way of threading our way through the complexities of the ecosystems within which we work. In an early Section (3000) a scheme was proposed for classifying techniques in terms of whether the information they produced came from the occurrence, utilization or response levels of either the environment or the animals concerned, or both. Emphasis has been placed throughout the text on first becoming clear on objectives, then arranging subsequent information-gathering or management activities in a fashion appropriate to the objectives (Section 2030). Almost always, for a wildlife research worker or wildlife manager, this involves obtaining two or more different kinds of information, as is elaborated in examples of problems mentioned earlier in this chapter (Section 8320).

The rules for making such field observations are simple, but if ignored serious difficulties may be encountered in interpreting the information obtained. *The gathering of different kinds of data should be planned to facilitate comparison*. This cannot be over-emphasized. Here are simple guidelines which may prove useful under two kinds of field situation.

1. If comparisons of one kind of information at different times of the year, or between years or between areas are to be made, *the golden rule is to be consistent*. The techniques should be applied in the same area or areas, in exactly the same way and preferably by the same persons making the earlier observations. If different observers are involved, special care must be taken to ensure that they produce the same or very similar results when making the same kinds of observations at the same place and at the same time.
2. If comparisons are to be made between two or more kinds of information in a study area, or between two or more study areas:
   (a) The first rule is to maintain consistency of application as mentioned above.
   (b) Equally important, in comparing different kinds of information, is the need to ensure that the observations and results are segregated geographically. As a simple example, if one wishes to learn to what extent different species, different densities, different sexes, or ages, or assessments of physical condition, or mortality, etc, are associated with different habitats, then the observations will obviously be classified and assessed separately for each habitat. Or, if it is desired to compare information from animals killed with information taken along transects located in the lower, middle and upper reaches of a watershed, then the information from animals shot in the lower parts is compared with the lower transects and animals from middle and upper areas are associated with middle and upper transects respectively. Likewise, if the worry is animal damage to vegetation, then associated observations will be taken in areas of heaviest damage, areas of no damage, areas where use occurs but where the level of use is still tolerated, etc.

TRANSECT No.: 4
DATE: 11 Aug. 1959

OBSERVERS: Mitchell, Riney
LOCATION: ·7 on 10 Mile Drive, Wankie.

POINT EVERY SECOND STEP						**MIL-ACRE PLOT SURROUNDING EVERY TENTH POINT**									**TREND**			
			Inches			Feet									BARE GROUND		GRASSES. VIGOUR	
NO.	Bare Ground	Litter	Grass Height	Annual	Per.	Shrub/Tree Height	Cattle	Sp. Hare	Hare	Horse	Wild	Eland	Zebra	Kudu	Impro	Down	+	−
1	✓	−																
2	✓	−																
3	✓	−																
4	✓	✓				1												
5	✓	−				4												
6	X	−	12"		✓	1												
7	✓	✓				5												
8	X	−	12"		✓	1												
9	✓	✓																
10	X	−	5"		✓	2	2	✓		1		1		✓	✓		✓	
11	X	−	24"		✓	2												
12	✓	✓																
13	X	−	5"		✓	11												
14	✓	−																
15	X	−	12"		✓	5												
⋮																		
100	✓	−				14				1					✓		✓	

(a)

DESCRIPTION OF TRANSECT.

Sample taken in fairly open Combretum woodland, highest trees being ±20ft. Most leaves were fallen. Litter consisted mainly of leaves and dead annual grass. There were three distinct vegetation tiers – up to 20ft (Av. 15ft); the shrubs up to 8 feet (Av. 4·6ft); and the grass & litter. Several trees encountered on transect had been pushed over by elephant.

MANAGEMENT HISTORY.

(b)

Burning Program   Early on alternate years for previous 7 years; before that, annually for about 20 years.

Cattle Grazing   Of about same numbers for previous seven years, before that, cattle were run, but comparative rate of stocking unknown.

VEGETATION PROFILE ✓   PHOTOGRAPH TAKEN ✓   20 PEDESTAL HEIGHTS AV. 2.3"

100 SHRUB/TREE IDENTIFICATIONS ✓   BUSH ENCROACHMENT Yes.

Following this practice, if there are significant relationships they will be revealed much quicker than if all data taken from shot animals, for example, were compared with data taken from all transects. It may of course be necessary to do this but much understanding will be lost if only a summary approach is used initially.

8341 *Multi-purpose Transects*

It is for the above reasons that the use of multi-purpose transects have been emphasized in several parts of the text. Their advantages are the speed with which they can be run, the comparative ease with which consistent results may be obtained and their flexibility in being able to accommodate different kinds of information taken along the same transect.

A sample form is shown in Figure 141, but this should in no way be used as a model for copying unless it happens to be relevant to the job in hand. The figure is included simply to demonstrate how information from the occurrence level (both animal and plant), the utilization level (plant) and the response level (both animal and from the environment) can all be incorporated in a single transect.

It is convenient to think of such transects as a string of points stretching approximately 100 m through one (as uniform as possible) habitat. Each point starts on the surface of the ground where various kinds of observations

Transect summary

Bare ground	90%
Litter	47%
Perr. grass	10%
Ann. grass	0
Grass + litter	57%
Ground covered by shrubs or trees	26%
Av. tree/shrub height	9.5 ft.
Av. grass height	8.9 in.

Total droppings

Cattle	5
Springhare	5
Hare	5
Kudu	1
Eland	1
Zebra	1

Conservation trend

Bare ground	10/10 downgrading (i.e., increasing)
Perennial grass	8/10 downgrading (i.e., decreasing)
Transect assessment	downgrading

Figure 141   An example of a multi-purpose transect observation form (a), description (b) and summary (c), from Riney, T. (1963a)

may be made. Since an environment is three-dimensional, it is possible to imagine an invisible line extending vertically upward from each sample point to the top of the tallest vegetation. Various other kinds of information may be added by recording various parts of the physical environment that intersect with this imaginary vertical line, such as height of grass, height of shrubs or trees, etc.

Along the same transects, at every other pace, or at every tenth pace (depending on study objectives and the density of the item measured) a circular plot may be inspected and records made of numbers of faecal pellets of various species, an estimate (or measure) of plant utilization recorded, assessments of conservation or vegetative trends may be made, evidence of recent fire described, number and size of erosion pavement stones recorded, or whatever other observations may be relevant and lend themselves to being included in a multi-purpose transect. Often two or more observations may be used in combination to form various indices to utilization or occurrence, such as the grass availability index described in Section 3535 (obtained by multiplying the percentage ground covered by the bases of perennial grass by the average height of grass along the same transect). In other words some kind of information taken at each of the three levels of an ecosystem (Occurrence, Utilization and Response) may be taken as an integral part of a multi-purpose transect.

Another feature of such transects is that with careful planning the information gathered may be designed to contribute to two or more separate studies. For example, the form shown in Figure 141 was designed to help with learning differences in habitat associations between several species of large wild mammal and extensively managed domestic animals; at the same time information taken along the same transects was used as a basis for describing differences in conservation status and in trends associated with several kinds of land use.

The use of these transects for particular purposes has been described in several places earlier in the text. Figure 142 illustrates their main characteristics. Appendix 3 is included as an example of kinds of arbitrary decisions that may have to be made to facilitate obtaining consistent results.

Each transect should be numbered, accurately located and the habitat it samples described, along with any relevant information concerning recent or past history, present use, presence of bedding areas or key feeding areas, etc. If a species is so scarce that defecations were seen along the transect but none were counted within the circular plots, this should be noted.

Each transect should be summarized as soon after running as possible, and as soon as possible comparisons should be made between transects. Even superficial comparisons of transect results will usually reveal differences in relationships, or raise questions which can only be answered by running additional transects in other similar habitats or in different areas. If one waits until leaving the field to make these preliminary summaries then much time and effort will be wasted and it is possible that much understanding will also be lost.

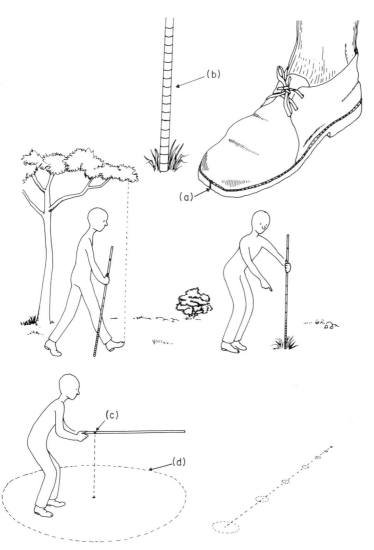

Figure 142   Framework of multi-purpose transects. Observations start immediately below a ballpoint pencil line on the toe of a boot (a) and extend vertically up in an imaginary line to the top of the tallest vegetation. A staff is marked in centimetres for estimating height of low vegetation (b). At periodic intervals, for example, around each tenth line point, other kinds of observations may be made within circular plots, whose radius is marked on the upper part of the staff (c), whose tip then delineates the edge of the circular plot (d). Points are recorded each time the right foot hits the ground. It is usually safe to start with 100 points, extending about 100 m in a straight line

If two persons are involved it is usually convenient for one person to enter the observations directly on the field form. If only one observer is involved it is often convenient to enter the observations, point by point and plot by plot, on a tape recorder and to transfer this information immediately afterwards to the form. If a tape recorder is used special care must be taken to ensure that no kinds of observations are missed as one proceeds along the transect. It is important to have a form in hand to help not only in remembering the various kinds of observations that are to be made at each point, but also to help in making them in the same sequence, which facilitates recording. Also helpful is a tally counter in one hand to press at every observation point, to remind one when the periodically observed circular plot is coming up and when the transect is finished.

8350   What To Do Next When You Don't Know What To Do Next

Normally an experienced biologist will have little difficulty in making a start towards clarifying a problem. However, even elderly biologists can find themselves in a situation where they wonder how to start making observations which are new to them and for which there is no ready guidance at hand. First, consider the normal options:

1. Consult the literature, suggestions from your boss, or other colleagues.
2. Re-examine the problem. Is it really different? Re-phrase the problem describing its elements as you would make them understandable to a curious 6-year-old child. This sometimes stimulates miraculous insight, for there are very few new problems.
3. Then of course there are always the animals and their habitats. It is consistent with the approach used in this book to fall back on field observations when in doubt, for always there are the basic animal–habitat relations with which we must eventually come to terms.

To elaborate, take a lunch and spend at least one 12-hour day in the field. As much of this time as possible should be spent in observing undisturbed animals within the area where the problem occurs. Keep the problem in mind while observing.

If after this you are still at a loss at to what to look for, try describing movement behaviour because this is simple and obviously relevant to animal–habitat relations. Take notes of your observations, thoughts, questions as you proceed. As you need them make appropriate notes on the spot. Look for answers to questions such as these: Where do the animals sleep? How much ground does an individual cover in a day? Map the pattern of movement—walking, running, resting, feeding. For each kind of movement what makes them move? (Unknown, contact with others of their kind, with man, with other species?) What is the rate of movement? If a group of animals is under observation how does the movement of the individual correspond with the movement of the group? How much time is spent in the various habitats? In short, throughout the day: where are the animals, what are they doing, and why?

Of course all species cannot be observed in this way and it may be appropriate to simply record each contact you make with individuals or groups and note their number, exactly what habitat they are seen in, what they were doing, and any specific details you may notice.

You may be dealing with such a wary, or scarce, or nocturnal animal that you may prefer to spend the day systematically combing through all different habitats and edges between habitats looking for trails, droppings, beds or other signs and recording them. Thus an indirect picture of animal occupancy of available habitats may be built up. Your observations may also reveal key areas of use as indicated by more intensive trail systems or other signs, concentrations of animals seen or by observing heavier use on plants. If this is the method chosen care must be taken to spend equal time in each of the available habitats and to make comparative observation in the same way, so you will be able to have some perspective of comparative animal–habitat occupance and use by the end of the day.

For any of these ways of observing, the problem of your concern should be kept firmly in your mind and, whatever your observations, examine them in terms of how relevant they are to answering your problem. Obviously the kinds of observations suggested above will neither define nor suggest answers to all animal problems (although it's not a bad way to start if your problem involves competition between two or more species). Regard the exercise as a trick if you like, a conspiracy to manoeuvre you into the field, looking at some real relation—any real relation will do—between animal and habitat. Then from this springboard of reality, what to do next about your problem usually resolves itself quickly. It is there, while observing, that questions will come (record them), there you will form theories (outline them). There you will get ideas for testing or proving the ideas. The next step is a working plan (Chapter 2000), and away you go.

The above should not be interpreted as *the* way to start on a problem if you are puzzled over how to start. This approach does not appeal to everyone; nor is it the best for everyone. Other approaches involve more intensive review of the literature, committee meetings of available biologists and administrators (a warning: it is amazing how many animal problems have been discovered in just such meetings, although non-existent in the environment under discussion). Or an investigator may wish to start with carefully controlled laboratory or field station experiments and then test his resulting theories in the field. The so called 'field approach' is not the only approach but it is obviously the relevant one to stress here, for this is what the handbook is all about. Finally, it is not surprising to find that field biologists are often capable of getting the appropriate evidence they need to solve their problems ... in the field, where the action, the excitement and the good life is.

470

8400   POLITICS AND WILDLIFE

There are many definitions associated with the words politics, politic, politician and political, and some of these definitions are likely to raise emotions unsuited to the spirit in which these pages are written. In considering the political aspects of wildlife I have thought especially of the influences by which individual government officers or private individuals seek to determine or control public policy because of vested interests; or who, for any reason, seek to control the appointments of, or actions of, those who manage affairs of the state that concern wildlife.

It is clear that any private or public organization or individual can get involved with politics and, in the sense used in this discussion can try to control or exert pressures on other organizations interested in wildlife to the end that policies or activities of the parks or game departments might change.

Thus special interests of groups or individuals can give rise to wildlife problems as we know them. The animals are commonly considered as 'problem animals', and the problem itself is often referred to as an 'animal problem', instead of what it really is—a human problem involving animals: cattle fences and zebra, crops and elephants, deer and trees, domestic stock and predators, cattle and the tsetse fly and so on and on. Such problems can of course be very real, or they can be non-existent, but animal problems almost always become political in some sense.

Although it may savour of over-simplification, it suits the happy spirit of this handbook to group political manoeuvres into two classes: the good and the bad. In any event this is a common way of reacting to the word politics so let us conform: first by considering bad politics and wildlife; secondly, good politics and wildlife.

8410   Bad Politics and Wildlife

Bad politics have one main characteristic by which the actions or manoeuvres can be easily recognized: they are usually based on unproven assumptions and they are sometimes end-products of a department trying to follow the laws of expanding gases and 'empire-build' or expand on something as unstable as 'hot air' (invented issues) rather than on a record of solid accomplishment. Those making decisions are often sincere; they simply bend with the blasts of pressure groups.

Another field man, the late Lord Wavell, included in his book of poems and other miscellaneous art forms called *Other Men's Flowers* (1944), a very accurate description of wildlife and bad politics, although I have not previously seen it referred to as such.

The following statement accurately describes a wildlife problem involving 'Birds, Bags, Bears and Buns'.

> The common cormorant or shag
> Lays eggs inside a paper bag.

The reason you will see, no doubt,
It is to keep the lightning out,
But what these unobservant birds
Have never noticed is that herds
Of wandering bears may come with buns
And steal the bags to hold the crumbs.

ANON.

This analysis of wildlife and bad politics will stand close inspection. Including it here may help in recognizing a familiar pattern of reasoning. Without over-labouring the comparison I would especially emphasize the first two lines, for they clearly illustrate a common characteristic of wildlife politics. The statement is clear enough. We know the bird well, and those of us suspicious, or having been subjected to some form of, technical training such as is now taking place, might wonder, for a start, what evidence there was for shags laying eggs inside paper bags. Arriving at conclusions with little or no relation to previous evidence or with no *relevant* evidence at all, is another feature.

Now rightly or wrongly a curious thing about political pressure is that the only thing that will satisfy the pressure group is some sort of action. And another curious thing about wildlife and politics is that often the action recommended to the department or organization concerned has nothing to do with the real animal–environment–human interest issues at stake, but 'something is being done'. That is all that seems to matter and political pressure recedes because 'something is being done', whether it is right or wrong or creating an even bigger problem.

As mentioned elsewhere, feelings sometimes run high when 'wildlife problems' are discussed and, as a trick to maintain the even temper of this discussion, examples mentioned will be real but anonymous so there can be no suspicion that anything other than principles is being considered.

To help you see the issues on an objective basis the following examples refer to the source of the problem that was publicized, the problem as given to the public, the solution that satisfied the public, the extent to which evidence confirmed the problem as a real one, and the effectiveness of the solution. Any resemblance between these examples and actual happenings is not coincidental. They are summarized accounts of real examples taken from several countries.

8411 *Fences*

Source of the problem publicized. A government department.

The problem as given to the public. A department interested in growing trees in a certain small area could not grow them because deer were eating the young trees, deforming them and preventing normal regeneration. Unless some action was taken the department would not be able to grow trees.

The solution that satisfied the public (and, initially, the department). Deer obviously had to be kept away from the trees and the two solutions suggested were: (1) to eliminate deer by shooting, or (2) to fence them out. Shooting was turned down because the area was adjacent to a very large unoccupied wilderness area and although the department heads were ambitious and the department expanding, the thought of exterminating deer from several hundred square miles of difficult country to protect trees in a study area of nearly 8 square miles was too much to consider, so the department built a 10-mile deer-proof fence. The fencing contractor made the fence up hill and down dale; made it strong with extra straining posts; made it 8 feet high, with eight strands, the bottom four close-spaced; made it with high-tensile steel wire and barbed wire, made it for a profit and charged a very large sum indeed. Just before the fence was completed, an attempt was made to drive the deer out before the fence was closed but most deer doubled back through the line of beaters and remained inside. The fence was closed anyway.

The fence was then complete, with great publicity, and the public accepted the whole business with considerable pride in how well the forests were being looked after and with considerable admiration for the ingenuity, the boldness of conception, and the sturdy evidence of action by public servants who were, after all, being paid from public taxes.

The effectiveness of the solution. Several months after the erection of the fence an inspection was made and good evidence of 72 crossings was found along the 10 miles of fencing. Both deer and pigs were freely crossing back and forth through the fence and the fence in fact provided a good opportunity for defining the areas of greatest local movement across the hillside. The fence was simply no barrier.

As a result of the inspection, and of repairs following appropriate recommendations, the fence became more of a barrier, but still not deer-proof, even with the considerable extra cost for battens, extra wires and labour.

Effort was then made to remove the deer from the forest by shooting. However, checks of deer numbers based on droppings counted along sample transects indicated that after 2 years' of shooting the population inside the fence was still increasing.

The extent to which evidence confirmed the problem as a real one. Further investigation inside the fence showed that the proportion of trees with their terminal shoots damaged to the extent that it would affect the quality of timber was negligible. The most interesting and final development was that a check on the numbers of desirable regenerating trees outside the fence, as compared with inside the fence, showed the location with the greatest number of regenerating trees was actually outside the fence where deer freely roamed, rather than inside.

Conclusion. The political tool that satisfied the public was the fence. Its ineffectiveness, and the fact that there was really no need to build it in the first place, was not publicized. A simple investigation to define the basic ecological problem in a way that would allow administrators to see for themselves what appropriate action they should take would have saved many thousands of dollars on this one project. This particular government did not then have an organization coping with this type of research. The saving to the government of the cost of this one fence would have been enough to employ a small but competent three-man research team for at least 2 years.

This is but one example. There are hundreds of examples of fences being built throughout the world as a result of political pressure of one sort or another. The reasons are always logical ones, the issues are always posed publicly as very grave ones, and the solution, the fence, is usually a very expensive proposition. The first issue to clarify, if we are to be objective about fencing, is: what evidence is there that the fence must be built in the first place? Humans usually don't have to have a reason not to do anything; we should be very clear about our reason for building fences, which cost much money. Relevant questions to ask at the time fences are first proposed as a solution to a management problem are, for example: How successful have other fences been in accomplishing the purpose for which the proposed fence is intended? Has any study been made of the basic, ecological, long-term trouble underlying this recent plea for fences? There will always be a use for certain types of fences under certain conditions, but when we start to ask questions like these I suspect that several million dollars a year on an international basis may be saved by reducing the frequency with which the fence is used not as an ecological tool but as a political tool.

8412 *Bounties*

The use of bounties on predators or other 'problem animals' (in the sense defined above) is a very old one. During the reign of Henry VIII, animals destroying crops, livestock, or game were declared 'vermin'. Not only could they be killed at any time but a reward, or bounty, was placed on them (Trefethen, 1964). Bounties are already well recognised as a political tool in wildlife circles and this is too well known to elaborate or to occupy space with numerous examples. It is enough to say that rarely do bounties accomplish the purpose for which they are intended, and all too frequently they create worse problems than the original ones, as when elimination of a predator stimulates an increase of an even less desirable prey species. Bounties usually stem from pressure groups outside the government and a senior government official in the wildlife service in the United States, several years ago, described government-supported bounty systems as: 'the most effective way of wasting government money that was ever invented'. The basic trouble with bounties is that as populations of animals (whose ears or tails are being collected for money) decline in number, the incentive to shoot them also drops and the bounty collector naturally moves to where the animals are more numerous, and his efforts therefore more profitable. This gives the area with recently lowered numbers of 'target' animals a chance to increase its population and so it goes with the effect of the bounty collectors being similar to that of a predator, taking animals in the way most suited to them, but never eliminating their prey. This is the secret of the perpetual drain on government resources. Bounties don't normally end because they have been successful; bounties normally go on forever if permitted. Why the initiation of a bounty is often successful in satisfying the pressure group's demand for action is one of the mysteries of human ecology, and considerably outside the scope of this handbook. A good public relations officer in a game department should be enough to all but eliminate the use of the bounty as a political tool within the area of his influence.

8413 *Shooting Campaigns*

In some countries a campaign to shoot animals can arise either from private groups ('outside pressure') or from representations within the government ('inside pressure'). The shooting usually stems from human concern with domestic animal diseases, with accelerated erosion or with regeneration of the forest trees or vegetative cover required to maintain productivity or to protect watersheds. For these problems wild animals have been held responsible: sometimes justly, sometimes unjustly.

When progress, in a shooting campaign, is measured by the consistency with which annual tallies of animals killed are increasing or maintaining their numbers then the shooting campaign becomes in effect very similar to a large-scale bounty system. The aim is to keep up high numbers of kills as evidence of action. This satisfies the political pressures that something is being done. Something is being done, right enough, but it usually results in a far more devastating and useless expenditure on governmental funds than the infamous, but more easily recognized as infamous, bounty system. If the shooting is heavy enough to reduce the population a third or more and then shooting slacks off, an eruption may be started (as described in Chapter 7000) and result in more animals than occurred in the area before the initial shooting commenced.

As with any other action programme, the reasons for starting the action of shooting should be perfectly clear before shooting starts.

I have been out on a shooting campaign to eliminate goats from an area where great damage was said to be taking place on one section of a mountain range. The most recent aerial photographs available, taken 6 months before, showed that the land was indeed very bare and badly eroding, and herds of goats could be seen in the picture. Publicity, designed to induce the government to kill the goats and solve the problem, had been going for 5 years. But when the shooting team of six men went out to shoot hundreds of goats they had to work hard for a week to shoot two goats. The real and original cause of trouble in this case had been fire and over-grazing by domestic animals, to the point where domestic animals could no longer exist. Feral goats spread in from adjacent areas and increased in numbers. However, by the time the government concerned sent its shooting party out to deal with the problem, the vegetation, far from being held back by goats, had grown up so fast—even in the absence of any control on the goat population—that in a space of 5 years the habitat was no longer suitable for goats, which had almost completely died out.

Sometimes shooting campaigns, starting from a legitimate concern in a specific small area, expand to become a countrywide campaign. One country was thus infected with a mass mania for shooting wild animals to reduce accelerated erosion and increase forest regeneration. A large organization was conceived, expanded on principles so aptly stated by C. N. Parkinson (1958) and continued for 20 years before a survey was made which indicated that over 80 per cent of all animals shot were taken from areas that were causing no worry whatever to any of the departments concerned with regenerating forests and accelerating erosion. A detailed study in one area regarded as most critical by these departments, and where wild animals were said to be the cause of serious erosion, was shown to parallel records on a nationwide basis, for where the wild animals were most numerous, there was the least trouble. Where domestic animals, in this case sheep, were present, either alone or in some combination with light populations of wild animals, there was always trouble which took the form of serious and accelerating erosion, the very problem publicly blamed on the deer.

This brings out another issue often associated with shooting campaigns, namely the scapegoat nature of the campaign against wild animals, as an action programme that satisfies the public while at the same time obscuring the real issue. All too often this real issue is simply bad land-management through unsuitable burning policies, or over-grazing by domestic animals. Especially where shooting campaigns are considered it is important to emphasize the need for investigations to define the real problems and to clearly determine the extent that wild animals are contributing to these problems before embarking on costly shooting campaigns. These investigations should be done at as early a stage as possible, for once a department builds up a large organization and has years of publicity justifying the high tallies of animals killed, it is most difficult to change policies without very careful attention to matters of saving face,

Another characteristic of these examples from various countries is the rather ironic twist that in addition to the fact that the bases for the campaigns are often on a false premise ecologically, the actual campaign, by virtue of the technique of shooting or poisoning, results (almost always quite unintentionally) in an increase of the very animals the campaign is designed to eliminate or control.

8420 Good Politics and Wildlife

This section is short, as a study of the examples given above under 'bad politics' will suggest obvious, more sensible, approaches. A department concerned with wildlife needs facts and a sound basis of understanding on which

476

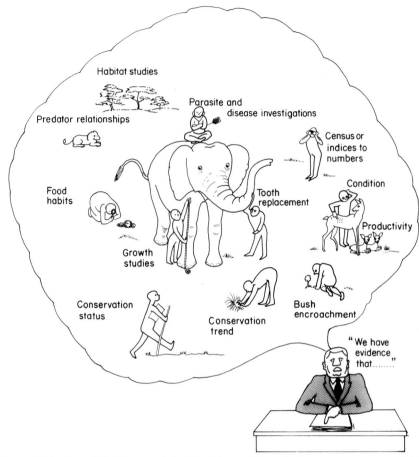

Figure 143   A politically sound (and safe) way for a Director to present wildlife problems to his government and to the public

to base their action programme. If a department is staffed with competent men who have this basic information, or with men who can obtain the necessary information, this is the best possible way in which to be free of, or at least to minimize, political pressures which have a way of insisting on the immediate solution of bird–bag–bear–bun problems.

The cartoon shown in Figure 143 is included as a suggestion that it is possible to do something about bad politics as it relates to wildlife. It is possible to obtain evidence to learn to what extent such potentially political tools as fences, bounties, shooting and poisoning campaigns or any other large-scale action programme is achieving its principal objective. Even more important it is possible to obtain evidence which clearly defines the problem in terms of the animals and their habitats and in terms of the extent to which they are in fact conflicting with human interests. A good clear definition of a problem is the first and most valuable step towards the solution of the problem. Since, in my opinion, it is impossible to over-emphasize this point, I have invented a truism:'Problems are easier to solve when we know what they are'.

We as field men should do what we can to encourage and develop the trend towards basing action programmes on appropriate and reliable evidence. This approach is simply applied common sense, and it can also be good politics. If we can do this we will be helping to speed the day when some sort of adjustment can be made between a stable landscape, our heritage of large mammals and those human interests that are and will always be behind any wildlife problem, real or imagined.

Chapter 9000

Wildlife conservation programmes in developing countries

Throughout the book emphasis has been on the importance of asking relevant questions and in developing approaches and techniques appropriate to answering these questions. The question of relevancy is usually within the context of the work of the game department—its present problems and long-term aims. These problems and aims will vary not only country by country, but within a country at different stages of its development. This chapter discusses the initiation of a conservation programme and the scope of a typical conservation department in a developing country.

9100 STEPS TO INITIATE A WILDLIFE CONSERVATION PROGRAMME

A country that initiates a wildlife conservation programme would logically take the following actions: (1) inventory; (2) selection and establishment of suitable areas as national parks or reserves to ensure the continued existence of at least one population of each species of the country's indigenous fauna in its natural surroundings; (3) introduction of measures to control hunting,

479

capture and trade of wild animals and their products in order to administer the exploitation of wildlife resources on a sustained-yield basis; (4) passage of legislation necessary to implement (2) and (3); (5) formation of a game department or other government organization to enforce that legislation and to carry out tasks (1), (2) and (3); (6) research into problems of wildlife; and (7) education of the public, at all levels, in the aesthetic and economic values of wildlife and the need for its conservation.

9110   The Inventory

The inventory stage consists of the listing of all forms of wildlife and the determination of its distribution and abundance. Although the completion of this work must depend on field surveys, a great deal of preliminary information can be obtained from existing publications and records (see Section 4200). Checklists of mammals and birds exist for almost all continents or zoological regions. The natural history museums can also be asked for information on the localities in which specimens have been collected in the region concerned (for elaboration, see Section 4300).

With even a general inventory in hand, it is then helpful to prepare a list of the most 'important' animals to be preserved; their 'importance' being judged on local rarity, the degree to which they are threatened by hunting, habitat destruction or other human activities, or the extent to which they are restricted in range completely or largely to the country in question. A convenient way to test international concern with the survival of a species is to consult the IUCN List of Endangered Species.[1]

9120   Establishment of National Parks and Reserves

In the following paragraphs the terms 'national park' and 'reserve' have both been used to denote areas where the object is to maintain all forms of indigenous fauna and flora in a natural state. The main difference between the two is that certain limited and controlled forms of land-use such as grazing, fishing and fuel-gathering may be allowed in a reserve but never in a national park. (For a more detailed definition of a national park, game reserve and sanctuary, see the guidelines prepared by representatives of African countries in Appendix 4.)

The creation of national parks and reserves is often the most effective first step to take to safeguard a country's wildlife from complete extinction. It is also the simplest, requiring initially only enough staff to patrol these areas. A generally accepted national objective is to preserve an example of each of the country's main zoogeographic regions, or most important ecosystems, in the natural state, with its native fauna and flora intact. Because few countries can afford to maintain more than a very limited number of such conservation

[1]Write to Survival Service Commission, IUCN, Morges (VD), Switzerland.

areas, great care must be taken in their selection to ensure that each contains the widest possible range of habitats and therefore animal species, typical of the region.

Some developing countries, particularly in Latin America, have added another dimension to the question of selecting parks. There, from large areas of wild, unused land, suitable areas are set aside as parks and reserves of the future, even before roads have reached the area or recreational demand has grown, as it will later, with increasing standards of living and increasing populations. The selection of parks and reserves can thus be an integral part of long-term planning. (Miller, K. R., 1980.)

When the initial allocation of national parks and reserves has been decided, it is likely that some animals on the 'important' (rare or endangered) list will occur in none of them. Special arrangements must therefore be made to safeguard such species, either by reintroduction into one of the national parks or reserves—if the species formerly occurred there and the habitat is still suitable, or by the creation of special reserves or sanctuaries especially for this purpose.

9130 Passage of Conservation Laws

Control of the killing or capturing of wild animals, and of trade in wild animal products, is difficult to introduce in a country whose inhabitants are used to traditional unrestricted hunting, or if hunting for trophies or meat has become commercialized. It is, however, an essential step if wildlife is to take a rightful place as a resource, yielding a sustained return for the benefit of the country as a whole, instead of being a means whereby a few people get rich quickly but destroy wildlife or its habitat in the process.

Effectiveness of control depends directly upon effective laws and on the number of enforcement staff in the field. Where a large staff is not immediately available, it is usually wise to introduce the regulation of hunting gradually, starting with species that are hunted for commercial trophies only (e.g. fur-bearers, crocodiles, elephants), and for which simple checks on possession, sale or export of trophies can remove the incentive to hunt illegally. Once this is accomplished, the staff can move gradually to the control of hunting of those animals that are killed for meat.

Two extreme views on the best plan for introducing wildlife legislation into countries previously without such legislation are often heard in regional international meetings by heads of parks and game departments. One group maintains that it is best to have the most advanced and comprehensive legislation passed as soon as possible. The other group advocates the plan of gradual development suggested in the preceding paragraph. The approach and philosophy is obviously a country's own decision.

Conservation laws, even in their simplest form, should include certain common provisions. These are: (1) the method of establishing national parks and other kinds of conservation areas; (2) a list of actions prohibited in such areas; (3) provisions for the control of hunting (see Section 9400); (4) provision for the control of the capture of live wild animals; (5) control of the sale, transfer or export of live animals or wild animal products; (6) appointment of a government agency responsible for enforcing the laws; (7) delegation of certain police powers to agency officers, such as the right to arrest offenders or to search property and seize incriminating articles where it is suspected that an offence has been committed; and (8) publicly stating the penalties to which convicted offenders are liable.

The importance of communication between game departments in adjacent countries should be emphasized because communication is prerequisite to mutually beneficial coordination. Although details of legislation will vary, a certain uniformity of legislation and effectiveness of control can avoid such problems as animals being poached illegally in one country then transported across a border to an adjacent country where the sale of that animal's skin and meat is perfectly legal.

9140   Formation of a Game Department (Wildlife Service)

Whether a separate *game department* or other government organization is given responsibility for the conservation of wildlife will depend on the potential value of the wildlife resource in the country concerned, and the funds immediately available for its conservation. Where circumstances permit, the creation of a separate game department with an adequate headquarters staff and field representation throughout the country is recommended. If this is not possible, wildlife conservation should be made the responsibility of an existing organization that is both conservation-oriented and has the widest possible coverage in the field. These criteria usually make the Forest Service the best choice, rather than the Tourist Department to which responsibility has been delegated in some countries.

Within a Forest Service, it is normal to establish a separate wildlife division to carry out the steps listed in Section 9100. For such services this would also imply (1) posting officers in non-forested parts of the country such as savannah or sub-desert areas, which may be of the greatest importance to wildlife; (2) developing a specially trained wildlife staff and (3) applying a national policy to both forested areas and other wildlife areas.

The actual number of persons employed and their duties will vary from country to country depending on local circumstances and also on the stage of development of the country.

9150   Research

A department responsible for wildlife conservation should include a small research unit of its own, or at least provide facilities for outside workers to assist with problems confronting that department.

The most important reason for research in a newly developing game department is that the Director needs a continuing supply of reliable information, facts and figures, about the animals, parks and reserves for which he is responsible to assist him in making practical day-by-day decisions and in forming specific objectives and long-term policy. Some countries have found that the appropriateness of research is improved if the research workers themselves are at the same time involved in some aspect of management.

Once a management policy is adopted it is necessary to know as much as possible about the consequences of this policy both before and after it is put into effect. Research and management of wildlife and park resources must go hand in hand.

Cooperating with other organizations who may assist with management-oriented research is an excellent policy. In most countries there is a museum or university in a position to carry out biological research. Close cooperation with such bodies, with research officers in other government departments and with visiting research workers from other countries, can facilitate the direction of a maximum research effort being concentrated on matters having application to national or regional wildlife management problems. A research coordinating committee commonly provides a basis for such coordination.

9160   Conservation Education

The future existence of wildlife depends in the long run on ordinary people accepting the aesthetic and economic value of wildlife. Before this can be achieved education is required at all levels, from the schoolchild to the adult. It necessitates the use of all media from books to films and museum exhibits and some countries have found sponsored visits to national parks and reserves of schoolchildren and specialized adult groups the most effective of all. It is unfortunate that few game departments, in their earlier stages of development, can afford an active part in an education programme of this sort. As an interim measure they should, however, help local conservation organizations such as wildlife conservation societies, museums and government tourist organizations in obtaining the materials they may require in the way of photographs or exhibits. In addition, the department might consider assisting in the preparation of conservation courses designed to be introduced into the curricula of both junior and senior schools.

Governments should consider the need for a special officer to deal with conservation extension even in the early days of the department.

9200   THE GAME DEPARTMENT

The staff necessary to administer the natural resources of wildlife, parks and reserves will have varied responsibilities, since the organization will advise government on matters of policy, both initiate and enforce legislation, select, inventory and manage parks and reserves.

9210   Staff

Although the specific arrangement of staff and responsibilities will naturally vary by country, it is useful to consider a basic minimum organization.

In the beginning and in its most simple form a conservation organization may have: (1) a person in charge at headquarters, supported by an appropriate small staff dealing with administration, accounts, stores and transportation; (2) an officer in charge of each of the game areas, reserves or parks over which the organization has jurisdiction and supported by staff appropriate to maintain the integrity of the area; and (3) a biologist or field officer to supply information useful for the officer in charge in making appropriate decisions regarding over-all policy, the drafting of legislation and the formation of specific objectives and goals. Initially this officer would assist the Chief Game Warden (or Director) by surveying the distribution and status of indigenous fauna, the selection of sites for national parks or reserves and in drawing up an outline plan for the control of hunting.

As conservation departments develop and mature, and as revenues from wildlife and parks increase, the increasing complexity will be reflected in a more elaborate staff structure. As an example, consider the organization of a game department headquarters in an East African country.

Five different divisions of work have been included: (1) A General Administrative and Accounts Section; (2) a Stores Section, which includes the issue of arms and ammunition, maintenance of vehicles, etc.; (3) a Licensing and Booking Office, responsible for the issue of game licences, controlled hunting area permits, certificates of ownership, export and sales permits, as well as for booking of controlled hunting areas and accommodation in those parks and reserves where the game department is responsible for the latter duty; (4) a Detection and Prosecutions Section to control trade in wild animals and wild animal products; (5) a Public Relations Section which interprets the work of the department to the public and in so doing initiates lectures, films, articles, organized visits to parks and reserves and other appropriate means to increase public awareness of the importance of the resource and the ongoing efforts being made to conserve and manage parks and wildlife.

Directors of game and park departments in developing countries have a special problem not only in recruiting suitable staff but in improving the capacity of existing staff to meet increasingly sophisticated problems. The two obvious ways of coping are by recruiting and through some form of in-service training.

9220   Parks and Reserves

The first need for permanent field staff comes with the setting aside of national parks and reserves. Here the normal practice is to put an officer in charge of each park or reserve. He should be responsible directly to the head of his organization and should be given the subordinate staff necessary to patrol the boundaries of the area, to make investigations necessary for the formulation of a management plan and to carry out the plan adopted.

Where possible, such officers should have received prior training in wildlife and national park management. If appropriate training cannot be found within the country the government may wish to send officers to other countries for training. Fellowships for training in wildlife and national parks management are available through UN Technical Assistance programmes, several bilateral aid programmes and private organizations.

9230   Control of Hunting and Capture

The need for a greatly increased field staff comes with introduction of control of hunting and capture of wild animals, and of trade in wild animal products. A great deal depends on how much help can be expected from the police in enforcing the law in rural areas, and from the customs authorities in controlling exports. Unless the conservation department can make its presence felt throughout the country, there is grave danger that the whole system of wildlife conservation laws will fall into contempt.

It is usually necessary to divide the country into a series of zones or controlled hunting areas and to appoint a game warden to each, with sufficient subordinate staff to ensure enforcement of hunting and associated regulations. Such areas will vary in size according to the amount of wildlife they contain, the prevalent forms of land-use, the ease of communication, and the extent of illegal hunting.

The names given to areas where hunting is allowed varies from country to country. In these discussions, as long as some form of management takes place they are referred to simply as controlled hunting areas. Hunting zones or hunting areas are areas where hunting legally or permissibly takes place but where no control is exerted. Hunting areas within managed forest areas are normally controlled hunting areas.

Similarly, the tactics used to enforce the law will vary from area to area, depending on the most prevalent form of offence. In some areas it may pay to keep most of the Ranger staff in outposts scattered throughout the region, whereas in others it may be more effective to keep a large part of the staff operating as a mobile force that can move out and take large bands of poachers, or descend on a part of the area and comb out numbers of offenders simultaneously. In countries where illegal trade in highly valued trophies, such as ivory, rhino horn or leopard skins is common, it is often necessary to appoint an additional warden with special detective training or abilities. He must be free to operate throughout the country, in order to break up chains of smugglers and middle men, and find leads to bulk buyers and exporters, who often provide the incentive for this form of lawbreaking.

Generally speaking, it is inadvisable to attempt full-scale control of hunting and sale of wild animal products until the game department is strong enough

486

to be represented throughout the country. Under exceptional circumstances, however, it may be possible to gradually introduce control, region by region. This is done by suspending application of the law to certain regions until the game department is in a position to staff them. But this approach should be avoided if possible, since it brings added difficulties to surrounding areas in which the law is applied, particularly if there is trade in wild animal products between uncontrolled and controlled regions.

9300   FINANCES

A problem commonly encountered by newly developing wildlife organizations is that as annual revenues from wildlife increase, departmental allocations do not increase proportionally. Developing countries that have established the principle of relating annual revenues from parks and wildlife to their annual budget requests have an obvious advantage in adequately financing their departmental activities.

The parks and wildlife organizations should not be expected to bear all costs incurred from their departmental budgets. The cost of administration and management should of course be borne by the department. However, the provision of public amenities, such as roads, night accommodation and restaurant facilities, is more properly the job of other departments (e.g. Public Works or Tourism) or of private concessionaires. Nevertheless, to minimize disturbance to wildlife and its habitat, and otherwise to insure that departmental policies and objectives are complied with, it is important that the Game and Parks Department has the last word on the siting of roads, paths and all buildings. Leases to private concessionaires should be short-term and renewable only if park laws and regulations have been strictly complied with.

9400   GENERAL HUNTING REGULATIONS

The wildlife conservation laws of almost all countries classify wild animals into the following categories: (1) totally protected species (not to be hunted), (2) partially protected or 'game' species (to be hunted only on licence), and (3) the remaining species (for the hunting of which no licence is required).

Of the partially protected or game species, some age or sex groups, such as immature animals, or obviously pregnant females, or females accompanied by their dependent young, are normally protected by law. The females of species that do not carry sporting trophies are also sometimes protected (e.g. elephant, lion, most species of deer, sheep or goats, and such antelopes as impala or kudu) especially when populations are low.

Wildlife conservation laws also deal with the different types of game licences that are available in the country. These vary greatly even in adjoining countries. Since the purpose of licensing is to apportion the allowable crop of game amongst would-be hunters in the fairest possible way, there is now a tendency to abolish or reduce the general game licence. This licence was formerly much used in most African countries, and entitled the holder to hunt large numbers of different species of animals. The recent trend is to require a hunter to take out a separate licence for each species of animal he wishes to hunt. Even in the case of game birds, where no practicable alternative to a general game bird licence exists, a daily or yearly bag limit is often imposed. If, as is usual, the law requires the return of a game licence on its expiry, with details of animals killed under its authority, this system has the added advantages of helping the game authorities to maintain better population statistics and of making the offence of shooting more than the licence allows easier to control in the field.

All wildlife conservation laws list various means of hunting game animals that are prohibited. They may prohibit, for instance: (1) the use of poisons; (2) the use of gins, traps, snares, pits, deadfalls, drop-spears, set guns, spears, bows and arrows; (3) the use of fire to surround or drive an animal; (4) driving an animal into water; (5) hunting an animal from a motor vehicle or aircraft, or using a motor vehicle or aircraft to drive an animal; (6) using an automatic or repeater rifle, or using a weapon of inadequate calibre or ballistics for the species of animal being hunted; (7) the use of any explosive projectile, or a projectile containing any drug or chemical agent having the property of anaesthetizing or paralysing an animal, or any bomb, grenade or explosive; (8) hunting an animal at night by means of a torch, spotlight or other artificial light.

Certain other methods of hunting game animals have been made illegal in some countries because they are considered unsporting, but are permitted elsewhere, where social conditions and traditions may differ. Examples of these are the use of dogs in hunting large mammals, baiting for carnivores and waiting to shoot animals at waterholes.

One other common feature of hunting legislation is the closed season. This can be applied to any species to protect it at a particular period of its annual cycle, such as when deer antlers are in the velvet, or when a fur-bearer's skin is worthless; most commonly it is used to protect a species during its reproductive season. In tropical regions, for species not showing a marked breeding peak, the closed season for reproduction is less effective, thus less often used in the tropics than it is in temperate regions.

All the controls on hunting so far mentioned are usually contained in the main body of a wildlife conservation law and they apply throughout the country, on private as well as on state lands. The owners of private land are, in some countries, free to make their own arrangements regarding hunting rights over their property, provided those arrangements comply with the provisions of the law. However, the exploitation of the hunting potential on public or state land is usually entrusted to the Game Department.

Comparative studies of wildlife and national park legislation have been made by FAO for the African, Latin American and Far Eastern Regions and include examples of model legislation. Officers interested in more details of the formation of wildlife legislation are advised to consult these documents or write directly to FAO, Rome. One such model, *Guidelines for the Preparation of Legislation*, prepared by African heads of game and parks departments, is included as Appendix 4.

Examples of ageing by tooth replacement in large mammals

Letters in the following formulas indicate incisors (I), canines (C), premolars (PM), molars (M), milk (m), permanent (P), erupting (e), absent (0).

Impala (*Aepyceros melampus*), after Child, G., 1964.

Adult tooth formula: $I\frac{0}{3}$, $C\frac{0}{1}$, $P\frac{3}{3}$, $M\frac{3}{3}$.

Defassa waterbuck (*Kobus Defassa Ugandae*, Neumann), Spinage, C. A., 1967

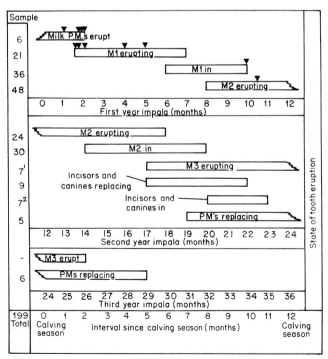

Figure A1   Maximum spread in relation to the calving season of specimens of impala in various stages of tooth eruption.▼ indicates known aged specimens; P. M., premolar.
(After Child, G., 1964)

Adult tooth formula: $I\frac{0}{3}$, $C\frac{0}{1}$, $P\frac{3}{3}$, $M\frac{3}{3}$.

African buffalo (*Syncerus caffer*, Sparrman)

Adult tooth formula: $I\frac{0}{3}$, $C\frac{0}{1}$, $P\frac{3}{3}$, $M\frac{3}{3}$.

Ageing by tooth replacement, following J. J. R. Grimsdell (1973), who formed the following eruption stages for the lower jaw only:

Table A1 Tooth eruption stages and age groups in Defassa waterbuck (after Spinage, C. A., 1967)

Stage											Estimate of age
	I^1	I^2	I^3	C	P^2	P^3	P^4	M^1	M^2	M^3	
1	$\frac{0}{e\text{-}m}$	$\frac{0}{e\text{-}m}$	$\frac{0}{e\text{-}m}$	$\frac{0}{e\text{-}m}$	$\frac{m}{m}$	$\frac{m}{m}$	$\frac{m}{m}$	$\frac{0}{0}$	$\frac{0}{0}$	$\frac{0}{0}$	Birth–1 year
2	$\frac{0}{m}$	$\frac{0}{m}$	$\frac{0}{m}$	$\frac{0}{m}$	$\frac{m}{m}$	$\frac{m}{m}$	$\frac{m}{m}$	$\frac{P}{P}$	$\frac{0}{0}$	$\frac{0}{0}$	1–2 years
3	$\frac{0}{m}$	$\frac{0}{m}$	$\frac{0}{m}$	$\frac{0}{m}$	$\frac{m}{m}$	$\frac{m}{m}$	$\frac{m}{m}$	$\frac{P}{P}$	$\frac{P}{P}$	$\frac{0}{0}$	2–3 years
4	$\frac{0}{P}$	$\frac{0}{m}$	$\frac{0}{m}$	$\frac{0}{m}$	$\frac{m}{m}$	$\frac{m}{m}$	$\frac{m}{m}$	$\frac{P}{P}$	$\frac{P}{P}$	$\frac{P}{P}$	3–3½ years
5	$\frac{0}{P}$	$\frac{0}{P}$	$\frac{0}{P}$	$\frac{0}{P}$	$\frac{P}{P}$	$\frac{P}{P}$	$\frac{P}{P}$	$\frac{P}{P}$	$\frac{P}{P}$	$\frac{P}{P}$	3½ years and older*

Tooth eruption classes

*Figures not clear but estimated all permanent teeth by 3½ years.

Grimsdell also found that counts of cementum rings on buffalo teeth were useful as ageing criteria and that the number or rings in M^1 correlated well with the height of the crown of this tooth above the gum line. This information is not included because the great variability in rate of tooth wear makes it necessary to develop such correlations area by area.

Red deer (*Cervus elaphus*) after Riney, T. (1957d).

Adult tooth formula: $I\frac{3}{3}$, $C\frac{1}{1}$, $P\frac{3}{3}$, $M\frac{3}{3}$.

In this example information was presented as an elaborated figure caption and on a single printed sheet for use in training research assistants who were issued with .303 calibre rifles and drawn from the ranks of hunters employed by the government in controlling deer. Before age groups based on tooth replacement were assigned, growth curves taken from the same animals and based on total length, length of skull, lower jaw, hind leg and girth measurements were examined as well as records from a few deer of known age.

Table A2 African buffalo (*Syncerus caffer*, Sparrman) (after Grimsdell, J. J. R., 1973)

Tooth eruption stages (lower jaw)

Stage	I^1	I^2	I^3	C	P^2	P^3	P^4	M^1	M^2	M^3	Estimate of age
0	e	e	e	0	e	e	e	0	0	0	Within 1 week
1	m	m	m	m	m	m	m	0	0	0	By 4 months
2	m	m	m	m	m	m	m	e	0	0	By 9 months
3	m	m	m	m	m	m	m	P	0	0	By 1 year
4	m	m	m	m	m	m	m	P	e	0	By 1 year 6 months–1 year 9 months
5	m	m	m	m	m	m	m	P	P	e	By 2 years
6	P	m	m	m	m	m	m	P	P	e	By 2 years 9 months
7	P	m	m	m	m	m	m	P	P	P	By 3 years
8	P	m *or* P	m	m	e	e	m	P	P	P	By 3 years 6 months–3 years 8 months
9	P	P	m *or* P	m	P	P	e	P	P	P	By 4 years
10	P	P	P	P	P	P	P	P	P	P	By 4 years 5 months
11	P	P	P	P	P	P	P	P	P	P	By 5 years

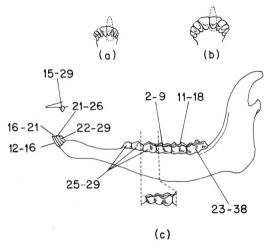

(c)

Figure A2 Numbers indicate the approximate age of red deer in months, when the front teeth (incisors and canines) and the first cheek teeth (three premolars) are replaced by permanent teeth, and when the last three cheek teeth (molars) erupt. (a) and (b) show milk and permanent front teeth respectively. The width of the .303 bullet is slightly over 7.5 mm; (c) illustrates the larger milk P⁴ with its three lateral folds
(Riney, 1957d)

In Figure A2, numbers in boxes indicate the approximate age in months when the incisors, canines and the three premolars are replaced by permanent teeth, and when the molars erupt.

The first year. Calves during the first year have all milk incisors. A .303 calibre bullet (just over 7.5 mm) hides, or nearly hides one of the central incisors of a calf—see insert (a). The first molar erupts between 2 and 9 months of age.

The second year. In the second year the milk incisors are replaced by permanent incisors. The central permanent incisors, larger in size as shown in insert (b), appear normally between 12 and 16 months of age, the second pair between 16 and 21 months, the third pair between 21 and 26 months, and the last pair usually erupts between 22 and 29 months, as the deer is turning 2 years old. The eruption period for the tusk (upper canine) varies between 15 and 29 months of age. Also in this yearling period the second molar erupts some time between 11 and 18 months of age. The last molar is absent, and the three milk premolars are beginning to show some signs of wear as they approach their third year. The third premolar is distinctly three-ridged—see insert (c).

The third year. Normally the first replacement activity in the third year is the eruption and replacement of the premolars, between 25 and 29 months of age. The third adult cheek tooth (P⁴) has but two lateral ridges. The last tooth to erupt is normally the last molar (M³) which has the most variable eruption time as it can appear between 23 and 38 months of age.

Evaluation. I am not as confident in using these criteria today as I was in 1957, although for calculating the proportions of animals that fall into the first, second or third year they may still be useful. For more precise ageing the timing of tooth eruption should be cross-checked in a particular study area and a critical examination of the spread of eruption times should be made following the procedures used by G. Caughley (1965).

White-tailed deer (*Odocoileus virginianus*), after Severinghaus, C. W. (1949).

Adult tooth formula: $I\frac{0}{3}$, $C\frac{0}{1}$, $P\frac{3}{3}$, $M\frac{3}{3}$ (see Table A3).

Table A3   White-tailed deer (*Odocoileus virginianus*) (modified from Severinghaus, C. W., 1949)

Class	I^1	I^2	I^3	C	P^2	P^3	P^4	M^1	M^2	M^3	Estimate of age in months
1	$\frac{0}{e}$	$\frac{0}{e}$	$\frac{0}{e}$	$\frac{0}{e}$	$\frac{0}{0}$	$\frac{0}{0}$	$\frac{0}{0}$	$\frac{0}{0}$	$\frac{0}{0}$	$\frac{0}{0}$	$0-\frac{1}{2}$
2	$\frac{0}{m}$	$\frac{0}{m}$	$\frac{0}{m}$	$\frac{0}{m}$	$\frac{0}{e}$	$\frac{0}{e}$	$\frac{0}{e}$	$\frac{0}{0}$	$\frac{0}{0}$	$\frac{0}{0}$	$\frac{1}{2}-2\frac{1}{2}$
3	$\frac{0}{m}$	$\frac{0}{m}$	$\frac{0}{m}$	$\frac{0}{m}$	$\frac{m}{m}$	$\frac{m}{m}$	$\frac{m}{m}$	$\frac{e}{e}$	$\frac{0}{0}$	$\frac{0}{0}$	$2\frac{1}{2}-5$
4	$\frac{0}{e}$	$\frac{0}{m}$	$\frac{0}{m}$	$\frac{0}{m}$	$\frac{m}{m}$	$\frac{m}{m}$	$\frac{m}{m}$	$\frac{e}{e}$	$\frac{0}{0}$	$\frac{0}{0}$	$5-7$
5	$\frac{0}{P}$	$\frac{0}{m}$	$\frac{0}{m}$	$\frac{0}{m}$	$\frac{m}{m}$	$\frac{m}{m}$	$\frac{m}{m}$	$\frac{e}{e}$	$\frac{0}{0}$	$\frac{0}{0}$	$6-7$
6	$\frac{0}{P}$	$\frac{0}{m}$	$\frac{0}{m}$	$\frac{0}{m}$	$\frac{m}{m}$	$\frac{m}{m}$	$\frac{m}{m}$	$\frac{e}{e}$	$\frac{e}{e}$	$\frac{0}{0}$	$7-9\frac{1}{2}$
7	$\frac{0}{P}$	$\frac{0}{m\ or\ e}$	$\frac{0}{m\ or\ e}$	$\frac{0}{m}$	$\frac{m}{m}$	$\frac{m}{m}$	$\frac{m}{m}$	$\frac{e}{e}$	$\frac{e}{e}$	$\frac{0}{0}$	$9\frac{1}{2}-11$
8	$\frac{0}{P}$	$\frac{0}{P}$	$\frac{0}{P}$	$\frac{0}{P}$	$\frac{m}{m}$	$\frac{m}{m}$	$\frac{m}{m}$	$\frac{P}{P}$	$\frac{e}{e}$	$\frac{0}{0\ or\ e}$	$11-13$
9	$\frac{0}{P}$	$\frac{0}{P}$	$\frac{0}{P}$	$\frac{0}{P}$	$\frac{m}{m}$	$\frac{m}{m}$	$\frac{m}{m}$	$\frac{P}{P}$	$\frac{P}{P}$	$\frac{0\ or\ e}{0\ or\ e}$	$13-17$
10	$\frac{0}{P}$	$\frac{0}{P}$	$\frac{0}{P}$	$\frac{0}{P}$	$\frac{m-P}{m-P}$	$\frac{m-P}{m-P}$	$\frac{m-P}{m-P}$	$\frac{P}{P}$	$\frac{P}{P}$	$\frac{P}{P}$	$17-20$
11	$\frac{0}{P}$	$\frac{0}{P}$	$\frac{0}{P}$	$\frac{0}{P}$	$\frac{P}{P}$	$\frac{P}{P}$	$\frac{P}{P}$	$\frac{P}{P}$	$\frac{P}{P}$	$\frac{P}{P}$	$20-24$

*The variation in timing of the eruption of these teeth is smaller than would be expected from comparing other species of similar size, even those closely related (see Figure 33). Before depending too heavily on these estimates of age it is suggested that eruption classes be recorded for a few animals of known age in your own study area, particularly deer over six months of age.

Plains zebra (*Equus quagga boehmi*, Matschei), based on Klingel, H. (1965).
Adult tooth formula: $I\frac{3}{3}$, $C\frac{1}{1}$, $P\frac{3}{3}$, $M\frac{3}{3}$.

The first milk premolar is small or absent and is disregarded for ageing purposes. The adult canine is rudimentary in mares. Table A4 shows the sub-adult classes of tooth eruption and ageing recognized by Klingel.

Klingel continues with further age classes, based on the extent of wear on teeth and especially on observations of the grooves or folds on the incisors which disappear as the incisors wear towards the gum. All signs of the groove

Table A4   Tooth eruption stages and age groups in Plains zebra (*Equus quagga*) (after Klingel, H., 1965)

Stage	I^1	I^2	I^3	C	P^2	P^3	P^4	M^1	M^2	M^3	Estimate of age
1	0/0	0/0	0/0	0/0	e/e	e/e	e/e	0/0	0/0	0/0	Newborn to a few days. Milk incisors not visible but felt through gums
2	m/m	0/0	0/0	0/0	m/m	m/m	m/m	0/0	0/0	0/0	1 to a few weeks
3	m/m	m/m	0/0	0/0	m/m	m/m	m/m	0/0	0/0	0/0	1 to a few months
4	m/m	m/m	e/e	0/0	m/m	m/m	m/m	0/0	0/0	0/0	6–9 months
5	m/m	m/m	m/m	0/0	m/m	m/m	m/m	e/e	0/0	0/0	1 year
6	m/m	m/m	m/m	0/0	m/m	m/m	m/m	P/P	0/0	0/0	$1\frac{1}{2}$ years
7	m/m	m/m	m/m	0/0	m/m	m/m	m/m	P/P	e/e	0/0	2 years
8	e/e	m/m	m/m	0/0	m/m	m/m	m/m	P/P	P/P	0/0	$2\frac{1}{2}$ years
9	P/P	m/m	m/m	0/0	P/P	P/P	m/m	P/P	P/P	0/0	About 3 years
10	P/P	e/e	m/m	0/0	P/P	P/P	m/m	P/P	P/P	e/e	About $3\frac{1}{4}$ years
11	P/P	P/P	m/m	e/e	P/P	P/P	P/P	P/P	P/P	P/P	About $3\frac{1}{2}$ years
12	P/P	P/P	P/P	P/P	P/P	P/P	P/P	P/P	P/P	P/P	About 4 years, complete set

disappeared on I^1 at 12–14 years, on I^2 at 13–16 years and on I^3 at 15–18 years.

Since the disappearance of incisor grooves is associated with tooth wear this is subject to great variation, as mentioned in the text. Ageing of young animals up to 4 years could probably be used for most management purposes.

Warthog (*Phacochoerus aethiopicus*, Pallas), after Child, G. *et al.*, 1965.

Milk tooth formula variable: $I\dfrac{1}{2\text{ or }3}$, $C\dfrac{1}{1}$, $P\dfrac{2,\,3\text{ or }4}{1,\,2\text{ or }3}$, and

Permanent tooth formula: $I\dfrac{1}{2\text{ or }3}$, $C\dfrac{1}{1}$, $P\dfrac{1,\,2\text{ or }3}{1\text{ or }2}$, $M\dfrac{3}{3}$.

Further variation in the permanent teeth resulted from the loss with age of incisors, premolars and the first and second molars. This loss was due to crowding through the prolonged growth of M^3 which reached a maximum of 70 mm.

In spite of this variation, the authors considered that ageing criteria based on the growth of canines and the last molar, combined with body growth measurements, could be useful in distinguishing warthog of up to 3 years of age.

The authors' recognition of differences in growth rates associated with

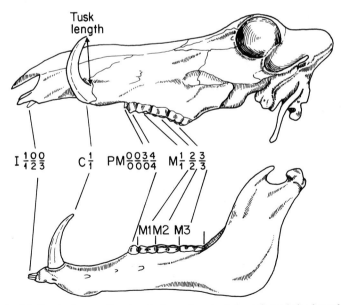

Figure A3   Method of recording tooth formulae, tusk length and the lengths of the 3 lower molars in warthog. Premolars are variable, the specimen shown above having only two upper and one lower permanent premolar (after Child, G. *et al.*, 1965)

496

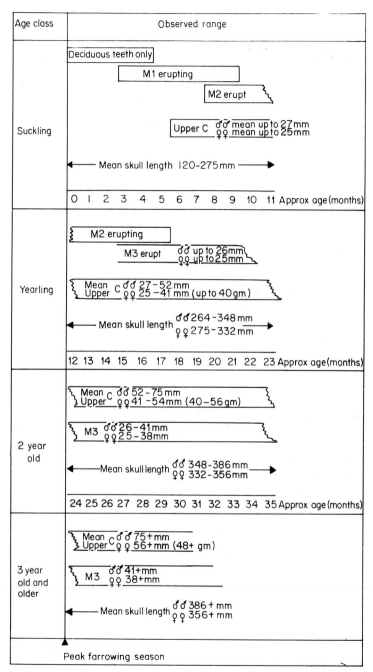

Figure A4   Use of the combination of several ageing criteria in warthog, a difficult species to age by a single criterion (after Child, G. *et al.*, 1965)

different planes of nutrition further emphasizes the need to develop ageing criteria in terms of a particular study or management area and not, at least with our present understanding, a general set of ageing criteria to apply to all wildlife areas in a single country.

Black rhinoceros (*Diceros bicornis* L.)

Tooth formula: I and C, vestigial or absent; $P \dfrac{3 \text{ or } 4}{3 \text{ or } 4}$; $M \dfrac{3}{3}$.

Age criteria: Two papers are available which describe the sequence of tooth eruption in black rhinoceros (Foster, J. B., 1965; Goddard, J., 1970). From these it is clear that the eruption of permanent teeth alone does not provide sufficient evidence for assigning age classes that will be of much use. This is mainly because of the slowness of eruption combined with variation in eruption times, although general age groups based on replacement alone are shown in Table A5.

J. Goddard used X-rays to locate the position of erupting premolars and molars when they were still beneath the bone, and his age classes also incor-

Table A5 Tooth replacement and age classes in Black rhinoceros (*Diceros bicornis* L.) (after J. Goddard, 1970)

Stage	P¹	P²	P³	P⁴	M¹	M²	M³	Estimate of age
1	$\frac{e}{e}$	$\frac{e}{e}$	$\frac{0}{0}$	$\frac{0}{0}$	$\frac{0}{0}$	$\frac{0}{0}$	$\frac{0}{0}$	Birth
2	$\frac{m}{m}$	$\frac{m}{m}$	$\frac{m}{m}$	$\frac{0}{0}$	$\frac{0}{0}$	$\frac{0}{0}$	$\frac{0}{0}$	½ year
3	$\frac{m}{m}$	$\frac{m}{m}$	$\frac{m}{m}$	$\frac{m}{m}$	$\frac{0}{0}$	$\frac{0}{0}$	$\frac{0}{0}$	1–1½ years
4	$\frac{m}{m}$	$\frac{m}{m}$	$\frac{m}{m}$	$\frac{m}{m}$	$\frac{P}{P}$	$\frac{0}{0}$	$\frac{0}{0}$	2–4 years
5*	$\frac{P}{P}$	$\frac{P}{P}$	$\frac{P}{P}$	$\frac{m}{m}$	$\frac{P}{P}$	$\frac{P}{P}$	$\frac{0}{0}$	5 years
6	$\frac{P}{P}$	$\frac{P}{P}$	$\frac{P}{P}$	$\frac{P}{P}$	$\frac{P}{P}$	$\frac{P}{P}$	$\frac{0}{0}$	6–7 years
7	$\frac{P}{P}$	$\frac{P}{P}$	$\frac{P}{P}$	$\frac{P}{P}$	$\frac{P}{P}$	$\frac{P}{P}$	$\frac{P}{P}$	8 years

*The P¹ is apparently lost at an early stage in one study (Goddard, J., 1970). In such cases a normal adult formula would be: $P \dfrac{0\ 2\ 3\ 4}{0\ 2\ 3\ 4}$; $M \dfrac{1\ 2\ 3}{1\ 2\ 3}$.

porate detailed observations of tooth wear. The criteria and age classes seem valid for his East African study area and workers in this region are well advised to consult Goddard's 1970 paper. But because tooth wear is an inseparable part of these criteria, the rate of wear must be cross-checked under differing conditions of soil and vegetation found in other areas.

Counting tooth rings proved an unsatisfactory means of ageing when Goddard tested this method.

The development of field criteria based on comparative size and shape, as shown in the text for elephant and buffalo, should prove useful for rhinoceros as well.

Elephant

Elephant teeth are remarkable, for the tusks that form the ivory of commerce represent the only incisors in large mammals that grow throughout life. Furthermore the only other teeth are the cheek teeth, of which there are six, three milk premolars and three molars in each jaw-half and the six teeth succeed each other, instead of occurring simultaneously as in other mammals. Elephant teeth are so large that there are never more than three, and normally but two cheek teeth above the gum line at any given time. The first premolar that appears gradually moves forward in the jaw, becomes worn and is succeeded by the second. The older teeth always move forward, becoming worn and being replaced by the next tooth in the series which in turn is replaced by the next and so on until the sixth and last tooth.

An animal of this size and with a potential lifespan similar to man naturally poses special problems in developing ageing criteria. Imagine, for example, the problem of locating known-age animals taken from areas similar to your study habitat. Fortunately a few people have been able to collect enough material from elephants shot on control or utilization campaigns in East Africa to greatly improve understanding of ageing in elephant. Most of the following information is based on the work of R. M. Laws (1966) who explored the uses and limitations of various ageing criteria based largely on a collection of 385 lower jaws of elephant which he supplemented with further material placed at his disposal by other workers.

Although only tooth eruption is mentioned in this appendix, and ageing by comparative size in the text, this by no means truly represents the work accomplished by Laws. Before assigning age classes he carefully compared and cross-checked with growth curves, counts of tooth rings for individual molars as they succeeded each other, exact sizes and shapes and numbers of ridges on the molars, weight and length of tusks and growth measurements on young elephant in captivity.

Ages assigned to erupting molars (premolars and molars being grouped together as part of a single series) are as follows: M^1 *is lost* at 2 years, M^2 at 6 years, M^3 at 13 to 15 years, M^4 at 28 years, M^5 at 43 years and M^6 *appears* at 30 years. Figure A5 (a), (b) and (c), illustrate this process. Although Laws recognizes thirty age groups, for purposes of assessing proportions of young

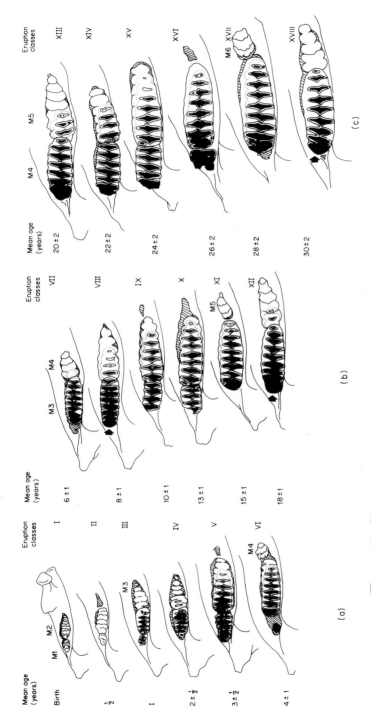

Figure A5 Cheek teeth in elephant, and their sequence of replacement and wear. Black areas show areas of wear. Arrows indicate that erosion is beginning (illustration taken from Laws, R. M., 1966)

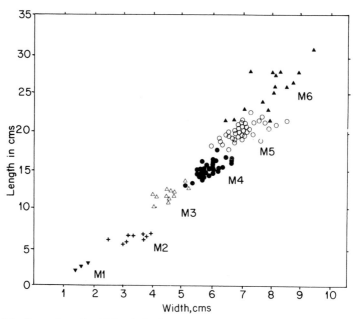

Figure A6   Length and width of elephant molars are used as an aid in identifying particular teeth (illustration taken from Laws, R. M., 1966)

and juvenile mortality, the first eleven groups are probably sufficient to recognize, for this carried the age estimates up to 13–15 years, that is to the disappearance of M^3.

Figure A6 plots length and width of molars as an aid in recognizing individual teeth. While this is doubtless appropriate for parts of East Africa, the validity of these measurements should be cross-checked before using in other regions.

Hippopotamus (*Hippopatamus amphibius* L.)

Milk tooth formula: $I\dfrac{2-3}{2-3}$, $C\dfrac{1}{1}$, $P\dfrac{4}{4}$,

Adult tooth formula: $I\dfrac{2}{2}$, $C\dfrac{1}{1}$, $P\dfrac{3}{3}$, $M\dfrac{3}{3}$.

R. M. Laws (1968) identified twenty-two age groups, the first ten of which include the tooth eruption sequence shown in Table A6. The remainder of his classes show different stages of tooth wear and are not included. For management purposes this seems justified as Laws' figures indicate that 53 per cent of the population was 8 years of age or under and 72 per cent was 17 years or younger. This leaves but 28 per cent hippo with a complete set of permanent dentition or older in his study area.

Table A6 Hippopotamus (*Hippopotamus amphibius* L.) (modified from Laws, R. M., 1968)

Stage	Tooth eruption classes										Estimate age of	Simplified age classes
	I^1	I^2	C	P^1	P^2	P^3	P^4	M^1	M^2	M^3		
1	e/e	e/e	e/e	m/m	0–e/0–e	m/m	0/0	0/0	0/0	0/0	Birth	
2	m/m	m/m	m/m	m/m	m/m	m/m	m/m	0/0	0/0	0/0	½ year	1
3	m/m	m/m	m/m	m/m	m/m	m/m	m/m	P/P	0/0	0/0	1 year	
4	P–e/P–e	P–e/P–e	P–e/P–e	m/m	m/m	m/m	m/m	P/P	0/0	0/0	3 years ± ½	2
5	P/P	P/P	P/P	m/m	m/m	m/m	m/m	P/P	0/0	0/0	4 years ± 1	3
6	P/P	P/P	P/P	m/0	m/0	m/m	m/m	P/P	e/0	0/0	7 years ± 1	
7	P/P	P/P	P/P	0/0	0/0	m/m	m/m	P/P	e/P	0/0	8 ± 2	
8	P/P	P/P	P/P	0/0	P/P	P/P	P/P	P/P	P/P	0/0	11 ± 2	4
9	P/P	P/P	P/P	0/0	P/P	P/P	P/P	P/P	P/P	e/e	15 ± 2	
10	P/P	P/P	P/P	0/0	P/P	P/P	P/P	P/P	P/P	P/P	17 ± 3	5
											Older	

As Laws indicated, the assignment of specific age to the eruption classes presents difficulties as most known-age specimens have been captive animals whose growth rates may differ from those of wild populations. A simplified grouping of those age classes, as for ages 0–1, 1–4, 4–8, 8–17 and over 17 may be easier to apply, as suggested in Table A6. As Laws suggests, such age groupings should be perfectly valid if used as comparative age classes rather than as a means of accurately determining age.

Newfoundland caribou (*Rangifer tarandus*)

Tooth formula: $I\dfrac{0}{3}$, $C\dfrac{0}{1}$, $P\dfrac{3}{3}$, $M\dfrac{3}{3}$.

Using lower jaws only, A. T. Bergerud (1970) estimates the age associated with permanent premolar and molar eruption as follows: M^1—3 months, M^2—13 months, $P^{2,3,4}$—24 months and M^3—25 months. All the permanent teeth were erupted by $27\frac{1}{2}$ months.

Although Bergerud noted a short calving season of only the first 2 weeks in June, similar estimates of caribou ages, from other regions, show considerable variation in P and M eruption dates, as shown in Table A7. The figures from Newfoundland may be due to individual variation. However, even if Canadian figures are disregarded there still remains considerable variation between regions, whether due to genetic or environmental influences or both. This suggests, once again, the need to develop or at least to cross-check published ageing criteria against local observations.

Table 7   Approximate age (months) of premolar and molar eruption in months for caribou (*Rangifer tarandus*) (after Bergerud, A. T., 1970)

Permanent tooth	Scandinavia	USSR	Newfoundland	Canada	Alaska
P^2	30	29	$24–27\frac{1}{2}$	17	24
P^3	30	29	$24–27\frac{1}{2}$	17	24
P^4	28	29	$24–27\frac{1}{2}$	17	24
M^1	4	$3\frac{1}{2}$	3	3	4
M^2	15	15	13	10	13
M^3	29	29	$25–27\frac{1}{2}$	17	25

Appendix 2

Field classification of African elephant and buffalo

In most large mammals of medium size there is little difficulty in separating calves or yearlings from older animals, shortly after a calving season. For larger animals separation even into rough age groups poses special problems. These two examples are given to illustrate how two biologists dealt with this problem in working with elephants (Laws, R. M., 1966) and buffalo (Grimsdell, J. R. R., 1973) respectively. In both cases much work had been devoted to developing ageing criteria using other kinds of evidence (see also Appendix 1) and in each example known-age animals were also involved.

Figure A7 illustrates a simple way of recognizing an elephant calf (b) and then relates comparative sizes and shapes to different age classes. Obviously there are still difficulties in applying these criteria in the field, but it is worth trying, for under many circumstances it may be the only choice. The following suggestions for using the Laws' elephant silhouettes in the field may prove useful.

1. Take a 3 × 5 in. card and draw a line representing the ground across its lower third.
2. Holding the card vertically and at reading distance, select a calf under 1 year old and mark its height on the card.
3. Then hold the card in front of Figure A7 so the marked height of the living calf corresponds to the height of Laws' calf silhouette and enter marks for other age groups.
4. The card can then be used to classify elephant into age groups entering the individual classifications on prepared forms.

A range finder is a great help in standardizing distances of observation, and time can be saved by preparing different cards to be used at varying distances. Once such cards are prepared groups without calves or solitary individuals can also be classified.

The considerations important for other kinds of classification, such as classification of undisturbed groups, apply equally to elephants.

Even with an accurate rangefinder this technique is difficult to use in the field. For following trends in a given area and for comparing study areas—for example, different parks, reserves or hunting areas—fewer categories can be used. For a start try using each of Laws' categories. If different observers

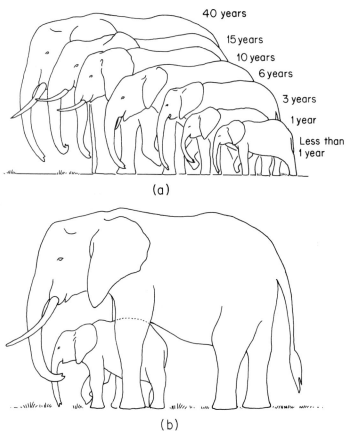

40 years
15 years
10 years
6 years
3 years
1 year
Less than
1 year

(a)

(b)

Figure A7 (a) Field age criteria for female African elephants in western Uganda. (b) Relation of average 1-year-old calf to average adult female (after Laws, R. M., 1966)

watching the same elephant group cannot produce the same results try reducing the categories from seven to five, for example: 0, 1–3 years, 3–6 years, 6–15 years and over 15 years. The cards used in classifying would be marked accordingly. For management purposes some reduction in categories would seem reasonable for R. M. Laws (1966) has estimated the average life expectancy of wild African elephant at 15 years.

Whatever criteria are adopted, record them. Ensure that contributing observers can duplicate each others' results and be consistent.

Equally practical are the outline drawings of African buffalo which Grimsdell prepared, based on photographs of animals of known age. As with the elephant silhouettes, one would not expect to distinguish all of these age groups in the field. However, using these drawings as a guide, it should be possible to at least recognize animals in their first year and most observers should be able also to classify buffalo that are well into their second year (Figure A8).

505

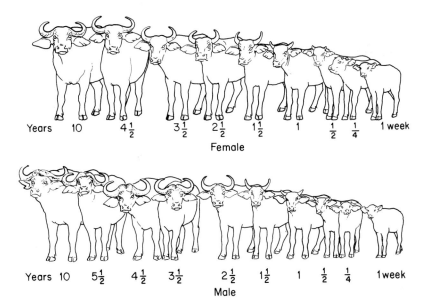

Years 10   4½   3½   2½   1½   1   ½ ¼   1 week

Female

Years 10   5½   4½   3½   2½   1½   1   ½ ¼   1week

Male

Figure A8 Drawings of female and male African buffalo of various ages, taken from photographs of animals whose approximate age was known. Buffalo in their first and second years should be consistently recognized (after Grimsdell, J. J. R., 1973)

Appendix 3

Arbitrary decisions associated with multi-purpose transects

An important prerequisite for two observers to closely duplicate each other's results is for them both to have a clear understanding of a set of arbitrary decisions which, when followed, contribute to a standardized field procedure. For comparison with data taken by different observers in other localities these arbitrary decisions should be recorded at least in summary form when the evidence is presented; otherwise it may be impossible for others to critically assess the information. Arbitrary decisions held in mind while gathering evidence are listed below. They are presented not as a guide to follow but to exemplify the kinds of detailed decisions needed to achieve consistent and comparable field results.

Every effort was made to keep a given transect in a given type of habitat and to record the location of the transect in such a way that the same area could be re-sampled in approximately the same place.

Transects were run in as straight a line as possible visually. Where the habitat was open a landmark was selected and a transect headed towards this landmark. Where the habitat was closed woodland it was sometimes more convenient to follow a compass bearing.

Where the habitat sampled was a narrow strip, or otherwise too small to accommodate a 200 yd transect, the line sometimes proceeded across the area, doubled back at a pre-determined angle, e.g. 30°, in another straight line, with another landmark or compass bearing until the 100 line points were completed by using as many such crosses as were necessary.

It was found that, particularly when running transects for the first time, effort had to be made not to look at the ground where the foot was about to strike and thus slightly weight the sample one way or another. With very little practice, however, it is possible to let the leg swing and the boot fall naturally and completely unselectively.

There are times when a straight walking course may be impossible, because of a low shrub or large tree-stump for example. The procedure adopted here was to pace one or two paces at right angles to the line of the transect, turn and continue parallel to the transect for the next pace or two as required, then return to the line of the transect with another one or two paces. If it was still impossible to re-join the line of the transect the measuring staff was thrust in at the approximate place where one's foot would have stepped. If the thrust is made without looking at the ground at the time of thrust this is as unselective as the rest of the sample points. A small mark on the measuring staff was used as the equivalent of the line on the boot and point observations were then recorded in the same manner.

When recording canopy height it is the canopy directly over the sample point that was recorded. Thus, although the tree may have been 50 ft in height, the canopy above the point may have been only 20 ft, and was recorded as 20 ft. Some sample points, particularly in heavy forest areas, may have several layers of canopy, for example, at 4, at 10, at 20 and at 50 ft. Each of these canopy layers (made up of different plants) was recorded and used in calculations of frequency of occurrence of different canopy height classes along the transect, but for calculation of percentage canopy cover for the entire transect, such a point would simply be recorded as canopy present, the same as if only one layer of canopy was present.

Litter was recorded as present even if the size was less than the $\pm\frac{1}{16}$ in. line on the boot, such as a very small scrap of grass or root. Fallen logs were counted as litter.

Grass was not counted unless the line marked on the boot came to or inside the basal diameter of the grass. Grass leaves, even green leaves that lay along the ground, were not counted as grass cover even though part of a living plant. Such leaves were included as grass canopy.

For stoloniferous grasses, grasses with rooting runners, basal diameter was interpreted as including either a hit on the main base, or on a rooted stolon. Stems and leaves in between were included as grass canopy.

As the foot pressed the ground, care was taken not to let the grasses pressed forward by the foot become involved in the interpretation of the sample point.

It requires somewhat more practice to become proficient at interpreting trend than to record observation of presence of, or absence associated with, status. This is natural because the final decision may involve observing and weighing several types of evidence. It is suggested that before trying the technique in an area in which it is desired to learn present trend, sample transects be placed in areas where the trend is known. For example, one can usually find places on a property where permanent water, or a dipping tank, has been established in the previous few years, or in the vicinity of a recently established village. Such an area, or any other type of situation known to be downgrading, that is characterized by an increasing amount of bare ground over the past few years, is useful to observe at an early stage in the use of this technique to become acquainted with one extreme. On the other hand it is usually possible to find areas that in the past have had heavy use and have become degraded in condition but which, within the past few years, have gone out of use; for example, where cattle have been removed and where burning has become less frequent or has stopped. This type of situation could be expected to show some signs of recovery on the bare patches of ground and should help the observer to become acquainted with this extreme as compared with the downgrading condition. Agricultural research stations, where burning or grazing experiments have been in progress for several years, are excellent sites for familiarizing oneself with local relevant evidence indicative of upgrading and downgrading areas.

Appendix 4

Guidelines in establishing national legislation on wildlife and national parks[1]

The purpose of these guidelines is to present in summary form a checklist of the basic and most important legislative provisions pertaining to wildlife and national parks. Because these provisions represent, on the whole, the minimum legal framework necessary to ensure sound wildlife management policies, they should, wherever possible, be included in a basic act, embracing both the protection and management of wildlife and various protected areas. It may be advisable to have detailed provisions on licences, such as permissible hunting areas, close seasons, bag limits and administration laid down in regulations under such a basic act.

The checklist is not intended to be exhaustive. Some provisions which are normally found in any basic act or regulation, such as those regarding detailed administrative provisions, penalties, etc. are omitted. These provisions are but guidelines. They are by their very nature dependent in their form and content on the socio-economic conditions of the country or region to which they apply.

A. Objectives of basic wildlife and national parks legislation

There should be, wherever possible, mention of the objectives and the policies which should be implemented by the legislation. These objectives and policies should include, for example, the protection of the wild fauna and flora; in particular, the preservation of those species, together with their habitat, which are faced with extinction; the rational utilization and management of the wild fauna and flora to ensure an optimum sustained yield within the framework of a national land-use policy; the protection and management of protected areas.

[1]These guidelines represent a modification and combination of guidelines jointly produced by 37 heads of African national parks and game departments in a series of FAO-sponsored biennial Working Parties on National Parks and Wildlife, and culminating in a final text produced at Fort Lamy, Tchad, in March 1967 (Anon, 1967). Based on this document, similar guidelines for establishing national legislation in wildlife and national parks have been produced and made available by FAO (Viale termi di Caracalla, Rome, Italy) for Asian and for Latin American Countries.

509

The selection of areas to be set aside as strict natural reserves, national parks, game reserves and special reserves, shall be carried out within the framework of national land-use planning, and of the rational utilization of land and water, and shall include as complete as possible a selection of unique, rare or representative natural habitats.

Areas set aside for the establishment of national parks shall as far as possible be of sufficient extent to constitute an ecological unit, and to enable the fauna to perform its seasonal movements within the park.

The Government shall ensure that land-use practices around strict natural reserves, national parks, game reserves and special reserves, do not lead to silting, flooding, drying-out of water courses or other harmful consequences inside such reserves and parks. It shall endeavour to prohibit any diversion and prevent any pollution of water courses upstream of such reserves and parks likely to be detrimental to the park or reserve. It shall also take the necessary measures to regulate and control fires in the neighbourhood of such reserves and parks.

The Government shall take advantage of the multiple values of its fauna, the management of which shall be an integral part of economic and social development planning, and of land-use policy. Management of wildlife and its habitat will complement existing types of exploitation where appropriate, and constitute a major form of use as well as a stabilizing force in some regions, and notably on marginal and wild lands.

Management of wildlife and its habitat shall in practice be based on selected and well-defined objectives. In national parks and game reserves an important aim of management will be to achieve and maintain a desired relationship between the fauna and its habitat. In special reserves management will particularly favour the animal or plant species which it is intended to protect by the establishment of such reserves. Within strict natural reserves, only those measures judged indispensable by the competent scientific authorities to safeguard the very existence of the reserve shall be taken. Outside strict natural reserves, national parks, game reserves and special reserves, one of the main aims of management will be to obtain, through licensed hunting and, where appropriate, through harvesting or domestication, an optimum sustained yield compatible with other forms of exploitation.

In order to attain the objectives enumerated in the preceding paragraph, the Government shall, in areas under its direct control, draft and implement management plans; in other areas they shall encourage the drafting and implementation of such plans by means of technical and financial assistance.

Management plans will be formulated on the basis of surveys and research, including an inventory of the resource, studies on rate of growth and renewal, on the behaviour of species and their relation to the environment, on economic and social aspects, and on the means of integrating the proposed management with land-use practices in adjacent areas.

Management plans, having indicated the long-term aims pursued within the ecological, economic and social context, will set out more immediate aims and the action necessary to achieve them. These plans will be flexible in their execution and will be subject to periodic revision.

B. Protected areas

Protected areas vary from country to country, though in the case of national parks definitions seem to be fairly uniform. Consideration should be given when providing a definition of these areas to economic, geographical, biological and other conditions in each country or region. In all probability, different objectives will be assigned to each of these areas and it might therefore seem difficult to suggest definitions of them. Although it may not be feasible to set aside in a given country all the various types of areas for which definitions are given below, it is felt that at least a need for some of them should be found in each country.

1. Strict natural reserve

The term 'strict natural reserve' should denote an area:

(a) placed under the control of the state, the boundaries of which must not be altered, nor any portion subject to alienation except by the competent legislative authority;
(b) set aside to permit the free interaction of natural ecological factors without any outside interference whatsoever, excepting that judged indispensable by the competent scientific authorities for the safeguarding of the very existence of the reserve;
(c) throughout which any form of hunting or fishing, any undertakings connected with forestry, agriculture or mining, any grazing, any excavations or prospecting, drilling, levelling of the ground or construction, any work involving the alteration of the configuration of the soil or the character of the vegetation, any water pollution, and, generally, any act likely to harm or disturb the fauna or flora, and the introduction of any exotic animal or plant species is strictly forbidden;
(d) where it is forbidden to reside, enter, traverse or camp and which it is forbidden to fly over, without a special written permit from the competent authorities, and in which scientific investigations may only be undertaken by permission of those authorities.

2. National park

The term 'national park' should denote an area:

(a) identical with (a)—'strict natural reserve'—above;
(b) set aside for the protection and conservation of outstanding natural animal and plant communities, geological formations and areas of natural scenic beauty, for the enjoyment of present and future generations in a

manner consistent with the management of the resources, and for scientific, cultural, educational and economic purposes. When occasionally outstanding historical and archaeological remains or sites are found within and forming part of the significance of the areas as defined in the previous sentence, these features or sites may be managed as part of the resources of the national park;

(c) in which hunting, killing or capturing of fauna, destruction or collection of flora and other objects is prohibited except for scientific and management purposes, and on condition that these purposes are pursued by, or under the control of, the park authorities.

The activities prohibited under 'strict natural reserves', (1) (c) and (d), should also be prohibited in national parks, except where they are necessary to enable the park authorities to carry out the provisions of paragraph (b) above, or to enable the general public to visit national parks.

The Government shall progressively reduce, and as soon as possible eliminate, such human settlements as still exist inside national parks.

The location and construction inside national parks and game reserves of roads, installations and buildings, for the use of park personnel and the public, shall be undertaken in such a way as to disturb as little as possible the fauna and flora, and so as to blend with the landscape.

3. Sanctuary

The term 'sanctuary' (wildlife sanctuary, game reserve, wildlife refuge, etc.) should denote an area:

(a) set aside for the conservation and management of wildlife and the protection and management of its habitat;
(b) within which hunting shall be prohibited except by, or under the direction or control of, the reserve authorities;
(c) where settlement and other human activities are restricted or prohibited.

4. Special reserve

The Government recognizes that it is important and urgent to accord a special protection to those animal and plant species that are threatened with extinction, or which may become so, and to the habitat necessary to their survival. Such animals shall benefit from total protection and the hunting, killing or capture of such animals, and the destruction or collection of such plants, may only take place with the express authorization of the Government; this authorization shall only be given in exceptional circumstances, either in order to safeguard the very existence of the species or for scientific purposes.

The protection of rare species will not prejudice the right of legitimate defence of life or property in case of immediate and absolute necessity, as may be provided for in national legislation.

The term 'special reserve' (nature monument, national monument, etc.) should denote an area;

(a) set aside to protect wild animal or plant species (special emphasis is given to species threatened with extinction) either as individuals or as populations and, in the case of wild animals, to ensure sufficient habitat for their survival;
(b) set aside for the protection of outstanding geological formations, historical and archaeological objects or sites (that sites or objects of historical and archaeological significance may also occasionally be included in national parks has been mentioned under national parks);
(c) in which all other interests and activities are subordinated to this end.

C. Acquisition of land for inclusion in protected areas

Rules should be laid down for the procedure when acquiring lands to be comprised in a protected area. Preferably such lands should be made state property so as to ensure adequate and uniform management of protected areas. Expropriation should accordingly be provided for.

D. Establishment of protected areas

The overall importance of strict nature reserves and national parks and the vast areas which are often comprised in them require their establishment by Act of Parliament, or the equivalent, while the establishment of other protected areas should normally be the responsibility of the Minister in charge of wildlife and national parks, normally the Minister of Agriculture, Ministry of Forestry or Minister of Natural Resources.

E. Demarcation and fees

To avoid as far as possible encroachments on protected areas, it is advisable that in places within, or on the boundaries of, and particularly on access roads to protected areas, provisions be made for notices to be erected indicating:

(a) the kind of protection afforded to the area;
(b) the extent of the area set aside (map);
(c) the penalties resulting from any contravention regarding prohibited activities within such an area.

The administrative body responsible for protected areas should be empowered to charge entrance and other user-fees where consistent with local policy and where necessary or desirable to help cover the costs of management, maintenance and development of these areas.

F. Buffer zones

It may sometimes be necessary to establish so-called buffer zones around the boundaries of protected areas. Such buffer zones may consist of a strip of public or private land of varying width adjacent to a protected area and managed in a manner compatible with such protected area and yet sometimes used for other land purposes such as timber production, agriculture and tourism. The buffer zone should be managed in such a way as to absorb fire, pollen, erosion, pollution, noise, free-ranging domestic animals or other elements considered detrimental to the integrity of the protected area.

National and local bodies, empowered with administrative and advisory functions, may be established. National bodies are commonly vested with an overall responsibility of the national park and wildlife resource, while local bodies may be charged with the responsibility for a specific protected area.

G. Officers

For the administration of the Act at the local level, officers having the necessary technical and educational qualifications for the administration and management of wildlife and national parks should be appointed. These officers should, *inter alia*, be empowered:

(a) to arrest any person committing, or suspected on reasonable grounds of having committed, an offence against legislation and detain him until his full name and place of residence are ascertained;

(b) to demand any person suspected of any act for which the holding of a licence is required to produce such a licence for inspection;

(c) to enter and search where there is reasonable grounds to suspect that an offence has been committed against legislation;

(d) to inspect and examine any animal or product thereof;

(e) to seize any animal or product thereof;

(f) to hold enquiries into offences.

H. Appointed officers deemed to be public servants

It may be advisable to give officers the status of public servants in order to give them an increased protection against all kinds of verbal or physical violence in the performance of their duties.

I. Protected animals and plants

It should be laid down that all wild fauna, protected or not, irrespective of whether found on public lands or private lands, belong to the state, which should be made responsible for its management and protection. Specifically designated animals and plants, the hunting, collection or destruction of which is forbidden owing to their rarity, should be listed as protected fauna and flora. The possession of such protected animals and plants should be considered an offence. (Occasionally it may be wiser to protect *all* wild fauna, except those animals listed as game animals.)

J. Game animals, close seasons, bag limits and minimum sizes

By the term 'game animals' it should be understood those wild animals listed in special Annexes to national legislative acts which may be hunted for sport under a licence.

In order to enable the responsible department to regulate the rational utilization of the resource at an acceptable level a system of close seasons, bag limits and minimum sizes should be established. During a close season, hunting is prohibited either generally or with regard to a particular species or family of species. Bag limits restrict the number of animals which must be taken during a certain time period (daily, weekly, monthly, yearly, seasonally). To avoid over-exploitation, it may sometimes be necessary to restrict the hunting of certain species below a certain minimum size or age.

K. Regulations for the implementation of the Act

It is of great importance to have a clause empowering the relevant Ministry to issue and, whenever necessary, revise regulations amending the lists of protected species, plants and game animals, close seasons, bag limits and minimum sizes. This makes it possible for the regulations to be re-examined and revised to keep pace with technical, social and economic changes by avoiding the complicated and time-consuming procedure of having to submit amendments to the legislation to the law-making body.

L. Sporting game licences

Game animals should be hunted only on the condition that a hunting game licence has first been issued by a specifically designated authority.

The applicant for a sporting game licence should state the locality where he intends to hunt, the period during which hunting is going to take place, the species to be shot and the response to be used.

A sporting game licence should be personal and valid not longer than 1 year. It should be issued only to persons above the age of 18 years (or that age considered consistent with prevailing conditions in the country concerned) in possession of a firearms licence. Limitations, restrictions, terms and conditions, such as close seasons, bag limits, minimum sizes and locality where the hunting may take place, may be prescribed.

For the purpose of controlling the actual game harvest it may be useful to lay down that the licences must keep a record of the game killed, which record should be submitted to the issuing authority after expiry of the licence.

Hunting should be exercised on public as well as on private lands. In the latter case the hunter should have to ask the private landowner for permission to hunt on his properties.

Hunting should under no conditions be carried out inside protected areas except by authorized personnel when necessary for management purposes.

In the context of licences, it would be advisable to define the term 'hunt'.

This could be defined as the killing, taking, trapping or capturing of any wildlife by any means, the pursuing, disturbing or molesting of the same, as well as every attempt to hunt or kill wildlife and every act of assistance of any other person.

M. Scientific hunting and capture licences

The granting of licences entitling individual scientists or individuals in biological institutions of repute to hunt or capture specified numbers of species of protected animals for the purpose of scientific research or collection for zoological gardens, museums or similar institutions, should rest with the relevant Ministry.

N. Commercial capture and trade licences

A licence to hunt and trade in animals (with the exception of protected animals) and their products should first be granted by the relevant authority to qualified persons before the latter engage in these activities. The licence should entitle the holder to hunt specified numbers and species of game animals and non-protected animals in specified areas outside protected areas.

The licensee should keep a record of the number and species of animals killed or captured, which record should be submitted annually to the relevant authority.

For the purpose of these guidelines, the term 'meat' shall mean fresh or preserved meat, fat and blood.

The Government shall prohibit throughout their territory any commerce or transport involving the meat of an animal belonging to a species listed in Annexes to the legislation as protected. All commerce or transport of meat of an animal belonging to such a species shall be prohibited except with written permission from the competent authorities.

The Government shall take the necessary measures to regulate and control, throughout its territory, commerce and transport of all game meat.

O. Firearms licences

A sporting game licence should be issued only to the holder of a firearms licence. The applicant for a firearms licence should be made to undergo a test as to his shooting ability and general knowledge of hunting wild animals. It is advisable to establish a minimum age limit for the possession of such a licence.

P. Suspension and revocation of licences

In addition to fines and other penalties, the possibility of suspending or revoking the above-mentioned licences should be provided for, if the licensee fails to comply with prescribed conditions or if he has contravened the provisions laid down in the Wildlife Act.

Q. Export and import of wild animals and trophies

It is essential that the export and import of wild animals and trophies be adequately regulated.

1. The Government shall prohibit the import, export and transit of live animals belonging to species listed in Annexes to the legislation except with the written authorization, and under the control, of the competent authorities.
2. The Government shall take the necessary measures to regulate and control the import, export and transit of all live wild animals.

For the purposes of the present guidelines, the expression 'trophy' shall denote any dead animal, belonging to a species listed in Annexes to the national legislation, and shall include any tooth, tusk, bone, horn, shell, claw, hoof, skin, hair, egg, feather, or any other durable portion of such animal, whether or not included in a manufactured or processed article, except articles of little value which, through a legitimate processing, have lost their original identity.

The Government shall take the necessary measures to control and regulate, throughout their territory, the manufacture of articles from trophies and the trade, import, export and transit of trophies and articles made therefrom.

The export of trophies shall be prohibited unless the exporter has been granted, by the competent authorities, a certificate or licence permitting export. Such a certificate shall only be issued where the trophies have been lawfully imported or lawfully obtained. In the case where export is attempted without a certificate having been granted, the trophy shall be confiscated, without prejudice to the application of other penalties.

The import of trophies shall be prohibited except on production of a certificate of export, delivered by the competent authorities of the country of origin, failing which the trophy shall be confiscated, without prejudice to the application of other penalties.

(i) Every trophy consisted of unprocessed rhinoceros horn or ivory exported in accordance with the provisions of these guidelines shall be identified by marks; a description of these marks, together with the weight of the trophy, shall be recorded in the certificate of export.
(ii) Similar marks shall, wherever possible, be placed on other trophies and the certificate of export shall include a description of such trophies permitting their identification.

These measures shall provide that found trophies, or trophies from animals killed accidentally or in defence of life or property, shall become property of the state.

R. Recognition of protected species

In order to render the prohibition on all dealing in protected species effective, appropriate illustrations of these animals should be prepared and circulated to Customs officers and others engaged in the trade of wild animals and their products; or the same persons should be otherwise instructed in the methods of identifying protected species and trophies derived therefrom.

S. Unlawful hunting methods

Below a list is made of those hunting methods which might be considered illegal (it is understood that it may not be necessary to legislate against all methods since some of them may not be applicable in a given country or region);

1. the use of motor vehicles, motor boats or aircraft, whether moving or stationary, either for the purposes of hunting, capturing or killing, animals, or wilfully to disturb, drive or stampede them for any purpose whatsoever except when these methods are employed by the competent authority or under its direction or control;
2. the use of fire to hunt, capture, or kill animals;
3. the use of military weapons and also of any firearms capable of firing more than one cartridge as a result of one pressure of the trigger, or reloading itself without further action by the operator, it being ensured that the approprite firearms are used which under normal conditions are capable of killing the animal outright.

Furthermore, hunting, capturing or killing of animals may be prohibited:

4. by night, with or without the assistance of lighting or dazzling equipment;
5. by use of drugs, poison, poisoned weapons and baits, or radioactive substances;
6. by use of nets, pits or enclosures, traps or snares, or set guns or explosives;
7. with the assistance of electrical sounding devices such as recording machines or other electronic equipment;
8. with the aid of any form of hide except where these methods are employed by the competent authority or under its direction or control.

T. Customary rights

Since customary rights—if excessively exercised—may be a threat to wildlife, it might be worth considering the restriction of these rights. (Preferably, customary hunting should be exercised under licence.) Also other types of customary rights, such as domestic-animal grazing or forest exploitation within protected areas, should as far as possible be prohibited or restricted.

When, exceptionally, customary rights may be exercised within a particular protected area, a statement to that effect should be incorporated in the declaration of that area.

The Government shall progressively extinguish such customary hunting and fishing rights as are an obstacle to the application of those legislative provisions which relate to the protection, conservation and management of wildlife.

U. Animals causing danger and damage

When occasionally certain species or individual animals constitute a danger or cause damage to life or property, authorization to destroy these animals should be provided for. The destruction should be controlled or carried out by officers so as to exclude any possible abuse of this right.

V. Legitimate defence

The fact that certain animals are protected animals or that certain hunting methods are prohibited should not prejudice the right of legitimate defence of life or property exercised in good faith in case of immediate and absolute necessity.

To prevent abuse of this right, it should be laid down that in all cases where an animal is found dead or is accidentally killed or killed in legitimate defence, the carcass and trophy should become property of the state.

Furthermore, if the animal is a protected animal, a report should be submitted immediately to the competent authority.

W. Breeding grounds

In some countries there may be a need for the establishment of so-called breeding grounds designed to allow breeding of wild animals in captivity for economic and industrial purposes, or for the propagation of protected species.

X. Education

Contracting Governments shall take all necessary measures so that:

1. the problems of wildlife conservation and management are given the required emphasis in their education programmes at all levels, and
2. the general public is made aware of the value of this essential part of the national heritage.

Y. Administrative responsibilities

The Government shall designate an organization to administer the wildlife and national parks resources. Included in its functions will be to follow up the application of the provisions in the national legislative acts relating to wildlife and national parks (see also Chapter 9000). Among others this will include the following terms of reference:

1. to revise periodically the lists of protected, partially protected and harvestable species annexed to the national legislative acts;
2. to coordinate research;
3. to ensure the necessary coordination with all measures relating to the purpose of the national wildlife and national park policies and relevant legislation which may be necessary with other Government organizations.

Appendix 5

Minimum requirements for commercial manufacture and the sale of dried and fresh meat[1]

A. General conditions

1. Game will be bled and eviscerated at the site of slaughter. In order to avoid unnecessary contamination it is desirable that means be provided for raising the carcass off the ground during the bleeding and evisceration. This may be either by a portable tripod with block and tackle or a garage breakdown van type of fitting, or such other method as may be proposed and is to the satisfaction of the Ministry of Health.

2. The game shall be transported to the processing factory with all expedition, and undue contamination from dust and inclement weather shall be avoided by satisfactory covering of the carcasses. The floor and sides of the vehicle shall be of metal and a roof covering shall be fitted.

3. Ramp and loading bay at the factory shall be of concrete.

4. The factory shall be subdivided for all practical purposes into (a) an 'Abattoir', (b) a 'salting and drying room' and (c) 'Packing and store room' and shall be fly-screened throughout.

Suggested minimum dimensions shall be (a) 20 × 15 ft; (b) 20 × 15 ft; (c) 10 × 15 ft.

5. Enamel baths, fibre glass or similar containers shall be used for the 24 hours salt soaking in lieu of galvanized iron containers.

6. Packing of prepared biltong shall be in well-constructed cardboard containers or similar but not hessian type of sacking.

B. General construction

1. As much ventilation and air movement as possible shall be provided.

2. Where the roof is of other than aluminium or asbestos in the slaughtering section, a ceiling shall be provided; in the drying rooms a ceiling shall be provided.

---

[1]These minimum requirements for commercial operators to obtain a licence were developed by the Public Health Department of Zimbabwe. It is included because it is more elaborate than most and should thus provide a useful check-list or guideline for examining and appropriately modifying when developing national requirements for the use of game meat for the first time.

522

3. Minimum height of ceiling shall be 10 ft.

4. Where gauze is used to provide lighting apertures, such gauze shall be fitted in a well-fitting type of frame in such a manner that gauze can be replaced as often as required with a minimum of disturbance that might cause a prevalence of dust or other contamination.

5. No gauze shall be used at a height where it is likely direct contamination from blood may occur.

6. All internal surface of wall shall be smooth steel trowel-finished and painted with oil paint.

7. Floors shall be of smooth concrete graded to open channels for drainage and all corners between walls and floor shall be curved.

C. Abattoir

1. Meat shall be off-loaded on to a concrete ramp or platform with direct access on to the hanging rail.

2. The hanging rail shall be over a graded and drained channel with a raised concrete platform at one side in order to allow an operator effectively to reach the top of the carcass.

3. The entrance shall be screened with self-closing fly-proof doors.

4. An adequate supply of pure water shall be made available at a point inside the abattoir and there shall also be a wash-hand basin with hot and cold water laid on, together with soap, nail brush and towel readily available.

5. All cutting-up tables shall be of close-jointed hardwood and replaced as often as necessary when the surface is no longer capable of adequate cleaning.

6. All knives shall be of good quality and capable of being readily cleaned and sterilized.

7. After cutting up, the meat shall be transported into the salt-processing area in such a manner as to avoid contamination, e.g. in bins.

8. All personnel engaged in handling of meat shall wear plastic aprons, head covering and rubber boots and they and their persons shall at all times be clean and tidy.

9. Adequate facilities shall exist for the disposal of bones, horns, hoofs and offal.

10. Procedure in relation to scraping and salting of hides shall be such that no nuisance results from the processing thereof and storage thereof.

11. A sink with hot and cold water shall be available.

D. Salt-processing and drying section

1. The premises shall be of an adequate size to permit the retention of the meat during the following periods:

| Salt and pepper processing | 24 hours |
| Drying into strips | 10–14 days |

2. Hanging rails and hooks shall at all times be free from rust or other contamination.

E. Packing room

The area shall be completely fly-proof and all contamination shall be avoided.
Tables with tops impervious to liquids shall be used.

F. Drainage

Drainage from the abattoir and drying-out sections shall be disposed of by means of a graded and sloping floor to a half-channel which in turn shall discharge to a grease gulley. Effluent shall be disposed of in French drains.

G. Changing rooms

Facilities shall exist for changing of clothes.
A shower unit and sanitation shall be provided for the use of the employees.

H. Fresh game marketing

When it is proposed that the supply of fresh or chilled meat be made available from retail sources, the following additional conditions shall apply:
1. Carcasses and offal shall be numbered on a duplicate basis which permits identification of the offal with the carcass.
2. Offal in the abattoir pending cold storage shall be stored on metal hooks to hang well clear of the wall and in such a position that meat inspection can take place.
3. Unless carcass and offal is to be delivered immediately to premises adjacent where cold room facilities exist, a cold room adequate for the number of carcasses to be dealt with, shall be provided at the place where skinning and dressing takes place.
4. Transport of all meat from slaughter/dressing premises to the wholesaler/retailer or municipal abattoir shall be in a satisfactorily hygienic manner.

I. General

No person suffering from any infectious or contagious disease shall be employed on or about the premises or handle any foodstuffs.

Appendix 6

Red data book categories
(After Goodwin, H. O. and Holloway, C. W., 1974)

The IUCN Red Data Book defines five categories of species (or subspecies or other taxonomic classifications) as follows.

1. Endangered species

In danger of extinction and whose survival is unlikely if the causal factors continue operating.

Included are species whose numbers have been reduced to a critical level or whose habitats have been so drastically reduced that they are deemed to be in immediate danger of extinction. Also included are forms that are possibly already extinct.

2. Vulnerable species

Believed likely to move into the endangered category in the near future if the causal factors continue operating.

Included are species of which most or all of the populations are decreasing because of over-exploitation, extensive destruction of habitat or other environmental disturbance; forms with populations that have been seriously *depleted* and whose ultimate security is not yet assured; and forms with populations that are still abundant but are *under threat* from serious adverse factors throughout their range.

3. Rare species

Species with small world populations that are not at present endangered or vulnerable, but are at risk.

Such forms are usually localized within restricted geographical areas or habitats or are thinly scattered over a more extensive range.

4. Out of danger

Species formerly included in one of the above categories, but which are now considered relatively secure because effective conservation measures have been taken or the previous threat to their survival has been removed.

5. Indeterminate

Species that are suspected of belonging to one of the first three categories but for which insufficient information is currently available.

N.B. In practice, endangered and vulnerable categories may include, temporarily, species whose populations are beginning to recover as a result of remedial action, but whose recovery is insufficient to justify their transfer to another category.

Convention on International Trade in Endangered Species of Wild Fauna and Flora (CITES Convention)

The idea of a convention was adopted by the UN Conference on the Human Environment held at Stockholm in June 1972 which recommended that a plenipotentiary conference be called as soon as possible to prepare and adopt a convention on export, import and transit of certain species of wild animals and plants. In November 1972 the United States Government extended an invitation to host such a conference. The Convention was adopted in February and March 1973, in Washington, DC.

The following extract of Article II of this Convention describes the fundamental principles used in forming subsequent regulations for trading in skins of endangered species.

ARTICLE II

Fundamental principles

1. Appendix 1 shall include all species threatened with extinction which are or may be affected by trade. Trade in specimens of these species must be subject to particularly strict regulation in order not to endanger further their survival and must only be authorized in exceptional circumstances.
2. Appendix II shall include:
   (a) all species which although not necessarily now threatened with extinction may become so unless trade in specimens of such species is subject to strict regulation in order to avoid utilization incompatible with their survival; and
   (b) other species which must be subject to regulation in order that trade in specimens of certain species referred to in sub-paragraph (a) of this paragraph may be brought under effective control.
3. Appendix III shall include all species which any party identifies as being subject to regulation within its jurisdiction for the purpose of preventing or restricting exploitation, and as needing the cooperation of other parties in the control of trade.
4. The Parties shall not allow trade in specimens of species included in Appendices I, II and III except in accordance with the provisions of the present Convention.

Appendix 7

Application for outside assistance

It has been pointed out in Chapter 9000 that a developing country can obtain, under UNDP Technical Assistance and OPAS Programmes, the assistance of experts both in the initial inventory stage and for the planning and development of national parks and reserves, and that fellowships for training in wildlife management can sometimes be obtained from the same source. Alternatively, a larger-scale wildlife conservation programme may be appropriate. Such a project normally involves a team of several experts, together with necessary equipment, to work for a number of years, and the project provides study fellowships for local staff. *Bearing these possibilities in mind, the UN country representative should be consulted on the matter.*

The decision to establish a UN-assisted project does not lie with the UN representative alone. The government which applies for assistance has to give the project a sufficiently high priority among other requests for aid in the agricultural and forestry field to place it within the country's development plan. The country's own authorities must therefore also be convinced of the benefits to be derived from the rational exploitation of its wildlife resources.

The fact that UN assistance normally requires the recipient government to provide some form of matching contribution often presents a further difficulty to those struggling to initiate a wildlife conservation programme. However, international organizations (such as the World Wildlife Fund and a number of private foundations and societies) exist that are prepared to give funds for specific conservation projects, and bilateral aid is also available from a number of countries. Funds to cover part of a programme can sometimes be obtained from these sources and can be used for all or part of the counterpart contribution for the programme for which UN assistance is sought.

Appendix 8

Guidelines for planning and managing national parks

The following guidelines have been taken from K. R. Miller (1980). They are included as examples of a kind of guideline that is often modified to conform to national or organizational practices. However, once particular sets of guidelines are adopted and followed, this can do much to improve the effectiveness of management and development in national parks.

Table A8.1   Sample work plan for the preparation of a management plan for a national park

Step	Officer in charge
Logistics	————
1. Basic information (office)	
(a) Objective and criteria	————
(b) Descriptive information	————
(c) Base map	————
(d) Future demands for area	————
(e) Factors requiring urgent attention	————
(f) Construction costs	————
(g) Expected budget for department	————
(h) Administration and personnel	————
(i) Institutional and political aspects	————
2. Inventory (field)	————
3. Limitations and constraints	————
4. Objectives of the park	————
5. Zoning	————
6. Boundaries	————
7. Management programmes	
(a) Interpretation and research	————
(b) Environmental management	————
(c) Administration and maintenance	————
etc. . . .	

Table A8.2   Suggested outline for presentation of the management concept

1. Objectives
    (a)
    (b)
    (c)
    (d)
    (e)
    etc.

2. Activities
    (a)
    (b)
    (c)
    (d)
    (e)
    etc.

3. Norms and guidelines:
    (a)
    (b)
    (c)
    (d)
    (e)
    etc.

4. Requirements:
    (a) Facilities (physical)
    (b) Equipment
    (c) Supplies (disposable)
    (d) Manpower

5. Timing:
    (schedule of when activities are to be implemented, and when requirements are to be available)

6. Use/value of the expected outputs:
    (benefits anticipated from research, protection, etc.)

Table A8.3   Suggested outline for presentation of the development concept for each development area

1. Name:

2. Objectives for management of the area:
   (a)
   (b)
   (c)
   etc. . . .

3. Architectural theme:

4. Physical structures and facilities to be constructed:

5. Infrastructure and special basic support installations:
   (a) Roads, parking areas, entrance-ways, bridges
   (b) Signs
   (c) Trails
   (d) Sewage
   (e) Water (potable or non-potable)
   (f) Electricity
   (g) Gas, salt water or other special requirement

Table A8.4   Sample presentation of the outputs expected from managing the national park

1. (number) ha. of the (name) biogeographic province, will be protected as a functional ecosystem in perpetuity. (Refer to the National Strategy for Conservation to note how this output fits into overall plan.)

2. (number) ha. of transition areas between habitat (name) and habitat (name), will be protected, along with lakeshore lands, wetlands, riverine areas, etc.

3. Protection of genetic materials will include: (name of species) important to agriculture, medicine, forestry, fisheries and research on human physiology; (name of species) which are endangered with extinction; (name of species) important to regulation of the environment, etc. Where known, state monetary values of particular species.

4. Maintenance of cultural values will include (name of structure), (name of sites), etc., which are of importance to national identity, to research on the past, past technologies, past land use, and for generating tourism. (Note foreign exchange potential of specific areas.)

5. Scenic beauty to be protected around (name) city, (name) highway, (name) cultural sites, (name) tourism and recreation areas. Outstanding physical resources of aesthetic importance to the nation will be protected (names). (Note any financial values, such as tourism, which relate to the scenery.)

6a. Educational opportunities will be provided for (number) of students to visit natural and cultural areas each year; facilities will be available for (name) graduate students to work on advanced studies in the natural and cultural environment to prepare for future careers as national scientists; facilities will be available for (number) of teachers to be trained in environmental education and interpretation of natural and cultural heritage of the nation. And the general public will be supplied with interpretative services, at a rate of approximately (number) per year.

6b. Research opportunities will be available for work on natural and cultural resources. Facilities will include a small research station with a laboratory (description), dormitory (number of beds) and (other). The area is particularly important for research (non-manipulative) on medicinal, agricultural, forest, fishery, climate, hydrology, and other critical questions facing eco-development. (Name examples of important research opportunities.)

6c. The area is particularly useful for monitoring weather, agricultural pests, plant succession, upstream relations to downstream fisheries, bird migrations, wild animal populations, and other natural resources of relevance to environmental management.

7. Recreation facilities (name types) will be available for (numbers) of visitors per year, serving (name) communities and cities. The monetary impact of recreation is expected to value $ per year. Tourism will also be promoted. Foreign exchange from non-national visitors can be expected to rise to $ per year. And the tourism potential of the area reinforces national policies towards the development of open international relations.

8a. The rural community will benefit from (number) of new jobs in such particular lines of work as (name). In addition, new roads and communications will provide added benefits for the transport of agricultural commodities valued at __$__ . A new clinic and school will be developed in the town of (name) to serve park staff and the local community. The expenditures derived from park activities and tourism are expected to total some __$__ per year, (x) per cent of which will be spent within the local community itself. This conforms with policies for supporting the redistribution of wealth to the rural areas of the country.

8b. The added economic acitivity in the rural area, and the relocation of individuals from lands to be included within the park, will reduce negative human impacts on fragile areas totalling (number) ha. Those individuals relocated will be provided with alternative opportunities in (describe).

9. The (name) watersheds will be included within the park thereby providing protection for the stable production of water. The total output of water is (number) litres/years, contributed to (name) city potable supplies, hydro-electric power totalling (number) kWh/year, navigation valued at (number) tons of cargo, fishery production valued at (number), tons (or $), etc. Potential values of the various rivers flowing from the park are estimated at __$__ in terms of future uses.

10. Specific sites (names) will be reclaimed by natural means to prevent erosion and the actual (or potential) destruction of natural or cultural resources (names), and downstream developments, food sources, etc. Note details of downstream investments actually or potentially affected by erosion from the park area.

Total $ value/year of quantifiable services to the
community to be expected from the park management: $ _____

Summary of services from park management not
exchanged on the monetary markets currently: _____
                                                     _____
                                                     _____

Table A8.5   Sample presentation of the inputs required for managing the national park

	Development cost ($)	Estimated annual cost ($)
1. Physical inputs include:		
(a) summary (list) of buildings showing sq. metres of contribution, and total cost	————	
(b) summary of roads and trails, showing (km) of construction, and total cost	————	
(c) summary (list) of other elements, such as, signs, landscaping, bridges, entrance ways, etc.	————	
(d) construction and development of interpretative and educational exhibits	————	
(e) communications (radio, telephone) internal and external to the park	————	
(f) electricity, gas and other energy systems	————	
(g) sewage and waste systems	————	
(h) vehicles by (types) and (numbers), total cost	————	
(i) equipment for the various activities of management summary (list	———— -	
(j) supplies		————
(k) maintenance on buildings, roads, equipment, and amortization		————
2. The development of human capacity includes:		
(a) (number) of personnel of (types) of categories, total cost of salaries and benefits		————
(b) training courses for (numbers) in (types) of areas of work		————
(c) regular annual training courses for (number) of personnel in (types) of materials		————
3. The management of the park requires that (name) legal measures be taken, that the (name) policy be revised, and that several (name) cooperative agreements be prepared. Legal capacity exists in the department, and its cost to the park project for these services will total	————	
4. Miscellaneous inputs: include special consultants, research activities, etc.	————	————
Total cost for managing the national park: Development cost	$ ————	
Estimated annual operating costs		$ ————

References cited

Adams, L. (1965) Progress in ecological biotelemetry. *Bioscience*, **15**, 83–6.
Alexander, M. M. (1959) The habitat map: a basis for wildlife management. *N. Y. Fish and Game J.*, **6**(1), 103–13.
Allee, W. C., Park, O., Emerson, A. E., Park, T. and Schmidt, K. P. (1949) *Principles of Animal Ecology*. W. B. Saunders Co., Philadelphia. 837 pp.
Andrewartha, H. G. and Birch, L. C. (1954) *The Distribution and Abundance of Animals*. University of Chicago Press. 782 pp.
Anon (1967) *Draft African Convention for the Conservation and Management of Wildlife*. Text adopted by the ad hoc working party on wildlife management at its second session (Fort Lamy, Chad, 6–11 February, 1967). FAO: AFC/WL-67/16 Annex 5. 11 pp.
Ansell, W. F. H. (1960) *Mammals of Northern Rhodeisa [Zambia]*. The Government Printer, Lusaka. 179 pp.
Asdell, S. A. (1946) *Patterns of Mammalian Reproduction* Comstock Publ. Co., NY. 437 pp.
Ashcraft, G. and Reese, D. (1957) An improved device for capturing deer. *Calif. Fish and Game*, **43**, 193–9.
Baranov, F. I. (1926) [On the question of the dynamics of the fishing industry] *Biull. Rybnovo Khoziaistra* (1925), **8**, 7–11.
Beale, D. M. (1966) A self-collaring device for pronghorn antelope *J. Wildl. Mgt.*, **30**(1), 209–11.
Beeson, R. W., Cronemiller, F. P., Deering, R. L., Fausett, E. P., Pitchlynn, P. and Show, S. B. (1940) *Handbook for Range Managers, Region 5* USDA Forest Service, Washington, DC. 212 pp.
Bergerud, A. T. (1970) Eruption of permanent premolars and molars for Newfoundland Caribou. *J. Wildl. Mgt.*, **34**(4), 962–3.
Bider, R. J. (1962) Dynamics and the temporo-spatial relations of a vertebrate community. *Ecology*, **43**(4), 634–46.
Bider, R. J. (1968) Animal activity in uncontrolled terrestrial communities as determined by a sand transect technique. *Ecol. Monog.*, **38**, 269–308.
Bleazard, S. R. (1973) Flight requirements in a multi-disciplinary project. FAO Luangwa Valley Conservation and Development Project, Working Document No. 6. FO:DP/ZAM/68/510. 17 pp.
Bradley, R. (1972) A photographic ageing technique used on warthog. *E. Afr. Wildl. J.*, **10**, 123–8.
Brody, S. (1945) *Bioenergetics and Growth*. Reinhold Publ. Co., Baltimore. 1023 pp.
Brozek, J. and Keys, A. (1950) Evaluation of leanness–fatness in man: a survey of methods. *Nutr. Abstr. Rev.*, **20**(2), 247–55.
Buechner, H. K. (1961) Territorial behaviour in Uganda kob. *Science*, **133**(3454), 698–9.
Caughley, G. (1965) Horn rings and tooth eruption as criteria of age in the Himalayan thar, *Hemitragus jemlahicus*. *N.Z. J. Sci.*, **8**(3), 333–51.
Caughley, G. (1970) Eruption of ungulate populations, with emphasis on Himalayan thar in New Zealand. *Ecology*, **51**, 53–72.

534

Caughley, G. (1973) Game Management (Part 2). Game Management and Habitat Manipulation, Luangwa Valley Conservation and Development Project: Zambia. UNDP, FAO: DP/ZAM/68/510 Working Document No. 1, Part 2, 50–157.

Caughley, G. (1976) Wildlife management and the dynamics of ungulate populations. *Applied Biology*, Academic Press, London, **1**, 183–246.

Caughley, G. (1977) *Analysis of Vertebrate Populations*. J. Wiley & Sons, Chichester, 234 pp.

Caughley, G. (1979) *Design for Aerial Censuses (In Aerial Surveys of Fauna Populations)* Austr. Nat. Pk. and Wildl. Serv. Spec. Publ. 1, 15–20.

Child, G. (1964) Growth and ageing criteria of impala, *Aepyceros melampus. Occ. Pap. Nat. Mus. S. Rhod.*, **27B**, 128–35.

Child, G. (1968) Behaviour of large mammals during the formation of Lake Kariba. *Trans. Natl. Mus. Rhodesia*. 123 pp.

Child, G., Sowls, L. and Mitchell, B. L. (1965) Variations in the dentition, ageing criteria and growth patterns in wart hog. *Arnoldia*, **1**(38), 1–23.

Child, G., Sowls, L. and Richardson, G. L. (1965) Uses and limitations of eye-lens weight for ageing wart hog. *Arnoldia*, **1**(39), 1–2.

Child, G. and Le Riche, J. D. (1969) Recent springbok treks (mass movements) in south-western Botswana. *Mammalia*, **33**(3), 499–504.

Clark, E. A. and McMeekan, C. P. (1952) New Zealand lamb and mutton: Part I—Anatomical characteristics of lamb and mutton carcasses. *N.Z. J. Sci. Tech., A*, **33**(5), 1–28.

Cronwright-Schreiner, S. C. (1925) *The Migratory Springbucks of South Africa*. T. Fisher Unwin Ltd, London. 140 pp.

Craighead, F. C. Jr. and Craighead, J. J. (1965) Tracking grizzly bears. *Bioscience*, **15**(2), 88–92.

Craighead, F. C. Jr, Craighead, J. J., Cote, C. E. and Beuchner, H. K. (1972) Satellite and ground tracking of elk. *NASA Sci. Publ.*, **262**, 99–111.

Craighead, J. J., Craighead, F. C. Jr and McCutchen, H. E. (1970) Age determination of grizzly bears from fourth premolar tooth sections. *J. Wildl. Mgt.*, **34**(2), 353–63.

Craighead, J. J., Craighead, F. C. Jr, Varney, J. R. and Cate, C. E. (1971) Satellite monitoring of black bear. *Bioscience*, **21**(24), 1206–12.

Crockford, J. A., Hayes, F. A., Jenkins, J. H. and Feurt, S. D. (1957) Field application of nicotine salicylate for capturing deer. *Trans. N. Amer. Wildl. Conf.*, **22**, 579–83.

Crook, J. H. (1966) Gelada baboon herd structure and movement: a comparative report. *Symp. Zool. Soc. London*, **18**, 237–58.

Croze, H. (1972) A modified photogrammetric technique for assessing age-structures of elephant populations and its use in Kidepo National Park. *E. Afr. Wildl. J.*, **10**, 91–115.

Cumming, D. H. M. (1971) Radio-tracking of warthog: some results and their bearing on studies of game–tsetse relationships. *Proc. Symp. on Biotelemetry*. S. Afr. Council for Sci. and Indust. Res., Pretoria, 109–29.

Dasmann, W. (1971) *If Deer Are To Survive*. Wildl. Mgt. Inst., Stackpole. 128 pp.

Davenport, L. A., Shapton, W. and Gower, W. C. (1944) A study of the carrying capacity of deer yards as determined by browse plots. *Trans. 9th. N. Amer. Wildl. Conf.*, pp. 144–8.

de Meillon, B., Davis, D. H. S. and Hardy, F. (1961) *Plague in Southern Africa. 1: Siphonoptera*. Government Printer, Pretoria, or CCTA/CSA Publ. (29), 1958.

de Nahlik, A. J. (1959) *Wild Deer*. Faber & Faber, London. 240 pp.

Dible, J. H. (1932) Fat mobilization in starvation. *J. Path. Bact.*, **35**, 451–66.

Dice, L. R. (1952) *Natural Communities*. University of Michigan Press. 547 pp.

Doman, E. R. and Rasmussen, D. I. (1944) Supplemental winter feeding of mule deer in northern Utah. *J. Wildl. Mgt.*, **8**(4), 317–88.

Dorst, J. and Dandelot, P. (1970) *A Field Guide to the Larger Mammals of Africa*. Collins, London. 287 pp.

Douglas-Hamilton, I. (1971) Radio tracking of elephants. *Proc. Symp. on Biotelemetry*, S. Afr. Council for Sci. and Industr. Res., Pretoria, 335–42.

Ellison, L., Croft, A. R. and Bailey, R. (1951) *Indicators of condition and trend on high range-watersheds of the Intermountain Region*. USDA Agricultural Handbook No. 19. 66 pp.

Errington, P. L. (1946) Predation and vertebrate populations *Q. Rev Biol.*, **21**, 144–77; 221–45.

Estes, R. D. (1968) Territorial behaviour of the wildebeest (*Connochaetes taurinus* Burchell, 1823), PhD Thesis, Cornell University, 1968. 151 pp.

Estes, R. D. (1974). Social organisation of the African Bovidae. *IUCN Publication*, New Series No. 24, 166–205.

Evans, F. C. (1956) Ecosystem as the basic unit in ecology. *Science*, **123**(3208), 1127–8.

Fairchild, M. (1926) *The Scientific Method*. Character Ed. Inst., Natl. Capital Press, Washington, DC. 4 pp.

Farrel, R. K., Kogger, L. M. and Winward, L. D. (1966) Freeze branding of cattle, dogs and cats for identification. *J. Amer. Med. Assoc.*, **149**(6), 745–52.

Fenton, P. F. and Dowling, M. T. (1953) Nutritional obesity in mice. *J. Nutr.*, **49**, 319–31.

Fisher, J., Simon, N. and Vincent, J. (1969) *The Red Book: Wildlife in Danger*. Collins, London. 308 pp.

Foster, J. B. (1965) Mortality and ageing of black rhinoceros in East Tsavo Park, Kenya. *E. Afr. Wildl. J.*, **3**, 118–19.

Funaioli, H. (1972) *The Mweka Taxidermy Field Guide*. Coll. of Afr. Wildl. Mgt., Mweka, Moshi, Tanzania. UNDP: FAO. 40 pp.

Giles, R. H. (ed.) (1969) *Wildlife Management Techniques* (3rd edn, revised) The Wildl. Soc. Washington, DC. 623 pp.

Goddard, J. (1970) Age criteria and vital statistics of a black rhinoceros population. *E. Afr. Wildl. J.*, **8**, 105–21.

Goodwin, H. O. and Holloway, C. W. (1972) *IUCN Red Data Book*, Vol. 1. *Mammalia* (looseleaf). From IUCN, Morges (VD), Switzerland.

Graham, M. (1935) Modern theory of exploiting a fishery, and application to North Sea trawling. *J. Conseil Expl. Mer.*, **10**, 264–72.

Greager, D. C. and Jenness, C. A. (1979) An acoustically sensitive transmitter for telemetering the activities of wild animals. *J. Wildl. Mgt.*, **43**(4) 1001–7.

Grimsdell, J. J. R. (1973) Age determination of the African buffalo, *Syncercus caffer* Sparrman. *E. Afr. Wildl. J.*, **11**, 31–53.

Grinnell, J. (1917) The niche-relationships of a California thrasher. *Auk*, **34**, 427–33.

Hall, T. C., Taft, E. B., Baker, W. H. and Aub, J. C. (1953) Preliminary report on the use of Flaxedil to produce paralysis in the white-tailed deer. *J. Wildl. Mgt.*, **17**, 516–20.

Hammond, J. (1942) British animal husbandry. *Endeavour*, **1**(4), 131–40.

Hanks, J. (1969) Techniques for marking large African mammals. *Puku*, (5), 65–86.

Hansen, C. G. (1964) A dye spraying device for marking desert bighorn sheep. *J. Wildl. Mgt.*, **28**(3), 584–7.

Harrison, J. (1964) *Mammals of Sabah*. Tien Wah Press, Singapore. 244 pp.

Hart, W. J. (1966) *A Systems Approach to Park Planning*. IUCN Publ., New Series, Suppl. Paper (4). 118 pp.

Harthoorn, A. M. (1976) *The Chemical Capture of Animals*. Baillière Tindall, London. 416 pp.

Heezen, K. L., and Tester, J. R. (1967) Evaluation of radio-tracking by triangulation with special reference to deer movements. *J. Wildl. Mgt.*, **31**, 124–141.

Hilditch, T. P. and Pedelty, W. H. (1940) The influence of prolonged starvation on the composition of pig fat deposits. *Biochem J.*, **34**(1), 40–7.

Hjort, J. G., Jahn and Ottestad, P. (1933) The optimum catch. *Hvalradets Skrifter*, **7**, 97–127.

Hodge, H. C., MacLachlan, P. L. and Bloor, W. R. (1941) Lipids of the fasting mouse: I—The relation between carcass lipids and liver lipids. *J. Biol. Chem.*, **139**(2), 897–915.

Hornocker, M. G. (1970) An analysis of mountain lion predation upon mule deer and elk in the Idaho primitive area. *Wildl. Monogr.* (21), 39 pp.

Howard, W. E. (1960) Innate and environmental dispersal of individual vertebrates. *The Amer. Midl. Nat.*, **63**(1), 152–61.

Howell, J. C. (1951) Roadside census as a method of measuring bird populations. *Auk*, **68**, 334–57.

Howell, R. and Neustein, S. A. (1965) *The Influence of Geomorphic Shelter on Exposure to Wind in Northern Britain*. U.K. Forestry Commission Annual Report for 1965, 201–3.

Huxley, J..S. (1931) The relative size of antlers of deer. *Proc. Zool. Soc. Lond.*, 819–64.

Jenkins, D. (1972) The status of red deer (*Cervus elaphus corsicanus*) in Sardinia in 1967. *Una Vita Per La Natura*, Camerino, Tipografic Succ. Savini-Mercuri, 173–95.

Keys, A., Brozek, J., Henschel, A., Mikelsen, O. and Taylor, H. L. (1950) *The Biology of Human Starvation*. University of Minnesota Press. 2 vols., 1385 pp.

Kindel, F. (1960) Use of dyes to mark ruminant feces. *J. Wildl. Mgt.*, **24**(4), 429.

Kireilis, R. W. and Cureton, T. K. (1947) The relationships of external fat to physical education activities and fitness tests *Res. Quart. Amer. Assn. Hlth. Phys. Educ.*, **18**(2), 123–4.

Klingel, H. (1965) Notes on tooth development and ageing criteria in the plains zebra *Equus quagga boehmi* Matschie. *East Afr. Wildl. J.*, **3**, 127–9.

Kolenosky, G. B. and Miller, R. S. (1962) Growth of the lens of the pronghorn antelope. *J. Wildl. Mgt.*, **26**(1), 112–13.

Krebs, C. J. (1972) *Ecology: The Experimental Analysis of Distribution and Abundance*. Harper & Row, NY. 694 pp.

Kruuk, H. (1966) A new view of the hyaena. *New Scientist*, **30**(502), 849–51.

Laws, R. M. (1952) A new method of age determination for mammals. *Nature*, **169**(4310), 972–3.

Laws, R. M. (1966) Age criteria for the African elephant, *Loxodonta A. Africana*. *E. Afr. Wildl. J.*, **4**, 1–37.

Laws, R. M. (1967) Eye lens weight and age in African elephants. *E. Afr. Wildl. J.*, **5**, 46–52.

Laws, R. M. (1968) Dentition and ageing of the hippopotamus. *E. Afr. Wildl. J.*, **6**, 19–52.

Le Munyan, C. D., White, W., Nyberg, E. and Christian, J. J. (1959) Design of a miniature radio transmitter for use in animal studies. *J. Wildl. Mgt.*, **23**, 107–10.

Leopold, A. (1937) *Game Management* Chas. Scribner's Sons, NY. 481 pp.

Leopold, A. S., Riney, T., McCain, R. and Tevis, L. Jr. (1951) The Jawbone deer herd. *Calif. Fish and Game Bull*, (4), 139 pp.

Leopold, L. B., Wolman, M. G. and Miller, J. P. (1964) *Fluvial Processes in Geomorphology*. W. H. Freeman & Co., SF. 522 pp.

Lines, R. and Howell, R. S. (1963) The use of flags to estimate the relative exposure of trial plantations. *Forest Record* No. 51, For. Comm. Lond.

Linhart, S. B. and Knowlton, F. F. (1967) Determining age of coyotes by tooth cementum layers. *J. Wildl. Mgt.*, **31**(2), 362–5.

Longhurst, W. M. (1964) Evaluation of the eye lens technique for ageing Columbian black-tailed deer. *J. Wildl. Mgt.*, **28**(4), 773–84.

Lord, R. D. (1959) The lens as an indicator of age in cottontail rabbits. *J. Wildl. Mgt.*, **23**(3), 358–60.

Lord, R. D. (1962) Ageing deer and determination of their nutritional status by the lens technique. *Proc. 1st Natl. White-tailed Deer Dis. Symp.*, 89–93.

Lowdermilk, W. C. (1934) Acceleration of erosion above geologic norms. *Trans. Amer. Geophys. Union 15th Ann. Mtg*, 505–9.

Lydekker, R. (1907) *The Game Animals of India, Burma, Malaya and Tibet*. Rowland Ward Ltd, London. 409 pp.

Mackay, R. S. (1970) *Bio-medical Telemetry* (2nd edn.). John Wiley & Sons, Inc. London and New York, 533 pp.

Madsen, R. M. (1967) *Age Determination of Wildlife: A Bibliography*. USDA Dept Library, Bibliog. (2). 111 pp.

Manville, R. H. (1949) Techniques for capture and marking of animals. *J. Mamm.*, **30**(1), 27–33.

Matthews, L. H. (1952) *British Mammals* Collins, London. 410 pp.

Merriam-Webster, A. (1946) *Webster's New International Dictionary of the English Language* (2nd edn, unabridged). G. & C. Merriam Co., New York. 3210 pp.

Miller, K. R. (1974) Guidelines for the management and development of national parks and reserves in the American humid tropics. *Proc. IUCN Mtg.*, *Caracas*, 1974, 94–5.

Miller, K. R. (1980) *Planificacion de Parques Nacionales Para el Econdesarrolo Fundacion Para la Ecologia y la Proteccion del Medio Ambiente*. Madrid. 500 pp.

Mitchell, B., Staines, B. W. and Welch, D. (1977) *Ecology of Red Deer: A Research Review Relevant to Their Management in Scotland*. Institute of Terrestrial Ecology. 74 pp.

Moffit, J. (1942) Apparatus for marking wild animals with colored dyes. *J. Wildl. Mgt.*, **6**, 312–18.

Morisawa, M. (1968) *Streams: Their Dynamics and Morphology*. McGraw-Hill Inc., New York, 175 pp.

Mosby, H. S. (ed.) (1963) Wildlife Investigation Techniques, 2nd edn. The Wildl. Soc., Washington DC. 146 pp.

Mossman, A. S., Johnstone, P. A., Savory, C. A. R. and Dasmann, R. F. (1963) Neck snare for live capture of African ungulates. *J. Wildl. Mgt.*, **27**(1), 132–5.

Moulton, C. R. (1920) Biochemical changes in the flesh of beef animals during under-feeding. *J. Biol. Chem.*, **43**, 67–78.

Murie, O. J. (1951) *The Elk of North America* Stackpole Co. and Wildl. Mgt. Inst. 376 pp.

Neal, B. J. (1959). Technique of trapping and tagging the collared peccary. *J. Wildl. Mgt*, **23**, 11–16.

Newgrain, K. and Horwitz, C. M. (1979) U.H.F. System for Radio-tracking wild animals. *Proc. 2nd Internat. Conf. on Wildl. Biotelemetry, Laramie, Wyo*. 187–92.

Norton-Griffiths, M. (1975) *Counting Animals*. Afr. Wildl. Leadership Found., Kenya, Publ. (1). 110 pp.

Odum, E. P. (1966) *Ecology*. Holt, Rinehart & Winston. 152 pp.

Oelofse, J. (1970) Plastic for game catching. *Oryx*, **10**, 306–8.

Orr, R. T. and Moore-Gilbert, S. M. (1964) Field immobilization of young wildebeest with succinylcholine chloride. *E. Afr. Wildl. J.*, **2**, 60–6.

Ottestad, P. (1933) A mathematical method for the study of growth. *Hvaldradets Skrifter*, (7), 30–54.

Overton, W. S. and Davis, D. E. (1969) Estimating the numbers of animals in wildlife populations. Chapter 21 in Giles, R. H. Jr. (ed.), *Wildlife Management Techniques*, 3rd edn, 403–55.

Owen-Smith, R. N. (1971) The contribution of radio-telemetry to a study of the white rhinoceros. *Proc. Symp. on Biotelemetry*. S. Afr. Council for Sci. & Industr. Res., Pretoria, 101–8.

Panwar, H. S. (1979) A note on tiger census technique. *Tigerpaper*, **6**(2), 16–18.

Parkinson, C. N. (1958) *Parkinson's Law*. John Murray, London. 122 pp.

538

Pienaar, U. de v. (1973) The capture and restraint of wild herbivores by mechanical methods. Chapter 9 in *Capture and Care of Wild Animals*. Human and Rousseau Publ., Capetown, 91–9.

Pitelka, F. A. (1959) Numbers, breeding schedule, and territoriality in pectoral sandpipers in northern Alaska. *Condor*, **61**, 233–64.

Riney, T. (1950) Home range and seasonal movement in a Sierra deer herd. Univ. Calif. (Berkeley), MA Thesis 42 pp.

Riney, T. (1951) Relationships between birds and deer. *Condor*, **53**(4), 178–85.

Riney, T. (1953) Notes on the Syrian lizard (*Acanthodactylus tristami orientalis*). *Copeia*, (1), 66–7.

Riney, T. (1955a) Evaluating condition of free-ranging red deer (*Cervus elaphus*), with special reference to New Zealand. *N.Z. J. Sci. and Tech. Sect. B*, **36**(5), 429–63.

Riney, T. (1955b) *Identification of Big Game Animals in New Zealand*. Dom. Mus. Handbook (4). 26 pp.

Riney, T. (1956a) Comparison of occurrence of introduced animals with critical conservation areas to determine priorities for control. *N.Z. J. Sci. and Tech.*, Sect. B, **38**(1), 1–18.

Riney, T. (1956b) A zooecological approach to the study of ecosystems that include tussock grassland and browsing and grazing animals. *N.Z. J. Sci. and Tech.*, Sect. B, **37**(4), 455–72.

Riney, T. (1957a) The use of faeces counts in studies of several free-ranging mammals in New Zealand. *N.Z. J. Sci and Tech.*, Sect. B, **38**(6), 507–32.

Riney, T. (1957b) Sambar (*Cervus unicolor*) in sand hill country. *Proc. N.Z. Ecol. Soc.*, (5), 26–7.

Riney, T. (1957c) *Minimum data required to record New Zealand Government shooting activities. N.Z.* Forest Serve. (cyclostyled), 18 Feb. 1957. 6 pp.

Riney, T. (1957d) *Milk tooth replacement in red deer*. N.Z. Forest Serv. 1 p.

Riney, T. (1960) A field technique for assessing physical condition of some ungulates. *J. Wildl. Mgt.*, **24**(1), 92–4.

Riney, T. (1962) *Utilization of Wildlife in the Transvaal, South Africa*. IUCN Publ., Ser. (1), pp. 303–5.

Riney, T. (1936a) A rapid field technique and its application in describing conservation status and trends in semi-arid pastoral land. *African Soils*, **8**(2), 159–258.

Riney, T. (1963b) Rare and threatened mammals in Africa. Proc. XVI Internat. Congress of Zool., Washington, DC., 20–7.

Riney, T. (1964) *The Impact of Introductions of Large Herbivores on the Tropical Environment*. IUCN Publ., New Series (4), 261–73.

Riney, T. (1967a) *Conservation and Management of African Wildlife*. FAO/IUCN African Spec. Proj. Stage III, FAO, Rome. 35 pp.

Riney, T. (1967b) *Ungulate Introductions as a Special Source of Research Opportunities*. IUCN Publ., New Series (9), 241–54.

Riney, T. (1970) Wildlife and nomadic stock in semi-arid lands. *World Rev. Anim. Prod. (Feed and Food)*, **26**, 82–92.

Riney, T. (1977) *Major Environmental Problems in Mozambique*. UNEP Consultancy Report, SSA No. 7-2600040. 133 pp.

Riney, T. and Caughley, G. (1959) A study of home range in a feral goat herd. *N.Z. J. Sci.*, **2**(2), 157–70.

Riney, T. and Child, G. (1962) Breeding season and ageing criteria for the common duiker (*Sylvicapra grimmia*). *Proc. 1st Fed. Sci. Conf., Rhodesia Assn. Sci.*, 291–9.

Riney, T. and Dunbar, G. A. (1956) *Criteria for Determining Status and Trend of High Country Grazing Lands in the South Island of New Zealand*. Soil Conserv. and Rivers Control Council, Wellington. 32 pp.

Riney, T. and Kettlitz, W. L. (1964) Management of large mammals in the Transvaal, *Mammalia*, **28**(2), 189–248.

Riney, T., Watson, S., Bassett, C., Turbott, G. and Howard, W. E. (1959) Lake Monk expedition: an ecological study in Southern Fiordland. *N.Z., D.S.I.R. Bull.*, (135), 75 pp.

Robinette, W. L. (1968) *Game Animal Surveys*. Proc. Joint FAO/WHO Expert Comm. on Afr. Trypanosomiasis, Geneva, 25–30 Nov. 1968, 3.3. 15 pp.

Robinette, W. L., Jones, D. A., Gashwiler, J. S. and Aldous, C. M. (1956) Further analysis of methods for censusing winter-lost deer. *J. Wildl. Mgt.*, **20**(1), 75–8.

Romanov, A. N. (1956) Automatic tagging of wild animals and prospects for its use. *Zool. J. Moscow*, **35**, 1902.

Scheffer, V. B. (1950) Growth layers on the teeth of Pinnipedia as an indicator of age. *Science*, **112**(2907), 309–11.

Schenkel, R. (1966) *Play, Exploration and Territoriality in the Wild Lion*. Symp. Zool. Soc. London, (18), 11–22.

Schwartz, C. W. and Schwartz, E. R. (1959) *The Wild Mammals of Missouri*. Conservation Commission (Mo.). 341 pp.

Seidensticker, J. C. IV, Hornocker, M. G., Wiles, W. V. and Messick, J. P. (1973) Mountain Lion Social Organization in the Idaho Primitive Area *Wildl. Monog.*, (35). 60 pp.

Severinghaus, C. W. (1949) Tooth development and wear as criteria of age in white-tailed deer. *J. Wildl. Mgt.*, **13**, 195–216.

Simpson, C. D. and Elder, W. H. (1968) Lens weights related to estimated age in greater kudu. *J. Wildl. Mgt.*, **32**(4), 764–8.

Slater, L. E. (ed.) (1963) *Bio-telemetry: The Use of Telemetry in Animal Behaviour and Physiology in Relation to Ecological Problems*. Macmillan Co., NY. 372 pp.

Smithers, R. H. N. (1966) *The Mammals of Rhodesia, Zambia and Malawi*. Collins, London. 159 pp.

Smithers, R. H. N. (1973) *Recording Data on Mammal Specimens*. Spec. Publ. Natl. Mus. and Mon., Rhodesia, Salisbury. 10 pp.

Spinage, C. A. (1967) Ageing the Uganda defassa waterbuck *Kobus defassa Ugandae*, Neumann. *E. Afr. Wildl. J.*, **5**, 1–15.

Spinage, C. A. (1973) A review of the age determination of mammals by means of teeth with especial reference to Africa. *E. Afr. Wildl. J.*, **11**(3 and 4), 165–87.

Springer, J. T. (1979) Some sources of bias and sampling error in radio triangulation. *J. Wildl. Mgt.*, **43**(4), 926–35.

Steenkamp, J. D. G. (1970) The effect of breed and nutritional plane on the chronology of teeth eruption in cattle. *Rhod. J. Agric. Res.*, **8**, 3–13.

Stevenson-Hamilton, J. (1947) *Wild Life in South Africa*. Cassell, London. 364 pp.

Stewart, D. R. M. (1967) Analysis of plant epidermis in faeces: a technique for studying the food preferences of grazing herbivores. *J. Appl. Ecol.*, **4**, 83–111.

Stewart, D. R. M. and Stewart, J. (1970) Food preference data by faecal analysis for African plains ungulates. *Zool. Afr.*, **5**(1), 115–29.

Stickel, L. F. (1946) Experimental analysis of methods for measuring small mammal populations. *J. Wildl. Mgt.*, **10**(2), 150–9.

Stoddart, L. A. and Rasmussen, D. I. (1945) Big game–range livestock competition on Western ranges. *Trans. 10th N. Amer. Wildl. Conf.*, 251–6.

Stoddart, L. A. and Smith, A. D. (1943) *Range Management*. McGraw-Hill, Inc. New York and London, 547 pp.

Stoneberg, R. P., and Jonkel, C. J. (1966) Age determination of black bears by cementum layers. *J. Wildl. Mgt.*, **30**(2), 411–4.

Strahler, A. N. (1956) *The Nature of Induced Erosion and Aggradation: Man's Role in Changing the Face of the Earth*. University of Chicago Press, 621–38.

540

Taber, R. D. and Cowan, I. McT. (1969) Capturing and marking wild animals. *Wildlife Management Techniques*, 3rd edn, revised. The Wildl. Soc., 277–317.

Taber, R. D., de Vos, A. and Altmann, M. (1956) Two marking devices for large land mammals. *J. Wildl. Mgt.*, **20**(4), 464–5.

Talbot, L. M. and McCulloch, J. S. G. (1965) Weight estimations for East African mammals from body measurements. *J. Wildl. Mgt.*, **29**(1), 84–9.

Talbot, L. M. and Talbot, M. H. (1963) The wildebeest in Western Masailand, East Africa. *Wildl. Monog.*, (12), 88 pp.

Taylor, W. P. (1948) *Outline for study of mammalian ecology and life histories*. US Fish and Wildl. Serv., Wildl. Leaflet (304). 26 pp.

Tester, J. R. (1971) Interpretation of ecological and behavioral data on wild animals obtained by telemetry with special reference to errors and uncertainties. *Proc. Symp. on Biotelemetry*. S. Afr. Council for Sci. and Industr. Res., Pretoria. 385–407.

Trefethen, J. B. (1964) *Wildlife Management and Conservation*. D. C. Heath & Co., Boston. 120 pp.

van den Brink, F. H. (1967) *A Field Guide to the Mammals of Britain and Europe*. Collins, London. 221 pp.

van der Walt, K. (1973) *The Capture and Care of Wild Animals* (Conclusion). Human & Rousseau Publ., Capetown, 208–11.

Wavell, A. P. (1944) *Other Men's Flowers*. Jonathan Cape Ltd. London. 447 pp.

Weeks, R. W., Long, F. M., Lindsay, J. E., Baily, R., Patula, D. and Green, M. (1977) Fish Tracking from the Air. *Proc. 1st Internat. Conf. on Wildl. Telemetry, Laramie, Wyo.*, 63–9.

Winters, S. R. (1921) Diagnosis by wireless. *Scient Amer.*, **124**, 465.

Young, E. (ed.) (1973) *The Capture and Care of Wild Animals*. Human & Rosseau Publ., Capetown. 224 pp.

Zwank, P. J. (1977) Monitoring mule deer mortality. *Proc. 1st. Internat. Conf. on Wildl. Telemetry, Laramie, Wyo.*, 131–40.

Index

546

552

Vehicles (*continued*)
  hunting from, 487
  maintenance, 484
Vermin, 473
Vested interests, 428, 440, 470, 473
Virus diseases, 291

Walkie-talkie radios, 402, 415, 418
Wandering, 42
Wariness, 102, 357, 397, 401, 403, 405, 430
Warthog, 29, 70, 123, 159, 160, 178, 309–312, 413
Water, 25, 32, 38, 70, 73, 108, 119, 123, 124, 137, 210, 221, 260, 267, 298, 308, 319, 351, 401, 427
  surface run-off, 251
Waterbuck, 38, 93, 94, 311, 413, 419
Waterfowl, 437
Weasel, 179
Weather, 298, 299

Weight, 172, 189–190, 191, 277
  limitations of, 190, 191
Wet season, 31, 97, 102, 309, 312
Wildebeest, 30, 31, 32, 38, 39, 52, 58, 59, 77, 84, 196, 311, 394, 410, 412, 413, 419, 420, 429, 434
Winter season, 96
Wet udder
  demonstration of, 277
  ratios, 162–163, 189, 196
Woolworth system, 331–332
Working plans, 3, 5–11, 12, 424, 444

Young plants, 206

Zebra, 29, 39, 47, 48, 55, 56, 59, 66, 77, 196, 291, 311, 403, 410, 413, 419
Zimbabwean trap, 414–420, 422
Zone of silence, 398–401
Zoning, 345, 405, 437
Zoological parks, 352